Research Reports in Physics

Research Reports in Physics

Nuclear Structure of the Zirconium Region
Editors: J. Eberth, R. A. Meyer, and K. Sistemich

Ecodynamics Contributions to Theoretical Ecology
Editors: W. Wolff, C.-J. Soeder, and F. R. Drepper

Nonlinear Waves 1 Dynamics and Evolution
Editors: A. V. Gaponov-Grekhov, M. I. Rabinovich, and J. Engelbrecht

Nonlinear Waves 2 Dynamics and Evolution
Editors: A. V. Gaponov-Grekhov, M. I. Rabinovich, and J. Engelbrecht

Nonlinear Waves 3 Physics and Astrophysics
Editors: A. V. Gaponov-Grekhov, M. I. Rabinovich, and J. Engelbrecht

Nuclear Astrophysics
Editors: M. Lozano, M. I. Gallardo, and J. M. Arias

Optimized LCAO Method and the Electronic Structure of Extended Systems
By H. Eschrig

Nonlinear Waves in Active Media
Editor: J. Engelbrecht

Problems of Modern Quantum Field Theory
Editors: A. A. Belavin, A. U. Klimyk, and A. B. Zamolodchikov

Fluctuational Superconductivity of Magnetic Systems
By M. A. Savchenko and A. V. Stefanovich

Nonlinear Evolution Equations and Dynamical Systems
Editors: S. Carillo and O. Ragnisco

Nonlinear Physics
Editors: Gu Chaohao, Li Yishen, and Tu Guizhang

Nonlinear Waves in Waveguides with Stratification
By S. B. Leble

Quark–Gluon Plasma
Editors: B. Sinha, S. Pal, and S. Raha

Symmetries and Singularity Structures Integrability and Chaos
in Nonlinear Dynamical Systems
Editors: M. Lakshmanan and M. Daniel

Modeling Air-Lake Interaction Physical Background
Editor: S. S. Zilitinkevich

Bikash Sinha
Santanu Pal
Sibaji Raha (Eds.)

Quark–Gluon Plasma

Invited Lectures of Winter School,
Puri, Orissa, India, December 5–16, 1989

With 164 Figures

Springer-Verlag

Berlin Heidelberg New York London
Paris Tokyo Hong Kong Barcelona

Dr. Bikash Sinha
Dr. Santanu Pal
Variable Energy Cyclotron Centre, 1/AF, Bidhan Nagar, 700 064 Calcutta, India

Professor Sibaji Raha
Fachbereich Physik der Philipps Universität Marburg, Mainzer Gasse 33,
D-3550 Marburg, Fed. Rep. of Germany

ISBN 3-540-51984-X Springer-Verlag Berlin Heidelberg New York
ISBN 0-387-51984-X Springer-Verlag New York Berlin Heidelberg

This work is subject to copyright. All rights are reserved, whether the whole or part of the material is concerned, specifically the rights of translation, reprinting, reuse of illustrations, recitation, broadcasting, reproduction on microfilms or in other ways, and storage in data banks. Duplication of this publication or parts thereof is only permitted under the provisions of the German Copyright Law of September 9, 1965, in its current version, and a copyright fee must always be paid. Violations fall under the prosecution act of the German Copyright Law.

© Springer-Verlag Berlin Heidelberg 1990
Printed in Germany

The use of registered names, trademarks, etc. in this publication does not imply, even in the absence of a specific statement, that such names are exempt from the relevant protectiv laws and regulations and therefore free for general use.

2157 / 3140-543210 – Printed on acid-free paper

Preface

This volume contains the notes of the main lectures presented at the Winter School on Quark–Gluon Plasma (QGP) held at the Toshali Sands resort near Puri, India, 5–16 December, 1989. The format of the volume can be considered as a comprehensive synthesis of introductory lectures dealing with the essential fundamentals of the field and the frontier level ideas, directly relevant for an in-depth understanding of the state of the art. The degree of enthusiasm and involvement displayed by the participants of the School convinced us that a permanent record in the form of this book would be appropriate.

This text begins with a discussion of experiments and experimental techniques: Roberto Salmeron relates the history of the search for signs of QGP in ultrarelativistic heavy ion collisions, after an exposition of relativistic kinematics. Michael J. Tannenbaum's contribution is primarily concerned with QGP activities at Brookhaven National Laboratory.

We continue with several chapters on the theoretical models and formalism. Jitendra Parikh covers Quantum Chromodynamics (QCD) and QCD plasma, in particular, perturbative QCD in many-body systems at finite temperatures, to study QGP thermodynamics. Frank Close discusses the properties of a single ground state nucleon in a nuclear medium, and in a volume of high energy density. This problem is of direct relevance to the phase transition from hadronic to QGP behaviour. A more detailed knowledge of the glue and its distribution inside hadrons is a pressing goal.

Narayan Rana and Charles Alcock present the very exciting developments in our understanding of the Universe a microsecond after its creation – an area of physics where QCD/QGP related ideas can be applied with some confidence. Narayan Rana focuses on the geometry and dynamics of the early universe going over to Big Bang nucleosynthesis; Charles Alcock covers the very recent work of possible changes in the relative abundances of various nuclei essentially due to the inhomogeneity introduced by QCD phase transition. The need to invoke dark matter, its possible origin and its relevance to QCD phase transition is discussed.

It is implicit that previous knowledge of QGP is not essential to understand the contents of this book. It is hoped that the material gathered here, like the lectures on which it is based, will inspire the reader to the same degree of enthusiasm as experienced by the participants of the Puri School.

Calcutta, Marburg *Bikash Sinha*
June, 1990 *Santanu Pal*
 Sibaji Raha

From the Editors

A little over two years ago, the International Conference on Physics and Astrophysics of Quark–Gluon Plasma (ICPA–QGP'88) organised by us at Tata Institute of Fundamental Research, Bombay, seemed to have enjoyed a satisfactory measure of success. To sustain the excitement and motivation generated in that conference and especially to introduce the young graduate students and fresh post-docs to this new and rapidly evolving discipline, we decided to organise a more intense school on the same subject, covering extensively, if not exhaustively, the experimental and theoretical scenario related to the physics of QGP, starting from the very early universe to the laboratory experiments of the present generation as well as the next generation. To many of us, the study of QGP is all the more stimulating, because it forges, as it were, a bridge between the microcosmos and the macrocosmos.

The Winter School on QGP, December 5–16, 1989, was organised at the gloriously enchanting and picturesque surroundings of Toshali Sands, a spot near Puri, a famous pilgrimage and sea resort in Orissa, situated in the eastern part of India. The setting and the surroundings helped a great deal in cementing a sense of intense involvement and eager anticipation both among the students and the lecturers.

The format of the school was neither entirely pedagogic nor synoptic but rather a synthesis of the two. Each lecturer started with an introduction to his subject accessible to a graduate student generally well prepared but without any prior expertise in the area of QGP. From there, he brought the students along to the very forefront of the discipline with remarkable clarity and lucidity. It was indeed a pleasure to witness at the end of the school the young participants hitherto unfamiliar with the field converse knowledgeably about the current issues and bubbling with enthusiasm to start contributing to this growing subject. It is our firm belief in editing these proceedings that a large share of that enthusiasm will filter through in the printed pages too, and that this volume will prove to be of immense help to the beginners and the practising experts in QGP.

On the experimental side, Roberto Salmeron covered single-handedly the entire area of searches for the signals of QGP in ultrarelativistic heavy ion collisions. His remarkable set of lectures started very appropriately with a lucid introduction to the relativistic kinematics of high energy collisions, concepts which are often glossed over in the usual curricula, leaving room for immense confusion in the minds of beginners. Salmeron thus rendered yeoman service not only to the students of the school but to the entire community through his clear enunciation of the concepts every theorist and experimentalist must know by heart. From there on, he covered in an inspiring set of lectures the entire state of the art of experimental

QGP searches – from direct photons to hadrons, from two particle interference (Bose-Einstein correlations) to intermittency, the resonance states of J/psi and the possible sources of its suppression, the changes in the slope of the average energy density deposition as a function of the average transverse momentum indicating a phase transition. Behind all these signals, however, there still remains the enigma of FRITIJOF, somehow managing to get things almost right, hinting that there are as yet no clear signals of QGP.

Robert Brockman, along with Salmeron, helped to "demystify" the mythology of the large scale detectors which are part and parcel of ultrarelativistic heavy ion physics research. He concentrated mainly on the CERN detectors, from TCP to ZDC, streamer chambers to nuclear emulsions, dimuon spectrometers, BGO crystals, massive on line data acquisition systems and so on. It was extremely useful for the Indian participants, especially as India is on the verge of becoming a full partner in one of the CERN experiments (WA80), where involvement with the hardware is a must. Is it then possible for us to build a calorimeter or indeed a TPC – how about drift chambers? It is indeed unusual to delve deep into the hardware of detectors in a school not solely dedicated to instrumentation, but Brockman's and Salmeron's attempts were instrumental in stimulating many young minds at the school to get their hands dirty while keeping the physics goals firmly in sight. For reasons beyond our control, however, we had to leave the write-up of Brockman's lectures out of this volume.

The other scheduled lecturer in the experimental section, Mike Tannenbaum, could not attend at the very last minute. He had planned to cover the features of the Brookhaven activities in this area, emphasising in particular the role of transverse energy measurements as a tool for QGP diagnostics. In his absence, Salmeron gave the students an introduction to this area too. But we are also including the write-up of Tannenbaum's contribution in this volume for the benefit of the readers.

On the theoretical side, Jitendra Parikh covered the essential area of QCD and QCD plasma. His discussion on how to avoid the pitfalls of perturbative QCD in many-body systems at finite temperature in order to study thermodynamics and other collective properties of QGP was incisively instructive. Quite clearly, the young (and often the initiated too) can embark on a rather elaborate and involved calculation using perturbative QCD, without fully appreciating the fact that unless one takes into account the collective modes semi-phenomenologically (at the present state of our understanding of strong interaction physics), one might end up throwing the baby away with the bath water. The essential foundation of thermodynamics (or the lack of its validity!) for studying the evolution of the reaction volume after the collision of two nuclei at ultrarelativistic energies was also discussed in detail.

Frank Close contributed greatly in bridging the two seemingly (but then again, why indeed?) uncorrelated areas, the properties of a single nucleon embedded in a nuclear medium in its ground state on the one hand and on the other hand, what happens to them when they are in a volume where sufficient energy density has been pumped in. The very important recent developments in our understanding of a baryon in a nucleus and exactly how important the glue distribution is for predicting the structure of a baryon was emphasised. Needless to say, these considerations can have profound consequences for our understanding of the QGP signals as well as the phase transition from the hadronic to QGP sector. It is felt that whether it is the EMC effect involving the hadronic structure in a nucleus in terms of the QCD degrees of freedom (quarks and gluons) or the study involving the QCD phase transition, a more detailed knowledge of the glue and its distribution inside hadrons or in the fireball is a central issue.

Thus, a more comprehensive study of the atomic nucleus revealing the quark structure of hadrons (through lepton–nucleus interactions) and its logical extension to high density/high temperature hadronic matter, eventually dissolving into a QGP, needs to be performed in a careful and self-consistent manner. The physics of CEBAF (Continuous Electron Beam Acceleration Facility) and the physics of RHIC (Relativistic Heavy Ion Collisions) should be mutually complementary, fitting into the wonderful goal of searching for a unified picture of the ultimate constituents of the hadronic world.

So what exactly is the message at this time for our search for QGP in the laboratory and the related theoretical questions? It is probably correct to say that no definitive signals are yet at hand, the central problem being the pollutants of the hot hadronic matter, evolving with time beyond the critical temperature of the phase transition and eventually to freeze-out. All the signals proposed so far fall short of the acid test, implying that most of the data can perhaps be explained without necessarily invoking the existence of a QGP, notwithstanding the unnatural assumptions often needed to make the mundane explanations work. But it is heartening to note that to date not a single set of data has been found which would rule out the formation of QGP. Perhaps the theorists have not yet succeeded in asking sufficiently discerning questions on QGP diagnostics, but there is always a hint for the future. Is there enough energy deposited? Is the statistics good enough? This last point is particularly relevant for the direct photon signals of QGP – a 30% effect will correspond to one direct photon for every thousand charged particles, not an easy task and one which certainly demands much longer running time. Strange and anti-strange mesons may help to an extent but so little is known about the process of hadronisation, final state interactions and/or the scattering of hadrons from the neighbouring hadrons in a hot dense medium! Indeed, while we are looking for a novel state of matter, it is imperative to realise that very little really is known about the properties of hadronic matter (the background, to wit) at high density and/or temperature. It is high time that the initiated take a harder look at this issue.

If signals relating to any one particular species of detected particles may not be truly clinching; what about their ratios? Yes, it may be difficult to do, involving the measurement of several different kinds of particles in the same experiment, but the dividends certainly justify the effort. The ratio of photons to dileptons is an apt example. Naively, direct photons from QGP arise from the annihilation of quarks and anti-quarks as well as from the so-called generalised Compton effect (q+g going to q+photon), a channel absent for diphotons in the lowest order. So the ratio will clearly provide an all-important "clue" to the elusive glue distribution. It is felt that looking for several signals in the same experiment and studying their ratios will clarify many cloudy issues.

Similarly for two particle interferometry! Pion interferometry has been traditionally done for the past 30 years, starting with the epoch-making discovery of the GGLP effect. As can be gleaned from the lectures of Salmeron, such measurements have now become a mainstay of RHIC processes, not to mention also electron–positron and hadronic collisions at high energy. The enormous importance of these measurements notwithstanding, one has to realise that extracting physically meaningful information from them is still not an unambiguous endeavour, what with the final state interactions, parametrisation dependence of the source distributions and so on. Moreover, pions being predominantly final state products, pion interferometry measures the size of the reaction volume at or near freeze-out, at any rate in its hadronic phase. It will thus be most important to devise interferometric experiments which could yield information about the volume of the initial stage of the evolution.

On the theoretical side, the study of non-equilibrium dynamics and its application to RHIC processes, especially at the early times of formation, must be addressed. The efforts of the past few years in unravelling the subtleties of a kinetic theory with QCD ingredients appear not to have borne the fruit they initially seemed to promise, the difficulties of formulating a gauge-invariant prescription being the root cause. We cannot help thinking that there is scope for a breakthrough there and perhaps some ingenious theorist will take a fresh look at the problem in the near future. What concerns the signals, a simultaneous analysis of the central rapidity regime and the fragmentation regime seems important to understand the role of the baryonic chemical potential, especially in suppressing the anti-quarks. The evaluation of the production rates of J/psi in hadron-nucleus and nucleus-nucleus collisions within the framework of a bound state equation (e.g. Bethe-Salpeter) is an urgent task. We believe that the experimental results obtained from the runs with sulphur will decide the course theoretical activities should take, with an eye to the forthcoming lead runs.

The message is quite clear. It is simply not possible to detect QGP by a short run without elaborate controlled experiments, be that for higher statistics, be that for scanning for the signals event by event, be that for the study of correlation experiments. The results obtained at CERN and AGS so far have proved beyond doubt that there is interesting physics to be learnt from the search for QGP and it is absolutely imperative to design more detailed experiments to iron out some of the prejudices and unclarities in our search for the truth.

Going over from the microcosmos to the macrocosmos, the most relevant issue is the possible consequences of a QCD phase transition on the evolution of the very early universe, approximately several microseconds after the Big Bang.

Narayan Rana introduced the students to the language, the techniques and the subtleties of the geometry and the dynamics of the early universe. In an excellent series of talks, both in terms of thoroughness and pedagogy, he elaborated from a classical point of view the issue of the observed abundances of the elements and its deep connection with the inhomogeneity produced in the early universe by the (presumably) first order QCD phase transition. The lectures covered the basic concepts of the Standard Big Bang cosmology; the question of the dark matter – whether it is necessary – was discussed. The students were given a taste of the intricacies involved in the study of primordial nucleosynthesis. Rana paved the way for the students to ask: What are the ingredients that go into the cosmic soup? How is the latent heat associated with the QCD phase transition taken away to the ambient hadronic universe? How many layers conducting heat away from one another exist in the early universe? What is the mechanism of this conduction? How sensitive are the results to the parameters of the phase transition, the critical temperature, for example?

Charles Alcock dealt with these questions at length, in a brilliant set of lectures. The prospect of the first order phase transition leading to a cosmic separation of the QGP and the hadronic phases, which eventually allow diffusion of baryons among them, as Alcock expounded, can have very important effects on primordial nucleosynthesis and on the overall picture of the evolution of the early universe. The stretching of the time scale associated with nucleosynthesis as a consequence of the QCD transition is a scenario decidedly absent in the Standard Big Bang cosmology.

The central issue of the dark matter came into focus again. As was argued by Witten some time ago, a QCD phase transition may perhaps produce stable configurations of strange quark matter (called strange nuggets in the literature) which would be opaque and thus be plausible baryonic candidates for dark matter. While in recent times the existence of nuggets has been called into question, it is undeniable that an understanding of the dark matter within the framework of known physics, without having to invoke exotics like axions or massive neutrinos, would indeed be a great step forward. Alcock told the students about the results of detailed calculations done by himself and collaborators on nucleosynthesis in the presence of a first order QCD transition, where the observed abundances of all light elements with the singular exception of ^{7}Li come out to be within acceptable limits even when one insists on a baryonically closed universe. The intriguing question about the reliability and the accuracy of the astrophysical observation of abundances of elements was also discussed, with the conclusion that the uncertainties are rather large.

Having argued that the QCD phase transition allows the universe to be closed by baryonic matter, Alcock then addressed the second part of the issue. If the dark matter is not to be the strange nuggets, could they be Jupiter-like planets? They would indeed be non-luminous; how can one search for them? Alcock gave the students an introduction to his group's future project, setting up optical telescopes in the southern hemisphere to look for the brightening of distant stars in the Magellanic cloud by the gravitational lensing action of these Jupiter-like objects. All of us look forward to the results of this experiment.

In addition to the lectures, there were also some specialised seminars delivered by more initiated participants. We decided not to include those write-ups in this volume, as the format of those talks would not match the style followed in the regular lectures where no previous knowledge about the subject was assumed.

The two weeks at Toshali Sands were a feast, a gourmet meal where each course was different from the others, yet complementing one another in a way we had not imagined possible. We are grateful to Springer-Verlag for letting the whole community savour that taste to some extent. The School was sponsored by the Department of Science and Technology and the Department of Atomic Energy, Government of India.

Contents

An Introduction to the Search for the Quark–Gluon Plasma
in Ultrarelativistic Heavy Ion Interactions
By R.A. Salmeron (With 81 Figures) 1

Hadron Production in Relativistic Heavy Ion Interactions
and the Search for the Quark–Gluon Plasma
By M.J. Tannenbaum (With 45 Figures) 108

Physics of the Quark–Gluon Plasma
By J.C. Parikh (With 17 Figures) 181

Quarks and Gluons in Hadrons and Nuclei
By F.E. Close (With 2 Figures) 233

Early Universe and Big Bang Nucleosynthesis
By N.C. Rana (With 8 Figures) 259

The Astrophysics and Cosmology of Quark–Gluon Plasma
By C. Alcock (With 11 Figures) 300

Subject Index 343

An Introduction to the Search for the Quark–Gluon Plasma in Ultrarelativistic Heavy Ion Interactions

R.A. Salmeron

Laboratoire de Physique Nucléaire des Hautes Energies,
Ecole Polytechnique/IN2P3-CNRS, F-91128 Palaiseau, France

Some notions about particle production in hadron-nucleon and hadron-nucleus collisions are initially recalled, before examining the transition from hadronic matter to a quark-gluon plasma. The essential characteristics of the experiments whose results are discussed in these lectures are outlined. An analysis is made of the main results obtained in the experiments performed at Brookhaven at 14.5 GeV/nucleon and at CERN at 60 and 200 GeV/nucleon in the search for the quark-gluon plasma under the assumptions of different signatures : transverse momentum distribution, direct photon production, the suppression of the J/ψ production, Bose-Einstein interferometry, fluctuations in rapidity and intermittency, and production of strange particles. A conclusion from the experimental situation is drawn.

1. INTRODUCTION

A great interest has arisen in the last years on the study of ultrarelativistic heavy ion interactions, due to the possibility of producing quark-gluon plasma if the energy density reached in those collisions is sufficiently high. These lectures are an introduction to this subject, with emphasis on the search for the quark-gluon plasma in experiments recently performed at Brookhaven and at CERN, and are intended to provide the reader with a background which would allow to follow the forthcoming litterature. They were organized to match the lectures delivered at this school by Alcock [1], Brockmann [2], Parikh [3] and Tannenbaum [4], and an effort was made to avoid repetitions. The lectures start in section 2 with a recall of fundamental variables, their relationships with detectors acceptance and performance, analysis of events in the laboratory and in the center of mass systems. In section 3 the transition from hadronic matter to quark-gluon plasma is examined. Section 4 is an outline of the main characteristics of detectors. Sections 5 to 11 deal with the experimental status of the proposed signatures of the quark-gluon plasma, transverse momentum distribution, direct photon production, the suppression of the J/ψ production, Bose-Einstein interferometry, fluctuations and intermittency, and the production of strange particles, respectively. Section 12 summarizes the conclusions.

Many references on this subject can be found in the Proceedings of the "Quark Matter" series of conferences [5], as well as in ref. [6-8]. Recent reviews have been made by Tannenbaum [9], Geist [10], Specht [11], and Schmidt and Gutbrod [12].

2. RECALL OF SOME CONCEPTS IN HADRON-NUCLEON AND HADRON-NUCLEUS INTERACTIONS

2.1 NOTATION

For the collision of a hadron-beam with a nucleon-target we shall use the following notation, in the laboratory system (LS) and center of mass system (CMS), in which, by definition, the total momentum is zero. (Fig. 1).

	LABORATORY SYSTEM		CENTER OF MASS SYSTEM	
	BEAM	TARGET	BEAM	TARGET
Mass	M	M_N	M	M_N
Momentum	$\vec{P}$	$\vec{P}_N$	$\vec{P}^*$	$-\vec{P}^*$
Energy	$E=(M^2+\vec{P}^2)^{1/2}$	$E_N=(M_N^2+\vec{P}_N^2)^{1/2}$	$E^*=(M^2+\vec{P}^{*2})^{1/2}$	$E_N^*=(M_N^2+\vec{P}^{*2})^{1/2}$
Four-momentum	$q=(E,\vec{P})$	$q_N=(E_N,\vec{P}_N)$	$q^*=(E^*,\vec{P}^*)$	$q_N^*=(E_N^*,-\vec{P}^*)$
Total momentum	$\vec{P}_{lab}=\vec{P}+\vec{P}_N$		$\vec{P}^*_{cms}=0$	
Total energy	$E_{lab}=E+E_N$		$E^*_{cms}=\sqrt{s}=E^*+E_N^*$	
Four momentum	$q_{lab}=(E_{lab},\vec{P}_{lab})$		$q^*_{cms}=(\sqrt{s},0)$	

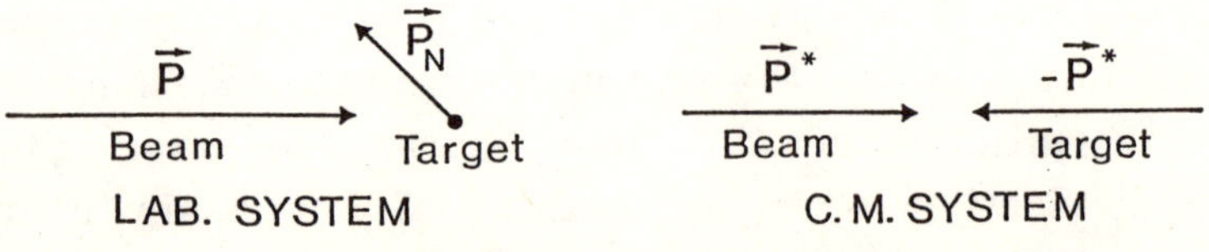

Fig.1. The momentum of beam and target in the LS and in the CMS.

In this notation, the velocity of the CMS in the LS is

$$\beta_c = \frac{|\vec{P}_{lab}|}{E_{lab}} = \frac{|\vec{P} + \vec{P}_N|}{E + E_N} \qquad (1)$$

the Lorentz γ-factor being

$$\gamma = \frac{E_{lab}}{\sqrt{s}} = (1 - \beta^2)^{-1/2} \qquad (2)$$

Let us assume that in a beam-target collision a particle is produced in the LS at an angle θ with momentum $\vec{p}$ and energy E, and in the CMS at an angle θ^* with momentum $\vec{p}^*$ and energy E^* (**Fig. 2**). We shall simplify the notation by writing $|\vec{p}| = p$ and $|\vec{p}^*| = p^*$. The beam direction will be the positive direction in both systems. The momenta $\vec{p}$ and $\vec{p}^*$ have longitudinal and transverse components, $\vec{p}_L$, $\vec{p}_L^*$, and $\vec{p}_T$, $\vec{p}_T^*$, respectively. The Lorentz transformations give:

$$p_T = p_T^* = p \sin \theta = p^* \sin \theta^*$$
$$p_L = \gamma_c(\beta_c E^* + p_L^*) = \gamma_c(\beta_c E^* + p^* \cos \theta^*) \qquad (3)$$
$$E = \gamma_c(E^* + \beta_c p_L^*) = \gamma_c(E^* + \beta_c p^* \cos \theta^*)$$

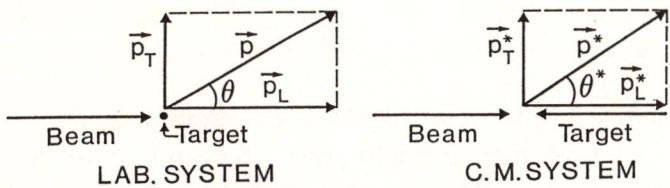

Fig.2. Longitudinal and transverse momentum and angle of emission of a particle in the LS and in the CMS.

The velocity of the LS in the CMS is $-\beta_c$, which gives $p_L^* = \gamma_c(-\beta_c E + p_L)$ and $E^* = \gamma_c(E - \beta_c p_L)$.

It is interesting to recall also the time transformation. If an event occurs in the LS at a time t at the space coordinate x, and in the CMS at a time t^* and space coordinate x^*, then:

$$t = \gamma_c(t^* + \beta_c x^*) \qquad (4)$$

Transverse mass. If the mass of a particle is m, by definition its **transverse mass is**:

$$m_T = (m^2 + p_T^2)^{1/2}$$

In some models of particle production the transverse mass is a suitable parameter to express the dependence of the differential cross section on p_T.

2.2 CASE OF A SINGLE PARTICLE

If a particle of mass m has in the LS a momentum $\vec{p}$ and energy $E = (m^2 + \vec{p}^2)^{1/2}$ its velocity in the LS is $\beta = p/E$. In its rest system its momentum is $\vec{p}^* = 0$ and its energy $E^* = m$. The Lorentz γ factor is given by $\gamma = E/E^* = E/m$. In summary,

$$\gamma = [1 - (\tfrac{p}{E})^2]^{-1/2} \qquad E = \gamma m \qquad p = \beta E = \beta \gamma m \qquad (5)$$

If the CMS is the rest frame of a single particle, then x^* is always equal to zero, $\beta_c = \beta$, $\gamma_c = \gamma$ and then

$$t = \gamma \, t^* \qquad (6)$$

This is an important relationship between the time t measured in the LS and the time t^* measured in the rest system of the particle. In particular, if t^* is the particle lifetime τ, the corresponding time in the laboratory is

$$t = \gamma \, \tau \qquad (7)$$

All properties of a particle, like the lifetime, the spin, the decay angular distribution, etc. must be measured in its rest system. For the lifetime this is obvious : from expr. (7), the time t measured in the LS depends on γ and therefore on the momentum $\vec{p}$, whereas τ is a constant, characteristic of the particle. For the decay angular distribution, assume the decay of a particle A into particles B and C : A $\rightarrow$ B + C. There must be energy and momentum conservation in all reference systems; as a consequence, in the LS the angle of emission of B or C might be limited. However, in the CMS, because the total momentum is zero, there is no preferential direction imposed by kinematics, and kinematically all angles of emission of B and C are allowed. If we find experimentally that in the CMS there is no isotropy, this is an indication that some physical process is imposing preferential directions for the emission of B and C. For example, in the decay $\pi \rightarrow \mu + \nu$, already at a momentum as low as 1 GeV/c the maximum angle of emission of the μ in the LS is only 2.3°, just because of energy and momentum conservation, whereas in the π rest frame all angles are allowed, from 0° to 180°.

Exercice 1 - *Deduce the expression of the maximum decay angle of the µ in the LS, in the π → µ + ν decay, as a function of the masses of the particles and the π momentum. Show that for a π momentum of 20 GeV/c this maximum angle is 0.1°. Take m_π = 0.140 GeV/c² and m_μ = 0.106 GeV/c².*

2.3 COLLISION OF TWO PARTICLES

The collision must be analysed in the CMS of the two colliding particles. The two main reasons for this are :

1 - In the CMS the energy has the smallest value of all reference systems. As consequence, it is in the CMS that we must compute the energies involved in the processes, particularly the threshold energies for the particle production. In the CMS the total energy $\sqrt{s}$ is $1/\gamma$ of the total energy in the LS : $\sqrt{s} = E_{lab}/\gamma$.

2 - Because in the CMS the total momentum is zero, when particles are created there are no preferential directions imposed by kinematics. If we find experimentally that some angles of emission are more probable than others, we can conclude that this is due to some physical process in the interaction. This is not the case in the LS, in which some angles may not be allowed just because energy and momentum would not be conserved. Let us take as an example the inclusive production of the J/ψ in proton-proton collision : p + p → J/ψ + ... From kinematics only, in the CMS of the p-p collision the J/ψ can be emitted at any angle, from 0° to 180°. In the LS, when the target proton is at rest the J/ψ can never be emitted at an angle greater than 17.7°, even if the energy of the incoming proton is infinite.

Exercise 2 - *Show that in the collision of a hadron with a nucleon at rest a J/ψ can never be emitted at an angle greater than 17.7° in the LS, even if the incoming hadron has infinite energy. Take M_ψ = 3.1 GeV/c2, $M_{nucleon}$ = 0.94 GeV/c².*

2.4 INCLUSIVE PRODUCTION OF A PARTICLE

Let us consider the inclusive production of a particle A, i.e., the collisions :

$$\text{beam} + \text{target} \rightarrow A + \text{anything} \quad (8)$$

The particle is emitted in the LS at an angle θ with momentum $\vec{p}$, which has transverse and longitudinal components $\vec{p}_T$ and $\vec{p}_L$, respectively.

We usally separate the cross section for the particle production into two factors : a function of p_T only, call $f(p_T)$, and a function of p_L only, call $g(p_L)$:

$$\sigma = f(p_T) \, g(p_L) \qquad (9)$$

This factorization, which has been used since the beginning of the study of particle production years ago, is empirical; there is no deep reason for it, but it is very convenient because each of those two factors has simple parametrizations which fit well the experimental data. A similar factorization of the differential cross sections is used:

$$\frac{d^3\sigma}{dp^3} = \frac{d^2\sigma}{dp_T^2} \frac{d\sigma}{dp_L} \qquad (10)$$

This factorization became currently used specially after Feynman's suggestion that the longitudinal component of the cross section, when measured in the CMS of the collision, would scale, i.e. <u>would not depend on the energy</u> $\sqrt{s}$, if we use as unit of p_L^* the maximum value of p_L^* in reactions (8). In other words, we define a new variable

$$x_F = \frac{p_L^*}{p_L^* \text{ maximum}} \qquad (11)$$

and measure $d\sigma/dx_F$ instead of $d\sigma/dp_L^*$ because $d\sigma/dx_F$ would not depend on the energy of the reaction. Feynman's assumption, which cannot be proven, was shown experimentally to be only approximately valid.

2.4.1 Differential cross section as a function of p_T^2 or p_T

It has been observed experimentally, about 20 years ago, that the Lorentz-invariant differential cross section, $E \, d^3\sigma/dp^3$, as well as the non-invariant cross section $d^2\sigma/dp_T^2$, when expressed as a function of p_T, show an exponential behaviour of the type:

$$E \frac{d^3\sigma}{dp^3} = A \exp(-a \, p_T) \qquad (12a)$$

$$\frac{d^2\sigma}{dp_T^2} = B \exp(-b \, p_T) \qquad (12b)$$

Sometimes we use an exponential function of the transverse mass, because this variable comes naturally in thermodynamical models of particle production :

$$E \frac{d^3\sigma}{dp^3} = G \exp(-\alpha\ m_T) \qquad (13a)$$

$$\frac{d^2\sigma}{dp_T^2} = H \exp(-\beta\ m_T) \qquad (13b)$$

In thermodynamical models, the coefficients a, b, α and β represent the inverse of a temperature.

We can write eq. (12b) as :

$$\frac{d\sigma}{dp_T^2} = \frac{d\sigma}{2p_T\ dp_T} = B \exp(-b\ p_T)$$

or

$$\frac{1}{p_T} \frac{d\sigma}{dp_T} = C \exp(-b\ p_T) \qquad (14)$$

(where C = 2B)

Fig. 3 shows an example of parametrization (12a) for π^+ and π^- production [13] and **Fig. 4** of parametrization (13a) for π^+ production [14], both at different energies at the CERN ISR. The lines in those two

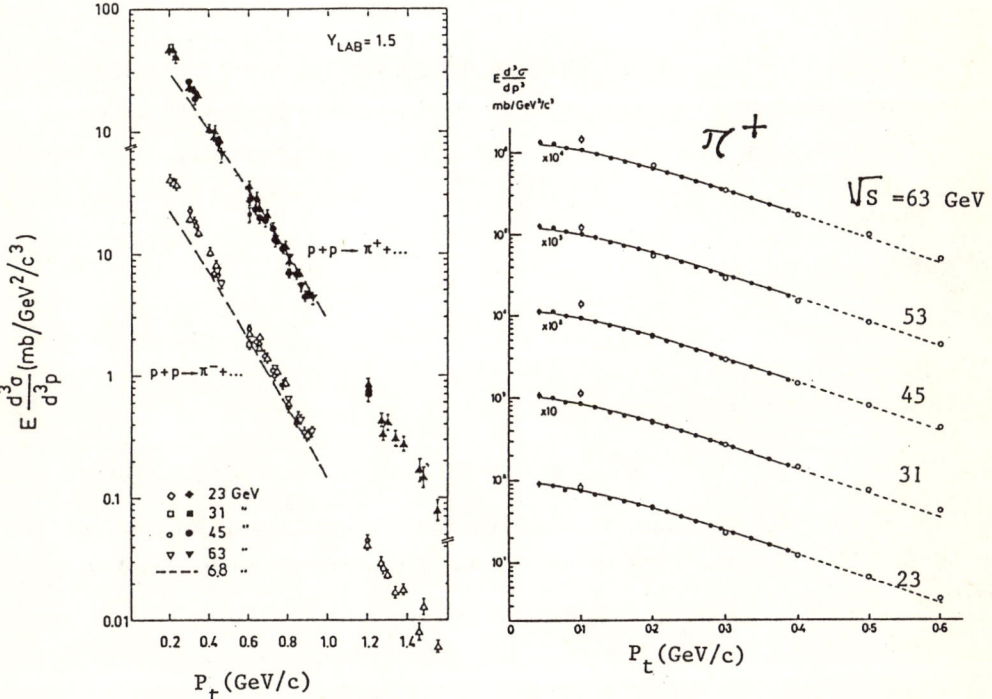

Fig. 3. Invariant differential cross-section versus P_T for π^+ and π^- production at the ISR [13]

Fig. 4. Invariant differential cross-section versus P_T for π^+ production at the ISR [14]

Figures are the fit of the experimental data to the functions (12a) and (13a), respectively.

When the P_T covers a wide range, of several GeV/c, the data usually cannot be fitted to a single function of the type (12) or (13). In this case the P_T range is divided into two or three intervals and a fit is done in each of them. For each P_T interval there is a value of the parameters a, b, α or β. We shall see examples of this in section 5.

2.4.2 Differential cross section as a function of p_L

It was explained above that we usually measure $d\sigma/dx_F$. The value of $p^*_{L\ max}$ which enters in the definition of x_F should be computed for a two-body process, but we usually make the approximation $p^*_{Lmax} \approx \sqrt{s}/2$, and use

$$x_F = \frac{2p^*_L}{\sqrt{s}} \qquad (15)$$

The differential cross section for the inclusive production of a particle is then written

$$\frac{d\sigma}{dx_F\, dp_T^2} = F(s,\, x_F,\, p_T^2)$$

Feynman's assumption that at high energies the function $F(s, x_F, p_T^2)$ becomes asymptotically independent of the energy means that:

$$\lim_{s \to \infty} F(s,\, x_F,\, p_T^2) = F(x_F, p_T^2) = f(p_T^2) g(x_F)$$

2.5 RAPIDITY

By definition, the rapidity of a particle is:

$$y = \frac{1}{2} \ln \frac{E + p_L}{E - p_L} \qquad (16)$$

It can also be expressed as

$$y = \sinh^{-1} \frac{p_L}{m_T} = \cosh^{-1} \frac{E}{m_T} \qquad (17)$$

The rapidity is connected with the longitudinal motion of the particle. The differential cross section for the particle production is then written:

$$\frac{d\sigma}{dy\, dp_T^2} = f(\sqrt{s},\, y,\, p_T^2)$$

and can be factorized into a function of y and a function of p_T^2:

$$\frac{d\sigma}{dy\, dp_T^2} = h(\sqrt{s}, y)\, k(\sqrt{s}, p_T^2)$$

2.5.1 Rapidity of the CMS in the LS

The total energy in the CMS is $\sqrt{s}$. The energy and momentum of the CMS in the LS are $\gamma_c \sqrt{s}$ and $\beta_c \gamma_c \sqrt{s}$, respectively. The rapidity of the CMS in the LS is:

$$Y_{cm} = \frac{1}{2} \ln \frac{\gamma_c \sqrt{s} + \beta_c \gamma_c \sqrt{s}}{\gamma_c \sqrt{s} - \beta_c \gamma_c \sqrt{s}}$$

or

$$Y_{cm} = \frac{1}{2} \ln \frac{1 + \beta_c}{1 - \beta_c} \qquad (18)$$

which is a constant for the same transformation.

2.5.2 Relationship between the rapidity of a particle in the LS and the rapidity in the CMS of the collision

The rapidities of a particle in the LS and in the CMS of the collision are, respectively

$$y = \frac{1}{2} \ln \frac{E + p_L}{E - p_L} \quad \text{and} \quad y^* = \frac{1}{2} \ln \frac{E^* + p_L^*}{E^* - p_L^*}$$

Making the Lorentz transformation on E and p_L :

$$y = \frac{1}{2} \ln \frac{\gamma_c(E^* + \beta_c p_L^*) + \gamma_c(\beta_c E^* + p_L^*)}{\gamma_c(E^* + \beta_c p_L^*) - \gamma_c(\beta_c E^* + p_L^*)} \qquad (19)$$

$$= \frac{1}{2} \ln \frac{E^* + p_L^*}{E^* - p_L^*} + \frac{1}{2} \ln \frac{1 + \beta_c}{1 - \beta_c}$$

or $\quad y = y^* + Y_{cm}$

the first term is the rapidity of the particle in the CMS and the second is the rapidity of the CMS in the LS. We see that the rapidity of a particle in the LS is equal to the sum of the rapidity of the particle in the CMS and the rapidity of the CMS in the LS.

This is a very important property of the rapidity. It means that the <u>shape</u> of the rapidity distribution is invariant under a Lorentz transformation. When we pass from the CMS to the LS the rapidity distribution is the same, with the y-scale displaced by an amount equal to y_{cm}. **Fig. 5** shows an example in which $y_{cm} = 1.5$; the rapidity y in the LS is related to the rapidity y^* in CMS by : $y = y^* + 1.5$.

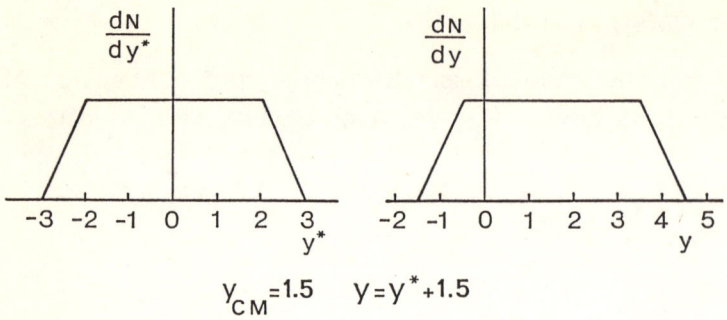

Fig. 5. Relationship between the rapidity distributions in the CMS and in the LS. See text.

2.5.3 Pseudorapidity

Let us assume that a particle is emitted at an angle θ in the LS (Fig. 2). The rapidity can be written:

$$y = \tfrac{1}{2} \ln \frac{E+p_L}{E-p_L} = \tfrac{1}{2} \ln \frac{\sqrt{m^2+p^2} + p \cos\theta}{\sqrt{m^2+p^2} - p \cos\theta}$$

For very high energies, $p \gg m$, and

$$y \approx \tfrac{1}{2} \ln \frac{p+p\cos\theta}{p-p\cos\theta} = \tfrac{1}{2} \ln \frac{2\cos^2\frac{\theta}{2}}{2\sin^2\frac{\theta}{2}}$$

or

$$y \approx -\ln \operatorname{tg} \tfrac{\theta}{2} \tag{20}$$

The function $-\ln \operatorname{tg} \theta/2$ is called <u>pseudorapidity</u> and is usually represented by η. We see that for very high energies the rapidity and the pseudorapidity are nearly equal:

$$y \approx \eta = -\ln \operatorname{tg} \tfrac{\theta}{2} \tag{21}$$

The pseudorapidity η is a very convenient variable, because it depends only on the angle of emission θ and is defined for any values of the mass and momentum of the particle and any value of the energy of the collision. We can measure η without knowing the momentum of the particle. For example, η can be measured for all particles in a detector without a magnetic field, like nuclear emulsions or a segmented calorimeter.

We see that the resolution $d\eta$ as a function of the precision $d\theta$ of the θ measurement is:

$$d\eta = -\frac{1}{\sin\theta} d\theta \tag{22}$$

The resolution is better at large angles θ. For a good resolution dη at small angles it is necessary to have a very small error dθ. In order to have a feeling it is interesting to see some numerical values of η and θ :

η	0	1	2	3	4	5	6	7	8
θ degree	90	40	15.4	5.7	2.1	0.8	0.3	0.1	0.04
θ milliradian	1581	698	269	99	37	13.4	4.9	1.8	0.7

Large value of η, greater than 5, are accessible with nuclear emulsions. **Figs. 6 and 7** are the pseudorapidity distribution, in the CMS of the collision, of two very high energy and very high multiplicity cosmic-ray events obtained by the JACEE Collaboration [15]. The event of **Fig. 6** was interpreted as a collision Ca + C (or 0), at the energy of ~ 100 TeV/nucleon; its multiplicity is He + (760 ± 30) Ns + > 300 photons, where N_s is the number of minimum ionizing particles. The event of **Fig.**

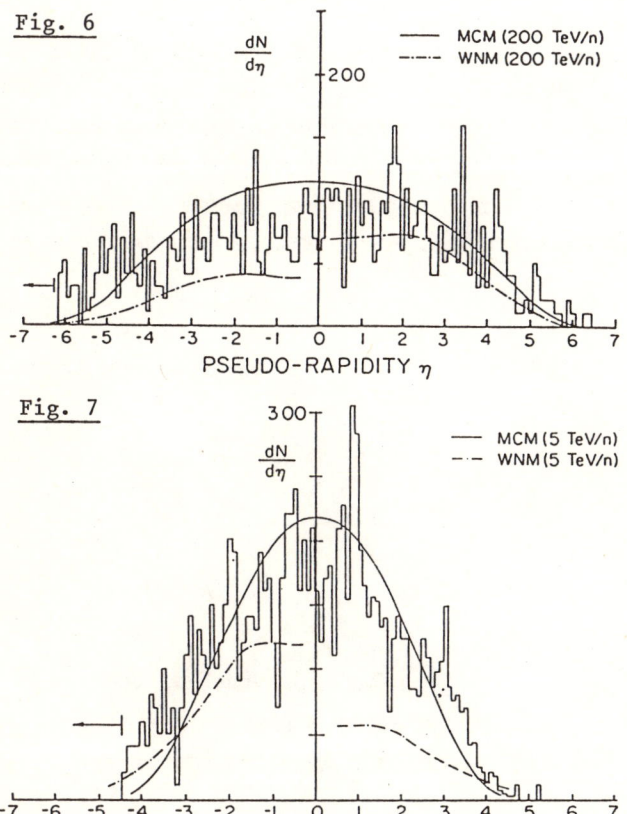

Figs. 6 and 7
CMS pseudorapidity distributions of charged particles in the Ca+C and Si+Ag Br cosmic ray events, respectively, found by the JACEE Collaboration [5].

7 was interpreted as a collision Si + Ag Br, at an energy of ~ 4 TeV/nucleon; its multiplicity is 5 N_h + (1010 ± 30) N_s + > 170 photons, N_h being the number of heavily ionizing particles.

Exercise 3 - Show that in the collision of a nucleon with momentum 200 GeV/c and a nucleon at rest, the rapidity of the CM in the LS is $y_{cm} = 3$.

Exercise 4 - Show that in the collision of a 200 GeV/nucleon oxygen nucleus (the 16 nucleons taken as one particle) with 50 nucleons of a heavy target (50 nucleons taken as one particle) the rapidity of the CM in the LS is $y_{cm} = 2.5$. Mass of the nucleon = 0.94 GeV/c^2.

Exercise 5 - Consider a detector with a conical symmetry around the beam axis, with its edges pointing to the target; a single particle to be detected must be emitted from the target at an θ between the values θ_{min} and θ_{max} as shown in *Fig. 8*. Assume that a particle A which is created in the target decays in a very short time inside the target into two muons : $A \rightarrow \mu_1 + \mu_2$. In the rest system of the parent particle A the two muons have equal and opposite momenta and one of them is emitted at an angle θ^* relative to a fixed axis, for instance the beam axis, the other is emitted at an angle $\theta^* + \pi$ *(Fig. 9)*. This angle θ^* follows some distribution law $f(\theta^*)$. The condition to detect particle A is to have both muons entering into the detector at angles between θ_{min} and θ_{max}. The geometrical acceptance of the detector is defined as the ratio of the number of particles A which are detected to the number of particles A which are produced and decay in the

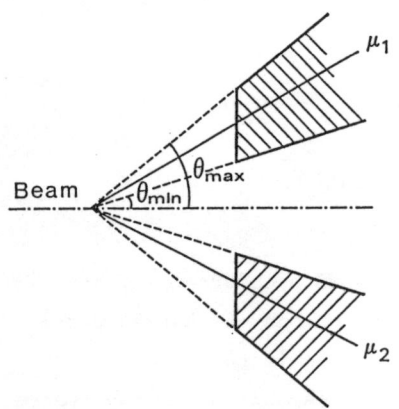

Fig. 8. Angular acceptance of the detector of exercise 5.

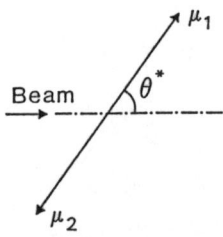

Fig. 9. Rest frame of the parent particle A which decays into two muons in exercise 5.

target. Compute the geometrical acceptance of a detector with $\theta_{min} = 100$ mrad, $\theta_{max} = 260$ mrad, for a particle A with mass $M = 3$ GeV/c² and momentum $P = 20$ GeV/c in the two cases :
   case 1 - angular distribution $f(\theta^*)d\theta^* = (1 + \cos^2\theta^*)d\theta^*$
   case 2 - angular distribution $f(\theta^*)d\theta^* = \sin\theta^* d\theta^*$.
(the answers are 0.17 and 0.44 in cases 1 and 2, respectively).

3. FROM HADRONIC MATTER TO QUARK-GLUON PLASMA

3.1 DENSITY OF NUCLEONS IN NORMAL NUCLEAR MATTER

Let us recall that the density of nucleons in normal nuclear matter is ρ_0 = A nucleons/nuclear volume = A nucleus/$(4/3 \pi R^3)$ fm³ where $R = r_0 A^{1/3}$ and $r_0 \approx 1.2$ fm. Therefore,

$$\rho_o \approx 0.14 \frac{\text{nucleons}}{\text{fm}^3} \qquad (23)$$

3.2 ENERGY DENSITY OF NORMAL NUCLEAR MATTER

The energy density of a nucleus of mass number A under normal conditions, ε_A, is given by the ratio of the rest mass of all nucleons to the nuclear volume :

$$\varepsilon_A = \frac{A \times \text{nucleon mass}}{\text{nucleus volume}} \approx \frac{A \times 940 \text{ MeV}}{\frac{4}{3} \pi r_o^3 A \text{ fm}^3}$$

or $\quad \varepsilon_A \approx 130 \frac{\text{MeV}}{\text{fm}^3} \qquad (24)$

Taking the radius of the proton as ≈ 0.8 fm, the energy density of a proton is $\varepsilon_p \approx 440$ MeV/fm³.

3.3 RELATIONSHIP BETWEEN ENERGY AND TEMPERATURE

This relationship is given by the Boltzmann constant, $K = 1.381 \times 10^{-16}$ erg/°K. Since 1 erg = 0.625×10^6 MeV, we have:

$$1 \text{ MeV} \approx 1.2 \times 10^{10} \text{ K}$$

3.4 TIME CORRESPONDING TO ONE FERMI

The time needed for a particle with the speed of light to travel the distance of 1 fermi is

$$t = \frac{1 \text{ fm}}{c} = \frac{10^{-13} \text{ cm}}{3 \times 10^{10} \text{ cm sec}^{-1}} = 3.3 \times 10^{-24} \text{sec}$$

Often one refers to this time as "one fermi".

3.5 DECONFINEMENT

We know that quarks and gluons are confined inside the hadrons. Within the standard model of strong interactions, which so far works very well, the interaction between two hadrons occurs via the interaction between the quarks of one and the quarks of the other, i.e., the exchange of gluons between the quarks. Since the quarks are confined, this means that the gluons must cross the confinement barrier, otherwise there would be no interaction.

Quantum chromodynamics predicts that at extreme high conditions of energy density and/or temperature there should be a deconfinement of quarks and gluons, and the hadrons should undergo a phase transition to a quark-gluon plasma (QGP). The QGP should occupy an extended volume relative to the nucleon volume during a short time, in the scale of the fermi. If a nucleus, which normally has a nucleon density $\rho_0 \approx 0.14$ nucleons/fm³ and an energy density $\varepsilon_A \approx 0.13$ GeV/fm³, is put into a state in which the nucleon density becomes $\rho \approx (10 \text{ to } 15) \rho_0$ and the energy density becomes $\varepsilon \approx (15 \text{ to } 25) \varepsilon_A \approx (2 \text{ to } 3)$ GeV/fm³, or into a state in which the temperature is higher than a critical value T_C, then deconfinement should occur and the hadronic matter should undergo a phase transition into a plasma of free quarks and gluons. This is illustrated in **Fig. 10**. Important theoretical considerations about the phase transition are given in the lectures by Alcock [1] and Parikh [3] at this school. See also Ref. [5, 16].

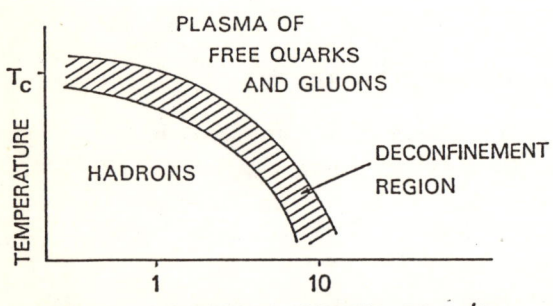

Fig. 10. Phase diagram of strongly interacting matter, with the transition from hadrons to QGP.

3.6 HOW TO OBTAIN EXPERIMENTALLY A HADRONIC STATE WITH HIGH ENERGY DENSITY OR HIGH TEMPERATURE ?

The only way we know today of obtaining hadronic states with high energy density and/or high temperature is with ultrarelativistic nucleus-nucleus collisions.

A first generation of experiments has been done in the last four years; three at Brookhaven, experiments E802, E810 and E814, and at CERN six large ones, NA34, NA35, NA36, NA38, WA80 and WA85, and eight emulsion experiments. At Brookhaven the AGS accelerates ^{32}S and ^{28}Si ions at 14.5 GeV/nucleon; at CERN the SPS accelerates ^{16}O and ^{32}S ions at 200 GeV/nucleon, which is about 14 times larger than the energy of the Brookhaven beams. In central collisions with heavy targets, energy densities of about 2 to 3 GeV/fm^3 should be achieved at CERN, which in principle should be about enough to produce the QGP.

Because of the nuclear transparency, the energy density achieved at Brookhaven is not 14 times smaller than at CERN. At 14.5 GeV/nucleon the stopping power is larger than at 200 GeV/nucleon and the efficiency to release energy in the collision is larger. The E814 Collaboration [17] estimated that an energy density of approximately 0.8 GeV/fm^3 is released in collisions of 10 GeV/nucleon Si ions with Al, Cu and Pb targets. The energy density obtained at Brookhaven is therefore about 1/3 to 1/4 of that reached at CERN, and not 1/14.

As we shall see in the course of these lectures, none of those experiments can clame to have detected unambiguously the QGP. However, since the field of nucleus-nucleus collisions at those high energies is quite a new one, very useful experimental information has been obtained, which may be used as guide lines for the future. Furthermore, some of the experiments have made quite a technological achievement by operating successfully in a very dense radiation environment. An example is given in **Fig. 11**, showing a streamer chamber picture of secondary particle trajectories of a ^{32}S+Au interaction at 200 GeV/nucleon, obtained by the NA35 Collaboration [18]. At this energy the average charged multiplicity, including electrons coming from π°'s via γ-rays, is about 500 particles, which is not easy to handle.

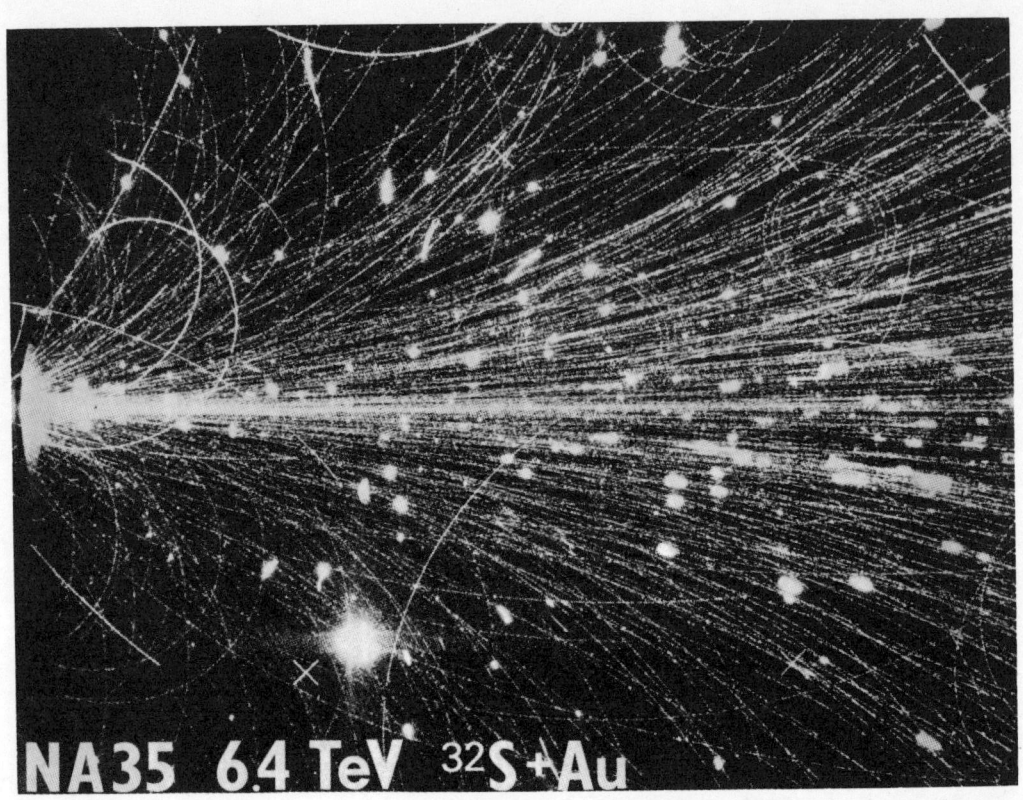

<u>Fig. 11</u>. NA35 Collaboration streamer chamber picture of the charged secondary tracks of a ^{32}S+Au collision at 200 GeV/nucleon.

3.7 CENTRAL AND PERIPHERAL COLLISIONS, CENTRAL AND FRAGMENTATION REGIONS

In the center of mass system of the two colliding nuclei they look like two Lorentz contracted pancakes. We distinguish between peripheral and central collisions. In a <u>peripheral</u> collision the two nuclei glance each other, with an impact parameter b which can be as large as about the sum of the two radii; some nucleons of both nuclei do participate in the interaction and others do not, remaining as spectators (Fig. 12a). In a <u>central</u> collision the impact parameter is zero or close to zero: if the two nuclei have different radii the smaller one makes a hole in the larger and all nucleons of the smaller can, in principle, participate in the interaction (Fig. 12b). In this case only the larger nucleus has spectator nucleons at the border.

In the collision the two nuclei overlap (partially in a peripheral collision, totally in a central collision) and a volume of hot hadronic

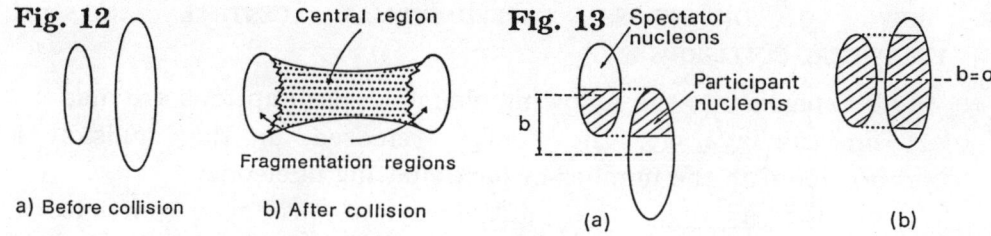

Fig. 12. a) Peripheral collision, b) central collision in the CMS of two colliding nuclei.

Fig. 13. a) Two Lorentz contracted nulclei before collision, b) the central and the fragmentation regions after the collision.

matter is created in which we usually consider two regions : a central region, made mainly of mesons created in the collision, and two fragmentation regions, which should be a mixture of fragments of the two initial nuclei with mesons, and therefore, rich in nucleons (Fig. 13). The idea of a fragmentation region was introduced by Yang, for the study of multiparticle production [19]. Some experimental support to the idea of central and fragmentation regions comes from the rapidity distribution of mesons in the C.M.S. of very high energy proton-proton collision, as at the CERN ISR, for example; the most probable rapidity of the pions produced in the interaction is close to zero and symmetrical around zero, suggesting that the pions come mainly from the central region. There is no sure experimental handle to help computing what can happen in the fragmentation regions.

There is an enormous lack of experimental information about what the central and the peripheral regions are made of. As examples, the average proportions of the different mesons and the particle density per unit volume are unknown in nucleus-nucleus collision. This lack of experimental information is a very serious difficulty which we have to face when making models about what particles should come off from nucleus-nucleus interactions. Striking examples will be given in section 8 with models of J/ψ absorption.

It is important to distinguish central and peripheral collisions because of the density of energy released. This density is small in a peripheral collision and is usually large in a central collision. However, due to the nuclear transparency there are central collisions in which a small energy is released. Because central collisions give the maximal possible enhancement to the temperature and energy density it is in these collisions that the search for the QGP is made.

3.8 HOW TO DISTINGUISH EXPERIMENTALLY CENTRAL AND PERIPHERAL COLLISIONS ?

In all experiments the following plausible assumptions are made:
1) on the <u>average</u>, the energy released in the collision is proportional to the number of participating nucleons;
2) on the <u>average</u>, the multiplicity of produced particles is proportional to the number of participating nucleons;
3) as a consequence of 1) and 2) on the <u>average</u> the multiplicity of produced particles is proportional to the energy released in the collision.

The last point has been verified in several experiments (NA36, WA80, WA85, E802).

We distinguish between central and peripheral collisions by measuring the energy released or the multiplicity of the secondary particles, or both.

3.9 COHERENT AND INCOHERENT INTERACTIONS

Some collisions of two nuclei are coherent : each nucleus interacts as a whole, there is a collective effect of the nucleons and the collision is not a series of independent nucleon-nucleon interactions. Other collisions are <u>incoherent</u> : there is no collective effect, the collision is a succession of independent nucleon-nucleon interactions.

In the experiments looking for the QGP it would be important to measure the degree of coherence of the collisions, i.e., the percentage of collisions which are coherent, because, for the same beam energy, the total energy in the center of the mass is not the same for coherent and incoherent collisions. In order to have an idea of the order of magnitude of the difference, let us assume a <u>central</u> collision of an ^{16}O nucleus with a ^{197}Au nucleus. The 16 nucleons of the ^{16}O will produce a hole in the center of the ^{197}Au nucleus in which there will be about 50 nucleons. If the collision is entirely incoherent there will be 16 center of mass systems of 16 independent nucleon-nucleon interactions. If the ^{16}O-beam has 200 GeV/nucleon, in each nucleon-nucleon CMS the energy, including the masses, will be $\sqrt{s} = 19.4$ GeV and the kinetic (available) energy will be 17.5 GeV; in the 16 of them the total energy will be $\sqrt{s} \sim 310$ GeV and the kinetic energy ~ 280 GeV. On the other hand, if the 16 nucleons of the ^{16}O interact coherently, as a single object, with the 50 nucleons of ^{197}Au also behaving as a single object, the total energy in the CMS of the 16+50 nucleons will be $\sqrt{s} \sim 550$ GeV and the kinetic energy ~ 488 GeV.

If the ^{197}Au nucleus interacts as a single object whose mass is 197 times the nucleon mass, $\sqrt{s} \sim 1120$ GeV and the kinetic energy ~ 920 GeV.

Exercice 6 - *Show that $\sqrt{s}$ and the kinetic energies in the CMS are indeed 310 and 280, 550 and 488, 1120 and 920 GeV, respectively, in the three cases given above. Mass of the nucleon = 0.94 GeV/c².*

3.10 THEORETICAL MODELS

There are models for particle production either in coherent or in incoherent collisions.

3.10.1 Coherent collisions.

The theoretical models for coherent collisions are either thermodynamical or hydrodynamical, or both. Usually thermodynamics takes into account the exchange of energy and temperature, whereas hydrodynamics takes into account mechanical motion of expansion and compression of the hadronic matter.

Only the models of coherent collisions allow an estimate of the energy density of the hot hadronic matter, under certain assumptions; the models of incoherent collisions do not allow. In the analysis of some experiments we sometimes face a hybrid situation : the energy density is estimated with a model of coherent collisions and the kinematics of the secondary particles are compared with models of incoherent collisions.

We do not know how to compute exactly the energy density in terms of quantities which are experimentally measurable. The most commonly used estimate is one proposed by Bjorken with a hydrodynamical model [20]. Other approaches have been made by Belenski and Landau in the past [21], and by Roberts recently [22].

<u>Bjorken's model</u>. In the collision of two <u>identical</u> nuclei of radius R let us consider a slice in the rapidity space of thickness Δy and let us call :

N = number of particles contained in this slab

$\langle E \rangle$ = average energy of the N particles.

The energy E contained in the slice is:

$$E = N \frac{d\langle E \rangle}{dy} \Delta y$$

Let us consider a volume V determined by Δy and the area of the nuclei transverse to the rapidity, at a proper time τ :

$$V = \tau \Delta y \pi R^2$$

The energy density is
$$\epsilon = \frac{E}{V}$$

or
$$\epsilon = N \frac{d\langle E\rangle}{dy} \frac{1}{\tau \pi R^2} \qquad (25)$$

which is Bjorken's estimate.

An interesting approximation can be made for the <u>central</u> region. In this region the longitudinal momenta are small and the following approximation holds:
$$N \frac{d\langle E\rangle}{dy} \approx \frac{dN}{dy} \langle M_T\rangle$$

where $\langle M_T\rangle$ is the average transverse mass of the secondary hadrons. The energy density is, therefore,
$$\epsilon \approx \frac{dN}{dy} \frac{\langle M_T\rangle}{\tau \pi R^2} \qquad (26)$$

which is Bjorken's estimate for the central region. It shows that the energy density in this region is proportional to the rapidity distribution of the secondary hadrons.

<u>Landau's model</u>. This model considers the central collision of two <u>identical</u> nuclei of mass M and radius R. The fundamental assumption of the model is that in the CMS of the collision the two nuclei would be <u>stopped</u>; all the kinetic energy would be used. It takes into account the relativistic contraction in the longitudinal direction. If E_{cm} is the kinetic energy (available energy) in the CMS, the energy density in this model is:

$$\epsilon = \frac{2\, E_{cm}^2}{\frac{4}{3}\pi R^3 M} \qquad (27)$$

For ultrarelativistic ions, the assumption that they stop is not physical. This leads to unrealistic, too high energy densities.

3.10.2 Incoherent collisions

The two most popular models of particle production in incoherent hadron-hadron collisions are the LUND MODEL [23] and the DUAL PARTON MODEL [24,25]. They originated from the old multiperipheral models. In both models the production of particles in hadron-hadron collisions is a two-step process:

step 1 - the two hadrons are excited;

step 2 - constituents of the two excited hadrons exchange a string of quarks-antiquarks, from which hadrons originate.

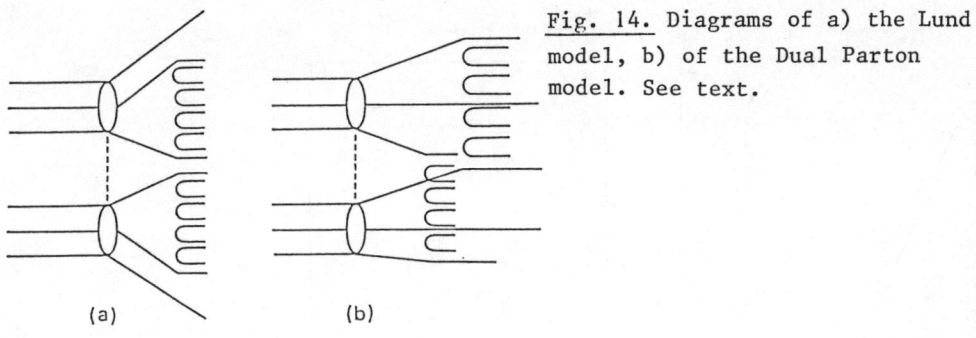

Fig. 14. Diagrams of a) the Lund model, b) of the Dual Parton model. See text.

There is a difference between the two models. In the Lund model the strings are formed between the constituents of the same hadron; therefore there is no exchange of colour between the colliding hadrons (**Fig. 14a**). In the Dual Parton model the string is formed between the constituents of the two colliding hadrons which, therefore, exchange colour (**Fig. 14b**).

It should be emphasized that both models were made for <u>soft interactions</u>, i.e. small P_T interactions, or in other words, for small momentum transfer between two successive components of the string. This is not always the case in ultrarelativistic nucleus-nucleus collisions.

Both models represent well experimental data obtained in soft interactions. As examples, in **Fig. 15** the invariant differential cross sections as a function of P_T for the production of negative and positive particles in proton-proton interactions at $\sqrt{s} = 53$ GeV, obtained at the CERN ISR [26], are compared with the prediction of the Lund model; the exponential decrease of the cross section is well represented by the model. In Fig. 16 the multiplicity distribution of negative particles in α-α interactions at $\sqrt{s} = 31$ GeV [27] is compared with the prediction of the Dual Parton model; the agreement is very good.

Both models are being adapted to take into account also hard scattering. As we shall see in the course of these lecture, experimental results obtained in high energy nucleus-nucleus collisions have been compared with both these models. We can ask, however, how surely can we extrapolate from nucleon-nucleon to nucleus-nucleus collisions. Both models have been put into Monte Carlo form. The Monte Carlo version of the Lund Model is called FRITIOF [28], of which there are several versions. There are three Monte Carlo versions of the Dual Parton model, called IRIS [29], VENUS [30] and DUAL MULTICHAIN MONTE CARLO [31].

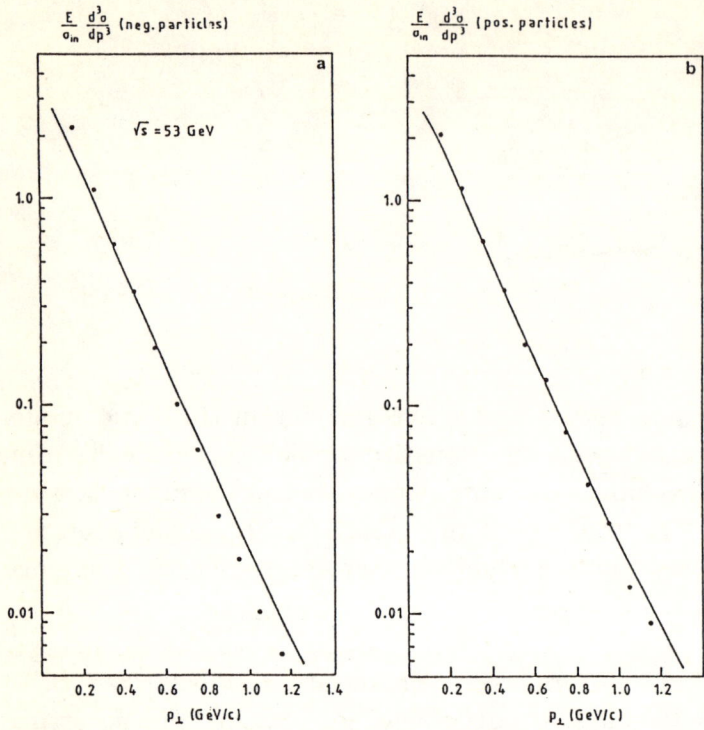

Fig. 15. Invariant differential cross sections versus P_T for the production of positive and negative particles in proton-proton interactions at $\sqrt{s}$ = 53 GeV [26], compared with the predictions from the Lund model, given by the solid lines.

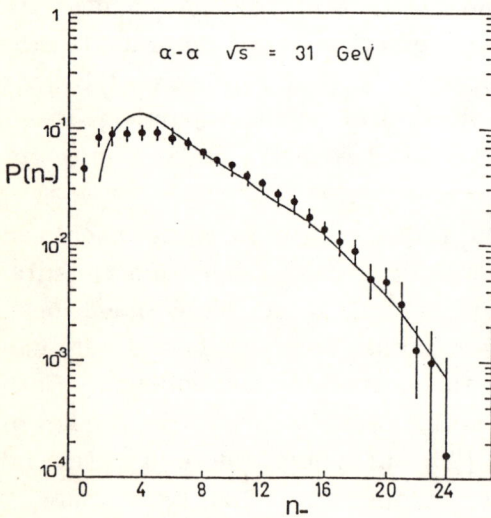

Fig. 16. Multiplicity distribution of negative particles produced in $\alpha - \alpha$ interactions at $\sqrt{s}$ = 31 GeV [27], compared with the prediction from the Dual Parton model, given by the solid film.

3.11 Signatures of the quark-gluon plasma

We shall review in sections 5 to 11 the experimental status of the proposed signatures of the quark-gluon plasma:

- P_T distributions
- Direct photons production
- Suppression of the J/ψ production
- Bose-Einstein interferometry
- Fluctuations in rapidity distribution and intermittency
- Production of strange particles.

4. THE EXPERIMENTS

We shall briefly outline in this section the main characteristics of the experiments whose results are discussed in these lectures. In order to make clear what the different detectors do, we shall present a summary of what can be measured in an experiment.

4.1 WHAT DO WE MEASURE IN AN EXPERIMENT ?

1 - <u>Energy</u>. In order to define the centrality of the interaction we measure either the electromagnetic or the hadronic energy with an electromagnetic or a hadronic calorimeter, respectively. All experiments performed at Brookhaven and at CERN have calorimeters.

As we have seen in section 3.10, the energy density of the interaction is related to the <u>transverse energy</u>. This quantity is extremely important because it is directly related to the violence of the collision. The highest transverse energies correspond to the most violent central collisions. It is in such events that we must look for the production of the QGP.

Calorimeters are usually divided into small modules, each module measuring a fraction of the energy. If a module i sees the target at an angle θ_i and measures the energy ΔE_i, the corresponding transverse energy is $\Delta E_{iT} = \Delta E_i \sin \theta_i$. The overall transverse energy is:

$$E_T = \sum_i \Delta E_i \sin \theta_i \qquad (28)$$

Tannenbaum [4,9] describes in detail the fundamental aspects of calorimetry.

2. <u>Multiplicity</u> of the secondary particles is measured either with visual detectors of tracks or with electronics devices. As examples of the

former there are streamer chambers, as in NA35 experiment [18, 32], or TPC's as in the NA36 [33] and E810 [34] experiments, or nuclear emulsions [35-38]. As examples of electronics devices there are tubular counters, like streamer tubes as in the WA80 experiment [39] or proportional tubes as in the E802 experiment [40], and silicon detectors, as in NA34 [41] and WA85 [42, 43] experiments.

3. <u>Particle momentum and sign of electrical charge.</u> This is usually measured with the trajectory of the particle in a magnetic field. If p is the particle momentum, Ze its electrical charge, B the field induction and R the radius of the trajectory, the Lorentz force on the charge gives :

$$p(GeV/c) = 0.3 \; B(Telsa) \; Z \; R(m)$$

4. <u>Particle trajectory</u>. It is usually given by some of the visual detectors of multiplicity mentioned above, or by a series of multiwire proportional chambers or drift chambers.

5. <u>Particle velocity</u>. It is usually measured with Čerenkov counters.

6. <u>Time of flight</u>. It is the time the particle takes to travel a distance between two points, where two fast counters are placed.

7. <u>Energy and direction of a single photon</u>. This can be measured with a segmented (modular) electromagnetic calorimeter. It was done in experiment WA80 [39].

A good survey of several of these techniques was made by Brockmann at this school [2].

4.2 THE E802 EXPERIMENT

The layout of the Brookhaven E802 experiment [40] is shown in **Fig. 17**. There is a target multiplicity array (TMA) made of tubular counters operated in proportional mode, a lead glass electromagnetic calorimeter (PBGL) and a zero-degree hadronic calorimeter (CAL); these three devices define central and peripheral collisions. Particle tracking is made with two sets of drift chambers (T_1, T_2) placed before and two sets (T_3, T_4) placed after a magnet. There is a powerful particle identification system, consisting of 160 slat plastic scintillator time of flight wall (TOF), a 96 segment aerogel Čerenkov counter (AEROČ), a 40 segment high pressure gas Čerenkov counter (GASČ) and an array of three gas Čerenkov counters (ČC) with their associated scintillators (S_1, S_2) and tracking chambers (T5-T7).

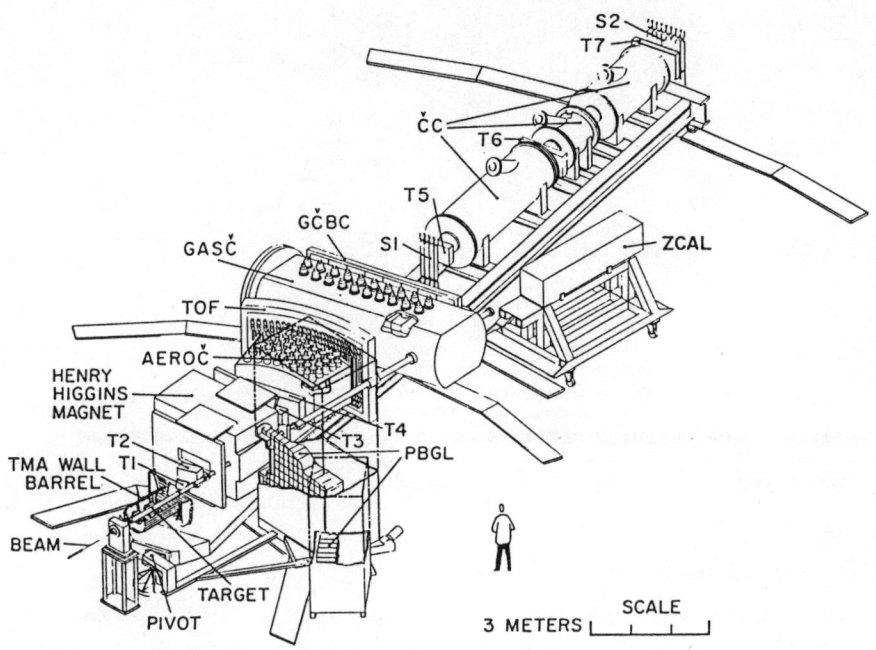

Fig.17. Layout of the Brookhaven E802 experiment.

4.3 THE WA80 EXPERIMENT

The layout of this detector is shown in **Fig. 18** [39]. It has a multi-array of 655 plastic scintillators named Plastic Ball, which allows to identify baryons and to measure their energy in the target region of $-1.7 \leq \eta \leq 1.3$. There are three calorimeters. A uranium-scintillators Zero-Degree Calorimeter (ZDC), placed at the end of the detector along the beam line, defines central and peripheral collisions; when the collision is central, little remains of the beam and the ZDC detects a small energy; when the collision is peripheral it measures the large energy which the beam still carries. An electromagnetic calorimeter called SAPHIR, made of 1280 lead glass modules measures the energy and direction of single photons. A hadron calorimeter called MIRAC (Mid Rapidity Calorimeter) with a large coverage measures the transverse energy flux. Multiplicities of charged particles are measured by sets of Iarocci streamer tubes, which cover a wide range of pseudorapidity $-1.7 \leq \eta \leq 4.4$, called LAM (Large Angle Multiplicity detector), SAM (Single Arm Multiplicity detector) and MIRAM (Mid Rapidity Multiplicity detector), respectively.

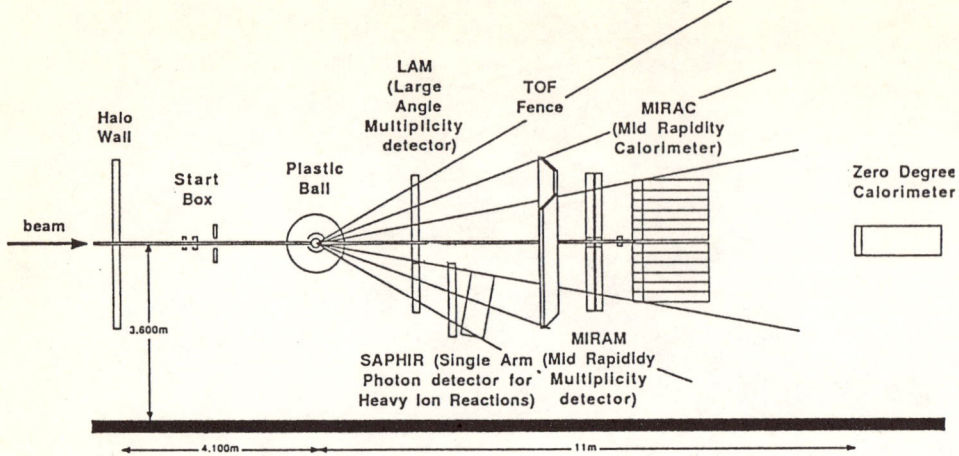

Fig.18. The CERN WA80 experimental setup.

4.4 THE WA85 EXPERIMENT

This experiment [42, 43] is installed at the Ω Spectrometer at CERN (**Fig. 19**). The target T is placed inside the Ω magnetic field. Tracks are measured by seven multiwire proportional chambers (MWPC) placed in the magnetic field and four others outside the field (MY1-MY4). The sensitive regions of the chambers and two hodoscopes HZ0 and HZ1 are matched to detect charged particles with transverse momentum above a certain minimum, which was chosen as ~ 0.6 GeV/c. There is a good detector of multiplicities, consisting of two arrays of 50 μm pitch silicon

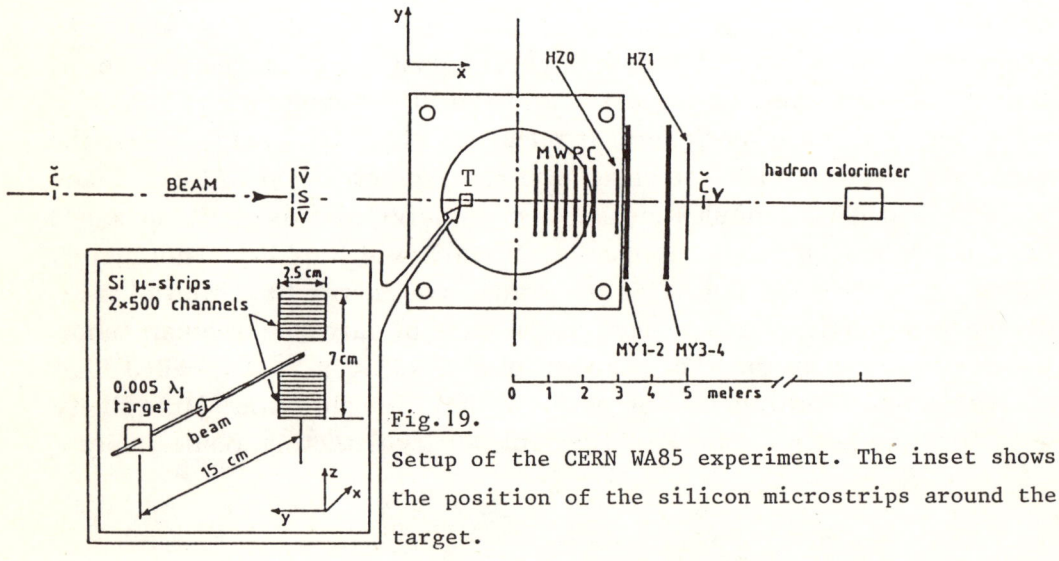

Fig.19. Setup of the CERN WA85 experiment. The inset shows the position of the silicon microstrips around the target.

microstrips, each with 512 channels, placed above and below the beam 15 cm downstream from the target; they measure multiplicities in the pseudorapidity range $2.1 \leq \eta \leq 3.4$. A hadron calorimeter placed along the beam line 25 m downstream of the target defines central and peripheral collisions (a zero-degree calorimeter).

4.5 THE NA34 EXPERIMENT

This detector [41] has two targets. One of them, target 1, is made of the wires of a multiwire proportional chamber; it is surrounded by a uranium-scintillators calorimeter (U/Sc) **(Fig. 20)**. There is along the beam line : a transition radiator detector (TRD) to identify electrons, target 2, a uranium liquid-argon calorimeter (U/LA), a uranium-scintillators calorimeter (U/Sc), followed by a Muon Spectrometer consisting of a magnet and multiwire proportional chambers (PC0 to PC6). Another part of the detector called External Spectrometer sees target 1 through a slit made in the front U/Sc calorimeter and was designed to detect photons. It consists of a multiwire proportional chamber followed by an iron plate for γ conversion (PC/Conv.) and two drift chambers (4 and 5) one on each side of a magnet. Photons produced in target 1, when crossing the iron plate, are converted into electron-pairs whose trajectories are curved in the magnetic field and detected by the drift chambers.

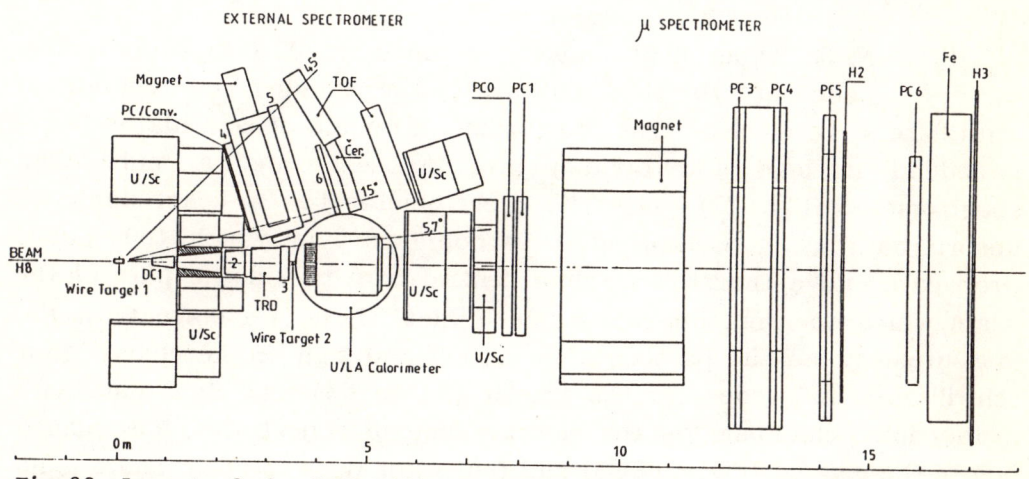

Fig.20. Layout of the CERN NA34 experiment.

4.6 THE NA35 EXPERIMENT

The NA35 detector [32] has a streamer chamber in a magnetic field, which allows the measurement of charged particles trajectories, momenta, sign of the charge and multiplicities (**Fig. 21**). It detects also secondary particles of some decays, among them the important Λ, $\bar{\Lambda}$ and K°_s decays. A veto calorimeter placed downstream along the beam line defines central and peripheral collisions (a zero-degree calorimeter). It has an electromagnetic calorimeter (PPD) and a hadronic calorimeter (Ring Cal + Intermediate Cal) both covering the pseudorapidity interval $2.2 \leq \eta \leq 3.9$.

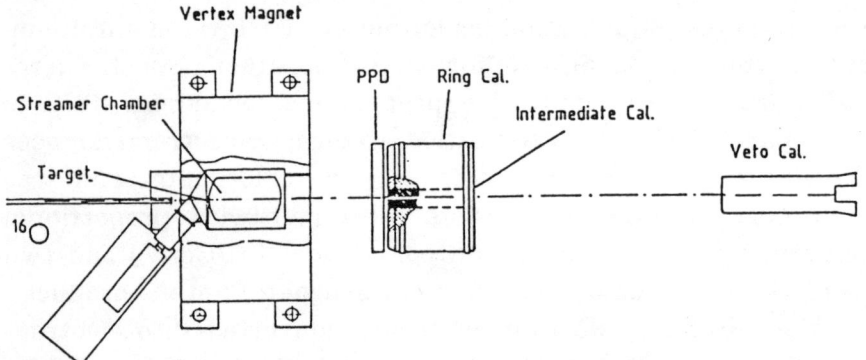

Fig.21. Experimental setup of the CERN NA35 experiment.

4.7 THE NA38 EXPERIMENT

The NA38 experiment consists essentially of a multiple active target made of 10 thin subtargets surrounded by cylindrical scintillators, an electromagnetic calorimeter made of scintillator fibers imbedded in lead and divided into 30 cells, and a multi-muon spectrometer (**Fig. 22**) [44]. The spectrometer has a beam dump to absorb hadrons, consisting of a carbon part in which the muons propagate and a tungsten-uranium central core to kill the part of the beam which does not interact in the target. There is a magnet for the measurement of the particle momentum and sign of the charge, four scintillators hodoscope for the trigger (R1 to R4) and eight multiwire proportional chambers for the measurement of trajectories, four placed before the magnet (PC1 to PC4) and four after (PC5 to PC8). There is no particle identification and no measurement of multiplicites. By definition a muon is a particle which leaves the target and reaches the hodoscope R4 at the far end.

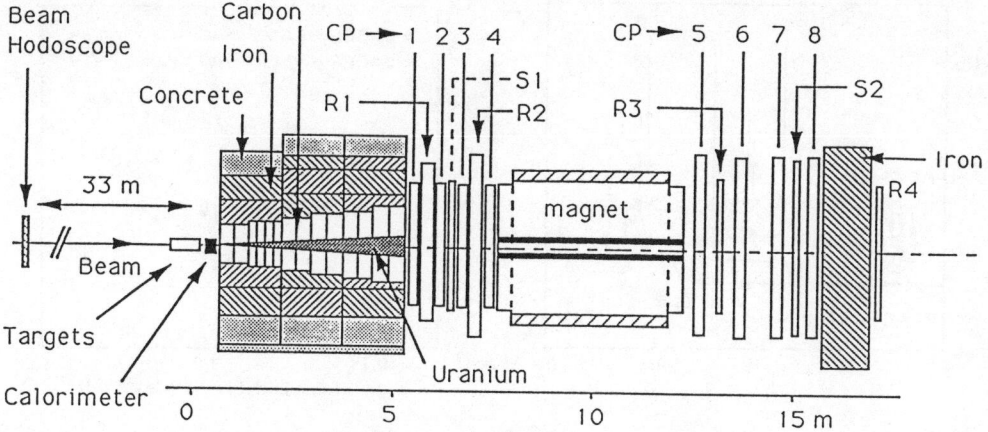

Fig.22. Layout of the CERN NA38 experiment.

5. TRANSVERSE MOMENTUM DISTRIBUTION AS A SIGNATURE OF QGP

In this section we shall start the review of the experimental status of the proposed signatures for the QGP.

5.1. THE AVERAGE P_T AT VERY HIGH ENERGIES

The JACEE Collaboration detected some very high energy cosmic-ray interactions with nuclear emulsions in which the <u>average</u> transverse momentum of the charged secondaries is significantly larger than the average values measured at lower energies [45]. Fig. 23 a) and b) show $\langle P_T \rangle$ versus the central rapidity density per unit colliding volume of nucleus-nucleus interactions $(dN/d\eta)/A^{2/3}$ and versus the energy density ε, respectively, of the JACEE events. Fig. 23a) shows also the results obtained by the UA1 Collaboration at the CERN antiproton-proton collider at $\sqrt{s}$ = 540 GeV. We see that, after a sort of a plateau, there is a rapid increase of $\langle P_T \rangle$ in the cosmic ray events at $\varepsilon \approx$ 2 to 3 GeV/fm^3. The average $\langle P_T \rangle$ does not go much above 1 GeV/c; but we should remember that most of the secondaries are pions, and for pions the average value 1 GeV/c is very high. The $\langle P_T \rangle$ for pions in most interactions is $\sim$ 0.34 GeV/c, as shown in Fig. 23a.

Van Hove [46] and Shuryak and Zhirov [47] have suggested that perhaps a rapid increase of $\langle P_T \rangle$ could be due to the formation of QGP. We should measure the average $\langle P_T \rangle$ as a function of the energy density, or some other quantity which should be proportional to the energy density, and see whether the trend sketched in Fig. 24 is

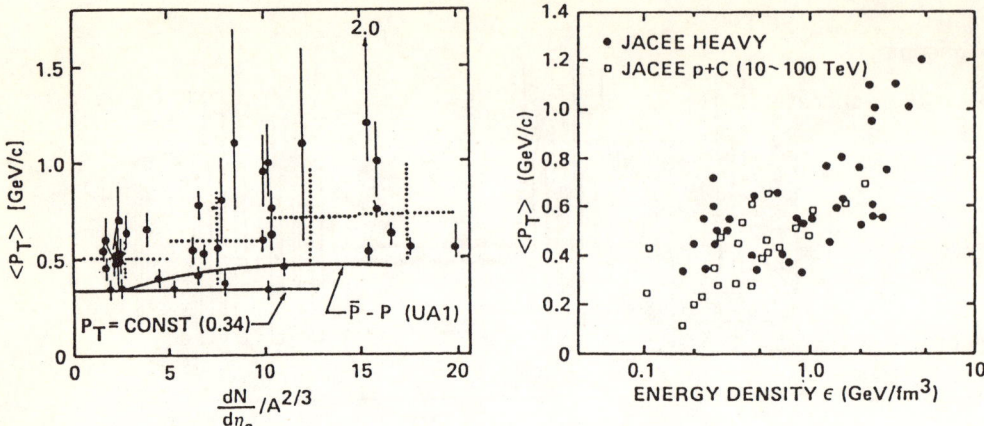

Fig. 23. Average transverse momentum of the charged secondaries obtained in the JACEE events [45]. See text.

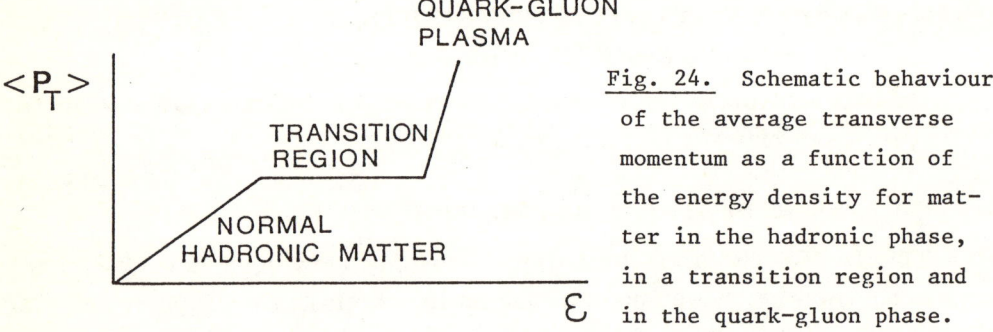

Fig. 24. Schematic behaviour of the average transverse momentum as a function of the energy density for matter in the hadronic phase, in a transition region and in the quark-gluon phase.

obtained; a small increase in $\langle P_T \rangle$ for small ε, corresponding to normal hadronic matter, a transition region in a sort of plateau and a rapid increase which would be due to the formation of the QGP.

Three experiments have results on $\langle P_T \rangle$: WA80, NA34 and NA38.

5.2 P_T DISTRIBUTION OF π°'s IN THE WA80 EXPERIMENT

The WA80 Collaboration measured the P_T distribution of π°'s produced in p + Au, O + C and O + Au interactions, in the three cases at 60 and 200 GeV/nucleon [48]. The direction and the energy of individual γ-rays were measured by the electromagnetic calorimeter SAPHIR (see **Fig. 17**). The number of π°'s was obtained from the γ-γ mass spectrum, as shown in **Figs. 25** for 200 GeV/nucleon ^{16}O+Au interactions and for 200 GeV p+Au interaction; the π° mass peak is clearly seen above a background. The interactions were divided into <u>central</u>, when the ratio of the energy measured by the zero-degree calorimeter to the beam

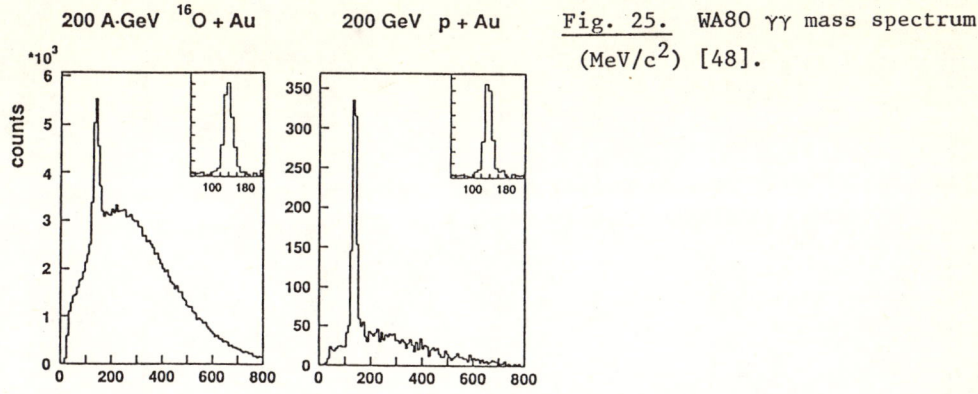

Fig. 25. WA80 γγ mass spectrum (MeV/c^2) [48].

energy was < 30%, and <u>peripheral</u> when that ratio was between 40% and 88%. The π°'s were always selected in the pseudorapidity range 1.5 ≤ η ≤ 2.1 in all the analyses described below. Besides this, a <u>minimum bias trigger</u> was defined by the requirements that less than 88% of the total beam energy should be measured by the zero-degree calorimeter and that at least one charged particle should be produced; such events are a mixture of central and peripheral collisions.

Two analyses have been made. In the first one, π° production in p+Au, 0+C and 0+Au collisions were compared at 60 and at 200 GeV/nucleon for minimum bias trigger events; the second was focused on p+Au and O+Au collisions at 200 GeV/nucleon. From the first analysis resulted the invariant differential cross section E d^{3}σ/dp^3 as a function of P_T which is shown in **Figs. 26** for 200 GeV/nucleon and for 60 GeV/nucleon. The data have been fitted to $(1/P_T) dN/dP_T \sim \exp(-P_T/T_0)$ in the transverse momentum range 0.8 GeV/c ≤ P_T ≤ 2 GeV/c. The resulting T_0 slope parameters are given in **Table 1**. We see that for these minimum bias events the values of T_0 are similar for protons and oxygen induced reactions and are not much sensitive to the target nucleus mass. However, T_0 obtained in 0+Au at 200 GeV/nucleon is at least 20% larger than the values obtained in proton-proton interactions at the

Table 1 - $T_0(MeV/c)$ obtained from the fit of the π° P_T spectrum for minimum bias trigger events by the WA80 Collaboration.

GeV/nucleon	p + Au	0 + C	0 + Au
60	179 ± 13	186 ± 8	200 ± 7
200	196 ± 8	190 ± 7	210 ± 5

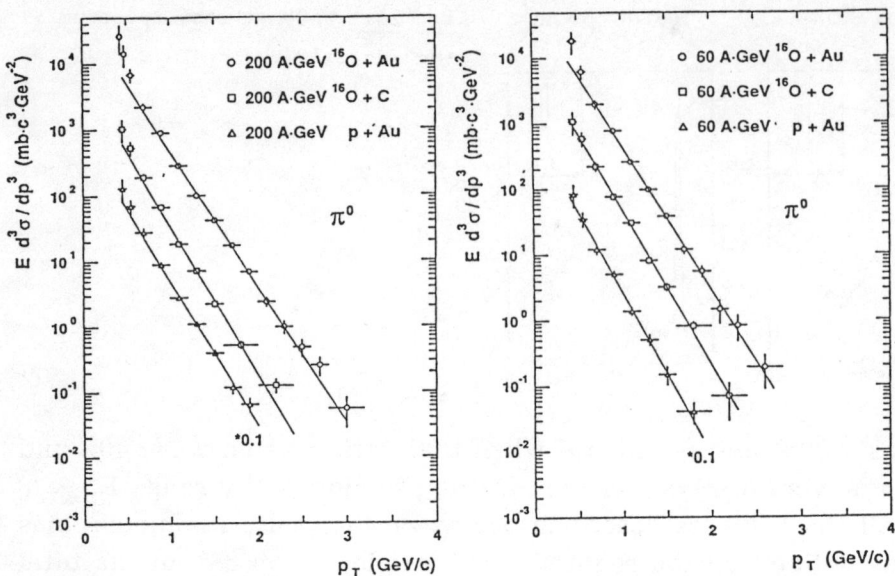

Fig. 26. WA80 invariant cross for π^0 production in O+Au, O+C and p+Au interactions at 200 GeV/nucleon and at 60 GeV/nucleon [48].

same energy and than the values predicted by FRITIOF. This might be due to imperfections in FRITIOF when proton-proton interactions are extrapolated to nucleus-nucleus collisions.

The P_T distributions of **Fig. 26** show a change in slope for $P_T \lesssim 0.8$ GeV/c, which is more pronounced for the heavy systems, being weakly indicated in the p+Au data. This change in slope is not new and is consistent with results from $\alpha + \alpha$ interactions at the CERN ISR [49] and with the results of the NA34 COllaboration which will be shown in the next section.

In the second analysis two comparisons of invariant differential cross sections were made, both at 200 GeV/nucleon. In one of them, shown in **Fig. 27**, the π^0 P_T spectrum of the <u>minimum bias events</u> in p+Au was compared to that of O+Au. The data were fitted to the same exponential function as in **Fig. 26**, in the same P_T range, 0.8 GeV/c $\leq P_T \leq 2$ GeV/c. The slope parameters for O+Au and p+Au are $T_0 = 210 \pm 3$ MeV/c and $T_0 = 196 \pm 4$ MeV/c, respectively, which are, of course, the same values obtained from **Fig. 26** (with slightly smaller errors). We can distinguish three regions in these spectra : one with $P_T \lesssim 0.8$ GeV/c, in which there is a change in the slope, already shown in **Fig. 26**; a second, in the range $0.8 \leq P_T \leq 2$ GeV/c where the fit was made; and a third with $P_T > 2$ GeV/c. In the third region the ratio of the cross sections

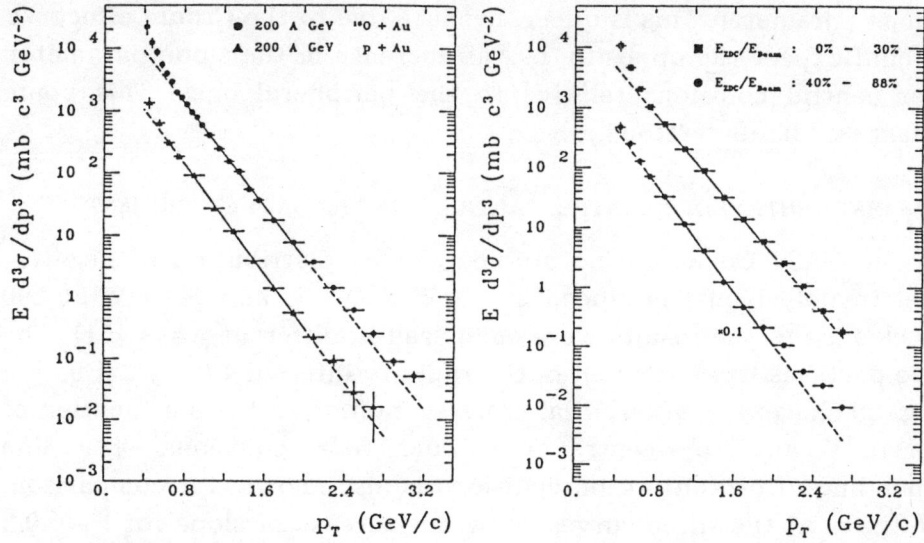

Fig. 27. WA80 invariant cross section for π^0 production in O+Au and p+Au collisions at 200 GeV/nucleon, for minimum bias events [48].

Fig. 28. WA80 invariant cross section for π^0 production in central (squares) and peripheral (dots) O+Au collisions at 200 GeV/nucleon [48].

O+Au to p+Au increases with P_T, reaching, for example, the value 1.6 at $P_T \sim 2.4$ GeV/c. This increase can be explained by the "CRONIN effect" [50] (an enhancement in the differential cross section as a function of P_T in p+nucleus compared to p+p interactions).

A second comparison at 200 GeV/nucleon was of <u>central</u> with <u>peripheral</u> collisions in O+Au events. This is shown in **Fig. 28**. We distinguish the same three regions in P_T as in the previous spectra. From a fit to the exponential function in the P_T range 0.8−2 GeV/c the slope parameters T_0 for central and peripheral collisions were found to be 220 ± 5 MeV/c and 189 ± 5 MeV/c. The conclusions from this comparison are :

a) The decreasing in slope parameter for $P_T \lesssim 0.8$ GeV/c is seen in the central collisions but not in the peripheral collisions data.

b) In the region 0.8 GeV/c $\leq P_T \leq$ 2 GeV/c the slope parameter T_0, measured with the small error of 5 MeV/c, is significantly larger for central than for peripheral collisions.

c) For $P_T \gtrsim 1.8$ Ge/Vc the spectrum for central collision events continues with the same slope as in the lower P_T region; whereas for peripheral collisions the spectrum shows a clear increase in the slope parameter. This is queer, because due to the Cronin effect we would expect the opposite, i.e., an increase in the slope parameter for central collisions relative to the peripheral ones. This point must still be understood.

5.3 P_T DISTRIBUTION OF NEGATIVE PARTICLES IN THE NA34 EXPERIMENT

The NA34 Collaboration studied the P_T distribution of negative particles (mainly pions) produced in p + W, ^{16}O + W and ^{32}S + W at 200 GeV/nucleon, and the results were compared in different ways [51]. The negative particles were selected in the rapidity interval $1.0 < y < 1.9$.

Fig. 29 shows the differential cross section, $d\sigma/dP_T^2$, as a function of P_T, for p-W and for <u>central</u> O+W and S+W collisions. The line corresponding to proton-proton collisions is also shown, for comparison. We see that : a) the three curves show an increase of slope for $P_T \lesssim 0.3$ GeV/c, analogous to what was observed by the WA80 Collaboration in

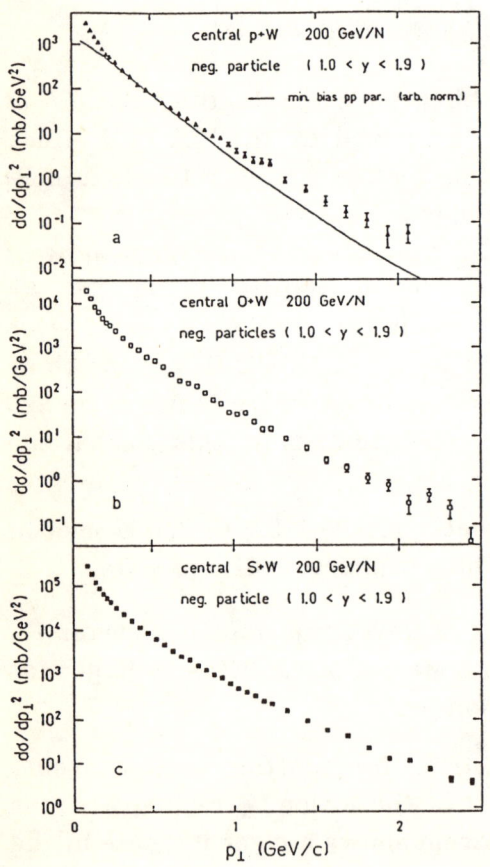

Fig. 29. NA34 differential cross section for negative particles produced in p+W and central O+W and S+W at 200 GeV/nucleon. The solid line in a) corresponds to p+p interactions (51).

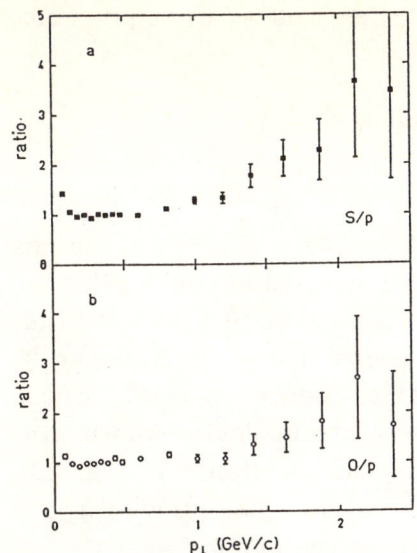

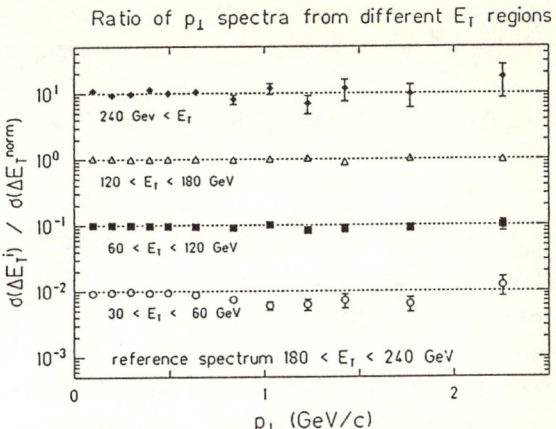

Fig. 30. NA34 ratios of the P_T distributions of negative particles in central S+W and O+W interactions to p+W interactions at 200 GeV/nucleon (51).

Fig. 31. NA34 ratios of the P_T spectra of negative particles in different E_T bands to the spectrum in the E_T reference band $180 < E_T < 240$ GeV (51).

the $\pi^°$ P_T-distribution for $P_T \lesssim 0.8$ GeV/c; b) the p+W data show an increase in the cross section relative to p+p for $P_T \gtrsim 1.0$ GeV/c, in agreement with the Cronin effect; c) the O+W and S+W data resemble closely to the p+W data.

Fig. 30 shows the ratios of the P_T distributions in <u>central</u> S+W and O+W interactions to p+W interactions, normalized to unity at small P_T. We see an enhancement for $P_T \gtrsim 1$ GeV/c; it could be due to the Cronin effect.

The P_T distribution was also studied as a function of the transverse energy of the interaction. The events were divided into transverse energy bands. Ratios of the P_T spectra at the different bands to the spectrum at one given band, arbitrarily taken as reference, were computed. **Fig. 31** shows the ratios as function of P_T, for S+W interactions, in which the reference E_T band is $180 < E_T < 240$ GeV. We see that the distributions of the ratios are flat; no E_T dependence of the P_T spectrum was detected.

The conclusions of the NA34 Collaboration study of the P_T distributions of negative particles produced in p+W, O+W and S+W interactions at 200 GeV/nucleon are that no abnormal effects were detected in the P_T spectra or in their ratios, as well as no dependence on the transverse energy of the interaction.

5.4 AVERAGE TRANSVERSE MOMENTUM OF THE J/ψ IN THE NA38 EXPERIMENT

The NA38 Collaboration studied the J/ψ production in O+U, O+Cu and S+U interactions at 200 GeV/nucleon. The J/ψ was detected in the pseudorapidity range $2.8 < \eta < 4$ in its $\mu^+\mu^-$ decay mode. We shall come back to this subject in detail in section 7. Here we shall just mention that the average transverse momentum of the J/ψ was calculated from the P_T distribution as a function of the transverse energy released in the collision. This energy was measured by the electromagnetic calorimeter in the pseudorapidity range $1.7 \leq \eta \leq 4.2$ (see section 4.7, **Fig. 22**). Preliminary result on $< P_T >$ of the J/ψ as a function of E_T for O-U collisions is shown in **Fig. 32** [52], without correction for acceptance. There is a slight enhancement of the average $< P_T >$ as E_T increases, as is usually seen in hadron production, corresponding to the first part of the curves shown in **Figs. 23 and 24**. The highest $< P_T >$ value, corresponding to $E_T \approx 100$ GeV is about 25% higher than the lowest at $E_T \approx 20$ GeV. There is no evidence for a rapid increase in $< P_T >$ of the type detected in the JACEE events.

The conclusion from this preliminary study of the NA38 Collaboration on the average transverse momentum of J/ψ is that nothing unusual was detected.

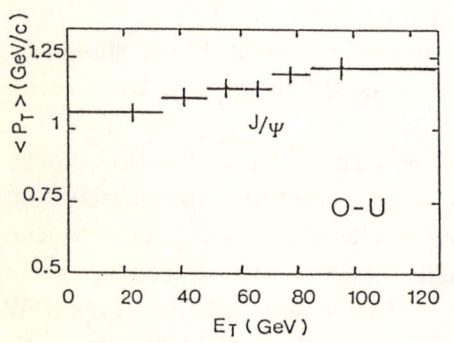

Fig. 32.
NA38 average $\langle P_T \rangle$ of J/ψ produced in 200 GeV/nucleon collisions as a function of the transverse enegy (52).

5.5 GENERAL CONCLUSIONS ON P_T DISTRIBUTIONS

The WA80 experiment found for π°'s produced in O+Au collisions at 200 GeV/nucleon a slope parameter larger for central than for peripheral collisions, and an increase in the slope parameter for peripheral collisions for $P_T \gtrsim 1.8$ GeV/c. The NA34 experiment detected no unusual features in the P_T distributions of negative particles. The NA38 experiment found nothying special in the $\langle P_T \rangle$ of the J/ψ.

The overall conclusion from these studies of transverse momentum distribution is that they provide important and useful information which was necessary to look for. However, concerning a possible signature of QGP nothing strikingly unusual was found.

6. DIRECT PHOTONS

6.1 DIRECT PHOTONS AS A SIGNATURE OF QGP

If a photon is produced in a QGP it leaves the hot plasma with a small probability of interacting in the outer freeze-out region (probability about 200 to 300 times smaller than that of a hadron). It would keep the memory of the temperature in which it was created and would therefore be a good signature of the QGP (**Fig. 33**). In order to see whether there are <u>directly</u> produced photons in the QGP we must study <u>inclusive</u> photon production and find out whether there are some kinematical differences between the sample of the selected inclusive photons and the photons which are known to originate from normal hadron decays. Direct photons as a signature of the QGP have been investigated by Hwa and Kajantie [53] and by Raha and Sinha [54].

Two experiments have studied inclusive photon production in heavy ion interactions : NA34 and WA80.

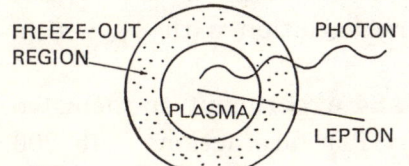

Fig. 33.
Photons as well as leptons produced in a hot quark-gluon plasma can cross the outer freeze-out region without interacting.

6.2 THE NA34 STUDY OF INCLUSIVE PHOTONS

The NA34 Collaboration studied inclusive photon production in p+W, O+W, S+W and S+Pt interactions at 200 GeV/nucleon [55]. The events were detected in the external spectrometer of the apparatus (section 4.5, **Fig. 20**); photons were converted into e^+e^- pairs in an iron plate and the e^+ and the e^- go through a magnetic field and drift chambers for the measuring of their momenta. The numbers of reconstructed photons were not large : 529 in p+W, 297 in 0+W, 850 in S+W and 105 in S+Pt the last two samples being put together for the analysis. The photon rapidity range was 1.0 < y < 1.9. The external spectrometer detects also π^+ and the π^-. The P_T distribution of the $\pi°$'s is assumed to be the same as that of π^+ and π^-. The calorimeter measures the transverse electromagnetic energy in the pseudorapidity range $-0.1 < \eta < 2.9$.

The basic philosophy of the analysis is the following. First a Monte Carlo programme has a photon generator designed to reproduce the kinematics of photons coming from normal hadron decays : $\pi° \rightarrow \gamma\gamma$; $\eta \rightarrow \gamma\gamma$, $3\pi°$, $\pi^+\pi^-\pi°$ and $\pi^+\pi^-\gamma$; $\eta' \rightarrow \eta\pi\pi$, $\rho°\gamma$, $\omega\gamma$ and $\gamma\gamma$; $\omega \rightarrow \pi^+\pi^-\pi°$ and $\pi°\gamma$. The numbers of η, η' and ω which are produced, relative to $\pi°$, were taken from proton-proton interactions at the ISR [56] :

$\eta/\pi° = 14.5\%$, $\eta'/\pi° = 6.3\%$ and $\omega/\pi° = 1.1\%$

Second, the P_T spectrum of the generated photons is compared with experimentally measured spectra. The results are given in **Fig. 34** for 200 GeV/nucleon interactions : a) p+W for $E_T > 20$ GeV; b) 0+W for $E_T > 60$ GeV; c) S+W (+Pt) for $E_T > 80$ GeV. The full lines are the P_T spectrum of photons coming from hadron decays obtained from Monte Carlo, normalized to the first experimental point in each case; there is a remarkable agreement with the experimental data. **Fig. 35** shows the ratio of dN_γ/dP_T as a function of P_T for 0+W to p+W in a) and for S+W to p+W in b). The ratios are consistent with 1 in the entire range of P_T. **Fig. 36** gives the ratio of number of inclusive γ's to the number of $\pi°$'s as a function of the transverse energy. There is no variation with E_T, within the errors.

CONCLUSIONS - The conclusions from the NA34 experiment on inclusive photons produced in p+W, 0+W and S+W (+Pt) interactions at 200 GeV/nucleon are, within the available statistics :

- The shapes of the P_T distributions agree well with that expected for photons coming from known meson decays.
- There is no difference between the spectra of inclusive photons produced in those three interactions.
- The ratio of number of γ's to the number of $\pi°$'s does not vary with E_T.

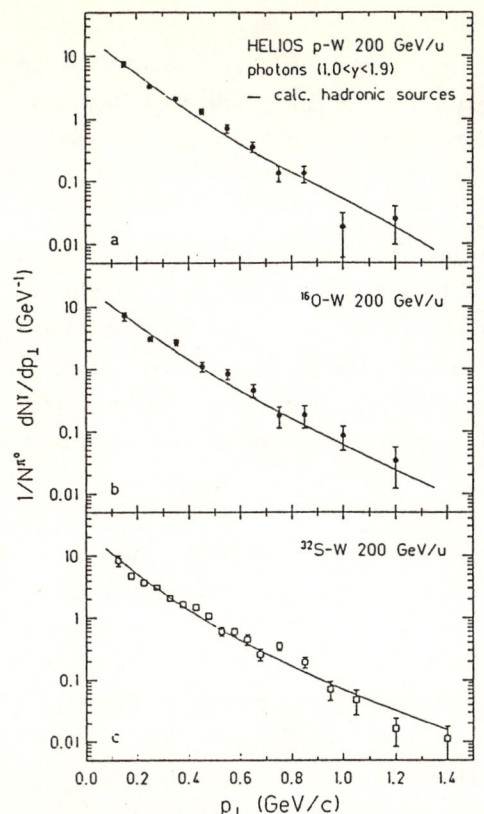

Fig. 34. NA34 P_T spectra of inclusive photons produced in the reactions p+W, O+W and S+W (Pt) at 200 GeV/nucleon. There is a good agreement with the spectra of photons coming from normal hadrons decays, given by the full lines.

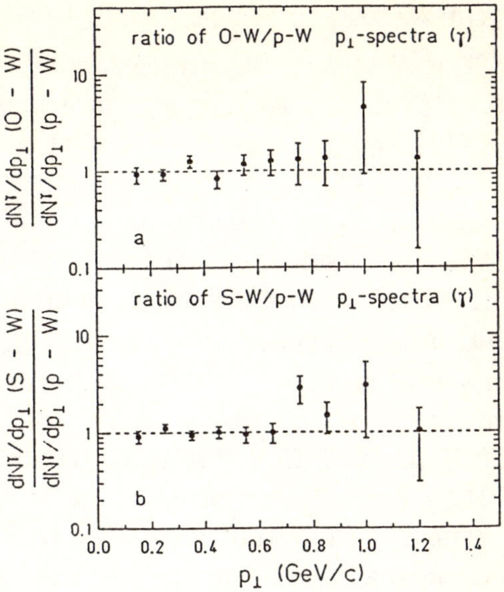

Fig. 35. NA34 ratios of the photon P_T spectra of Fig. 34 : a) O+W to p+W; b) S+W to p+W. They are consistent with 1 in the entire P_T range.

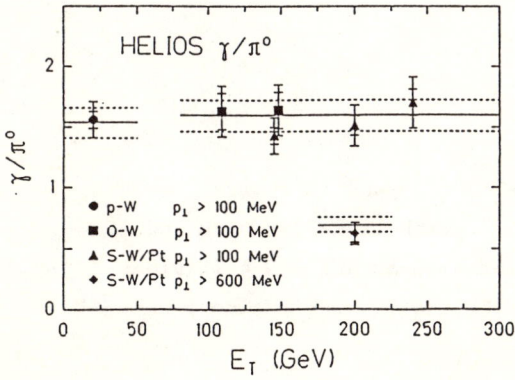

Fig. 36. NA34 ratios of the number of inclusive photons to the number of π^0's as a function of E_T. There is no variation with E_T.

6.3 THE WA80 STUDY OF INCLUSIVE PHOTONS

The WA80 Collaboration studied inclusive photons produced in p+Au, O+C and O+Au central interactions at 60 and 200 GeV/nucleon [48]. The energy measured by the zero-degree calorimeter allows a classification of the collisions into central and peripheral. Photons are detected by the electromagnetic calorimeter SAPHIR (see section 4.3, **Fig. 18**). The resolution of $\pi°$ mass obtained from two photons is 5 to 8% and gives an idea of the good quality of this calorimeter. **Fig. 37** shows the inclusive photon P_T-spectrum, $1/N_{event}$ dN/dP_T, for 200 GeV/nucleon O+Au, O+C and p+Au, and 60 GeV/nucleon O+Au and p+Au central interactions. In the same Figure the prediction from the Monte Carlo FRITIOF for 200 GeV/nucleon O+Au is given, showing a disagreement with the experimental data.

The experimental data were fitted to $1/P_T \, dN/dP_T \sim \exp(-P_T/T_{eff})$. The values of T_{eff} obtained from the fit are given in **Table 2**. They show that T_{eff} increases slightly with the target mass, the projectile mass and

Table 2 - T_{eff} *(MeV/c) obtained from the fit*

GeV/nucleon	p + Au	O + C	O + Au
60	198 ± 3	181 ± 2	215 ± 2
200	215 ± 4	193 ± 2	234 ± 2

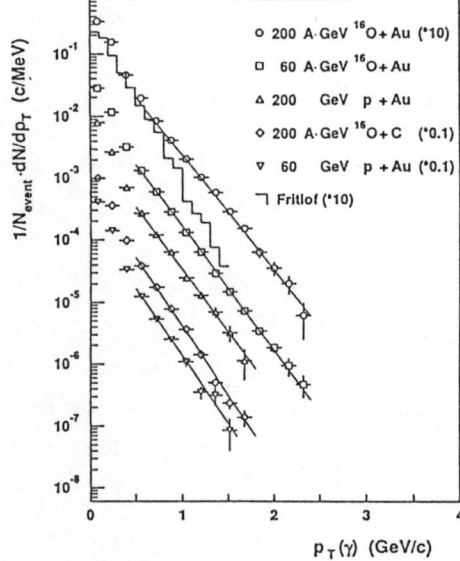

Fig. 37. WA80 inclusive photon P_T distributions for p+Au and central O+Au and O+C collisions at 60 and 200 GeV/nucleon.

the incident energy. The Teff predicted by Monte Carlo FRITIOF is about 20% smaller and does not show any dependence on the target mass. As we have pointed out before, this discrepancy might be due to imperfections in FRITIOF.

The experiment also studied the average $\langle P_T \rangle$ of the inclusive photons. A "truncated" $\langle P_T \rangle_\gamma$ is defined as a function of an arbitrary constant cut-off C, by :

$$\langle P_T \rangle_{\gamma, C} = \frac{\int_C^\infty P_T \frac{dN}{dP_T} dP_T}{\int_C^\infty \frac{dN}{dP_T} dP_T} - C \qquad (29)$$

$\langle P_T \rangle_\gamma$, C was studied for C = 320, 400 and 520 MeV/c as a function of the centrality (energy) of the collision and as a function of the entropy density of the collision, defined as $S \propto dN/d\eta \; A_{inc}^{-2/3}$. Here $dN/d\eta$ is taken in the central pseudorapidity distribution and A_{inc} is the number of incident nucleons which participate in the interaction. **Fig. 38** shows $\langle P_T \rangle_{\gamma, 400}$ for C = 400 MeV/c, as a function of the centrality of the collision, i.e. the energy released in the interaction, given by the ratio zero-degree calorimeter to beam energy. One notices a small increase of $\langle P_T \rangle_{\gamma,400}$ as the energy increases, which is the trend observed in high energy cosmic ray events and at the CERN $\bar{p}p$ collider **(Fig. 23)** as well as by NA38 for the J/ψ **(Fig. 32)**. **Fig. 39** shows $\langle P_T \rangle_{\gamma,400}$ as a function of the entropy of the interaction; it increases with S and tends to level off at high values of S.

<u>CONCLUSIONS</u> - The main results from the WA80 Collaboration on inclusive photons produced in p+Au, O+C and O+Au at 60 and 200 GeV/nucleon are :

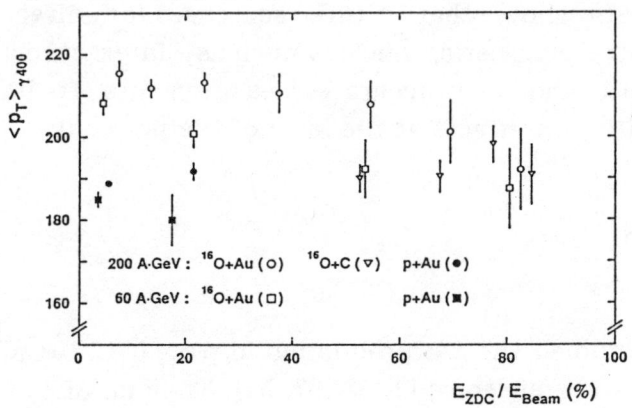

Fig. 38. WA80 experimental $\langle P_T \rangle_{\gamma,400}$ for inclusive photons from truncated P_T distributions (see text), as a function of the centrality of the collision.

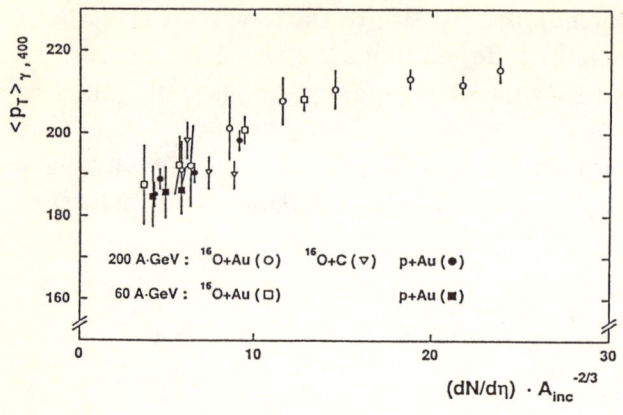

Fig. 39. WA80 experimental $\langle P_T \rangle_{\gamma,400}$ for inclusive photons from truncated P_T distributions (see text), as a function of the entropy density of the collision.

- T_{eff} obtained from the exponential fit for <u>central</u> collisions are $\sim$ 20% smaller in FRITIOF than the experimental values.
- $\langle P_T \rangle_{\gamma,400}$ increases of $\sim$ 15% from peripheral to central collisions, showing the trend observed in other experiments, like UA1 and high energy cosmic ray events **(Fig. 23)** or the $\langle P_T \rangle$ of the J/ψ measured by NA38 **(Fig. 32)**.
- $\langle P_T \rangle_{\gamma,400}$ also increases with the entropy, levelling off at some value of the entropy.

6.4 GENERAL CONCLUSIONS ON INCLUSIVE PHOTONS

WA80 detected the small effects summarized in the previous paragraph. Both, NA34 and WA80, made a very important search for direct photon production and provide fundamental information on inclusive photon production. However, concerning the QGP neither of the two experiments detected any strong effect which could be an evidence for <u>direct</u> photon production in the nucleus-nucleus interactions which they have studied.

These two experiments show that future searches for direct photon production in ultrarelativistic nucleus-nucleus interactions should require high statistics and very accurate measurements, to be able to distinguish among different effects at the level of the per cent.

7. THE J/Ψ SUPPRESSION

7.1 INTRODUCTION

The NA38 experiment studied the J/ψ production in p+U, 0+U, 0+Cu and S+U interactions at 200 GeV/nucleon [44, 52, 57, 58]. The aim of the experiment was to search for the quark-gluon plasma by studying the production of muon pairs and the A-dependence of the J/ψ production. The interest of muon pairs as a QGP signature comes from the fact that muons, like the photons, have no strong interactions. If muons are produced in the plasma they leave the hot plasma volume and the outer freeze-out region, where hadrons start to be formed again, without interacting. Therefore, they should keep memory of the conditions of the plasma from which they originate, like the temperature, for example, the same as direct photons should do (see section 6.1, **Fig. 33**). Independently of any experimental proposal and prior to any experimental data, Matsui and Satz predicted that the quark-gluon plasma should suppress the production of the J/ψ [16]. The reason is that the QGP has a strong colour field that should prevent the formation of the attractive potential which is necessary to bind the c and $\bar{c}$ quarks together in order to become the bound state which would be the J/ψ. We shall come back to this point in section 8. When the NA38 results became known several interpretations of the J/ψ suppression without the existence of the QGP have been proposed. We shall examine in this section the NA38 experimental results and in section 8 the present theoretical interpretations. An up to date account of the subject can be found in Ref. [59].

7.2 THE DETECTION OF THE J/Ψ AND THE MEASUREMENT OF THE TRANSVERSE ENERGY

The lay-out of the NA38 is described in section 4.7 and is shown in **Fig. 22**. The apparatus is triggered on muon pairs and measures the electromagnetic energy released in the interaction. The invariant mass of the muon pair and its transverse and longitudinal momenta are related to the energy of the interaction on an event per event basis. The J/ψ is detected via its decay J/ψ → $\mu^+ + \mu^-$, which has a branching ratio of (6.9 ± 0.9)%. The dimuon mass spectrum shows a continuum and a peak corresponding to the J/ψ. The J/ψ events have mass laying between 2.7 and 3.5 GeV/c^2 due to the resolution of the apparatus. We shall always refer to this particular mass interval as the "J/ψ region" or "J/ψ band" (see **Fig. 44** below).

7.3 THE ORIGIN OF THE J/ψ

In inclusive reactions of the type : hadron + hadron → J/ψ + anything, the J/ψ is not always prompt produced; in many cases it originates from the decay of χ-states and in some cases from the decay of ψ'. The only data on this question come from an experiment on π^--Berilyum interactions at 185 GeV/c [60] in which ~ 40% of the J/ψ's originate from the radiative decay of χ-states, χ → J/ψ + γ, and ~ 5% from the decay of the ψ' (about 50% of ψ's decay into J/ψ). The χ-states involved are mainly the χ_{c1} (3510) and the χ_{c2}(3555). The NA38 detector is not adapted to measure such decays. However due to the similarity of kinematical conditions at 200 GeV/nucleon it is plausible to assume that also in the NA38 experiment ~ 40% of the J/ψ's originate from radiative χ-decays and ~ 5% from ψ' decays. Many charmonium states are produced in 200 GeV hadronic interactions. If we call $N_{\eta c}$, N_χ and N_ψ the numbers of produced η_c, χ and J/ψ respectively, Ref. [60] gives the ratio

$$\rho = \frac{N_{\eta c} + N_\chi}{N_\psi} \sim 3\text{-}4$$

As the ISR energies the ratio ρ may reach ~ 10 [61].

7.4 TRANSVERSE AND LONGITUDINAL MOMENTA OF THE J/Ψ.

As an example, in order to show orders of magnitude, in **Fig. 40** is given the P_T-distribution of the J/ψ's detected in the NA38 experiment in O+U interactions at 200 GeV/nucleon, for all values of the transverse energy E_T, without correction for acceptance. We should notice that it is peaked at ~ 0.8 GeV/c, falls off as P_T increases and there are practically no events with P_T > 3 GeV/c. **Fig. 41** shows the longitudinal momentum

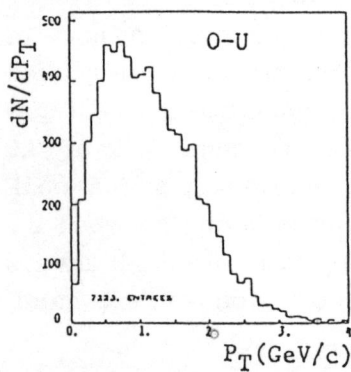

Fig.40. P_T distribution of J/Ψ produced in 200 GeV/nucleon O-U collision, not corrected for acceptance - NA38 experiment.

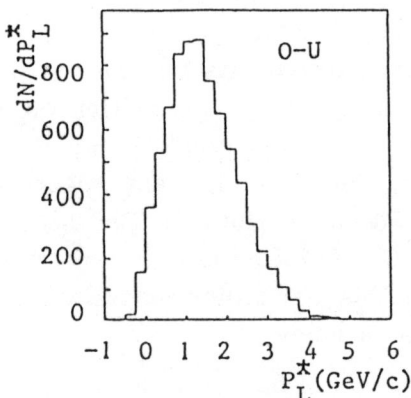

Fig. 41. P_L^* distribution of J/Ψ produced in 200 GeV/nucleon O-U collisions, not corrected for acceptance - NA38 experiment.

of the J/ψ, P_L^*, computed in the center of mass of a nucleon-nucleon collision, obtained in 0+U interactions at 200 GeV/nucleon, for all values of E_T, not corrected for acceptance. Notice that the distribution is peaked at ~ 1.5 GeV/c, falls off as P_L^* increases and there are no events with $P_L^* > 4$ GeV/c. The corrections for acceptance will not alter the trends of the P_T and P_L^* distributions.

7.5 THE TRANSVERSE ENERGY SPECTRUM

The detector is triggered on muon pairs irrespective of the electrical charges of the muons. As a consequence, not only the required $\mu^+\mu^-$, but also many $\mu^+\mu^+$ and $\mu^-\mu^-$ pairs are detected; they come mainly from the decay of $\pi^\pm$ and $K^\pm$ mesons. These pairs are also an evidence that, in the sample of $\mu^+\mu^-$ pairs, many are a background originated from those decays. We shall see below how this background is evaluated, to be subtracted from the overall, raw $\mu^+\mu^-$ spectra. **Fig. 42** shows the transverse energy spectra for 0+U and S+U interactions. They both show the raw, overall measured spectra for opposite-sign muon pairs, the like-sign muon pairs, and the signal of opposite-sign pairs after subtraction of the $\mu^+\mu^-$ background.

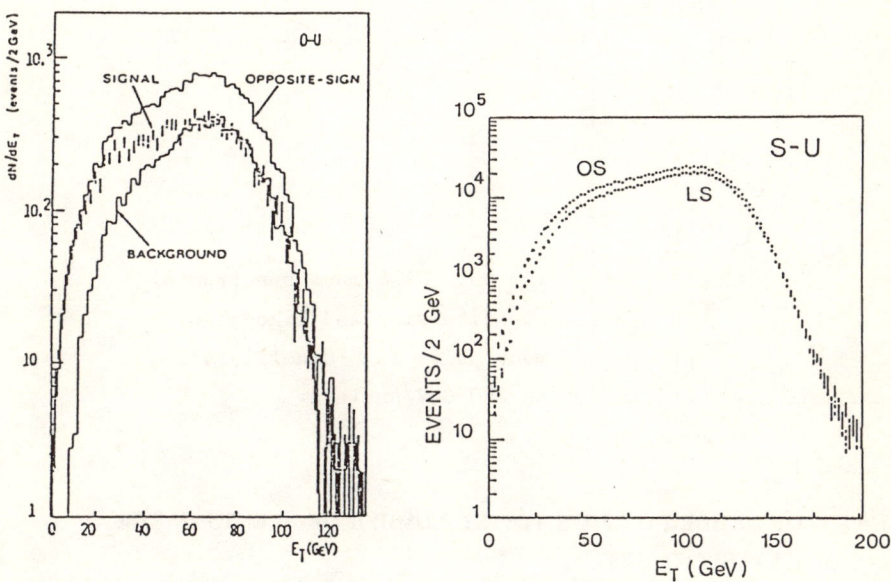

Fig. 42. NA38 transverse energy spectra measured in 0+U and S+U collisions at 200 GeV/nucleon.

7.6 THE DIMUON MASS SPECTRUM

Fig. 43 shows the mass spectrum of the same-sign and opposite-sign dimuons. We should notice that in the J/ψ mass band, $2.7 < M_{\mu\mu} < 3.5$ GeV/c^2, the fraction of same-sign dimuons is small relative to the opposite-sign. The opposite-sign background was estimated from the like-sign pairs with the expression

$$N_{+-\,back} = 2\sqrt{N_{++} \cdot N_{--}}$$

where $N_{+-\,back}$ is the background of $\mu^+\mu^-$, N_{++} and N_{--} are the numbers of $\mu^+\mu^+$ and $\mu^-\mu^-$, respectively. Calling $N_{+-\,raw}$ the overall measured number of $\mu^+\mu^-$, the signal is :

$$\text{Signal } \mu^+\mu^- = N_{+-\,raw} - N_{+-\,back}$$

The resulting $\mu^+\mu^-$ mass spectra for 0+U, S+U and 0+Cu events are shown in **Fig. 44** for all values of E_T.

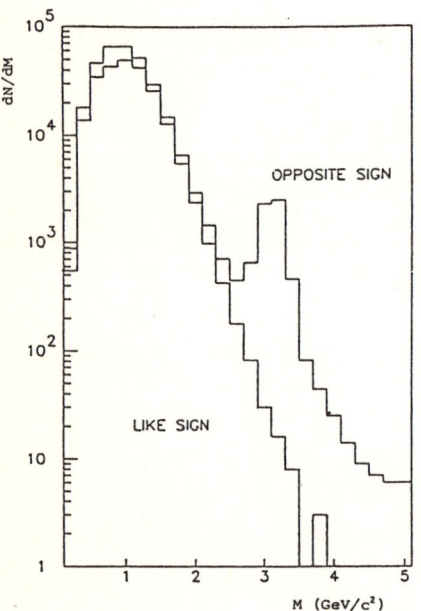

Fig. 43. NA38 mass spectrum of opposite-sign and like-sign muon pairs in 0+U collisions at 200 GeV/nucleon.

7.7 THE RATIO OF NUMBER OF J/Ψ'S TO THE NUMBER OF EVENTS IN THE CONTINUUM MASS SPECTRUM.

The rate of production of the J/ψ was evaluated by the ratio S of the number of J/ψ's, N_ϕ, to the number of events N_c of the continuum mass spectrum which are in the J/ψ mass region (2.7 - 3.5 GeV/c^2) :

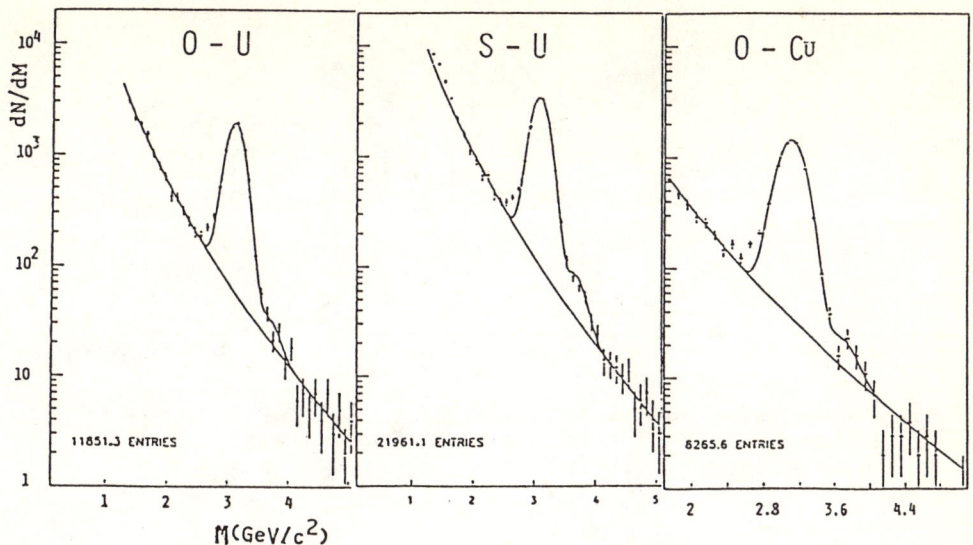

Fig. 44. NA38 mass spectrum of $\mu^+\mu^-$ pairs in O+U, S+U and O+Cu interactions, for all values of E_T.

$$S = \frac{N_\psi}{N_c} \qquad (30)$$

The choice of N_c in the J/ψ mass band is arbitrary. It was chosen because in this band the $\mu^+\mu^-$ background is very small, few per cent, as can be seen in **Fig. 43**. The events are mainly Drell-Yan and therefore N_c can be calculated in a theoretical work. It was checked, however, that the conclusions on the variation of S as a function of E_T (which we sall see below) do not depend on the interval of the continuum mass spectrum in which N_c is counted.

7.8 J/Ψ PRODUCTION AS A FUNCTION OF THE TRANSVERSE ENERGY

The main idea is to compare the rate of J/ψ production in central collisions with the rate in peripheral collisions. Since the centrality of the collision is related to the energy which is released, the J/ψ production rate was measured as a function of E_T. Large E_T events correspond to central collisions and a sample of small E_T events consists predominantly of peripheral collisions.

The events were divided into transverse energy bands. For the events of each E_T band the mass spectrum was obtained, analogous to the ones shown in **Fig. 44**, and a value of S, expr. [30] was calculated. **Fig. 45** shows : in a) the values of S as a function of E_T for

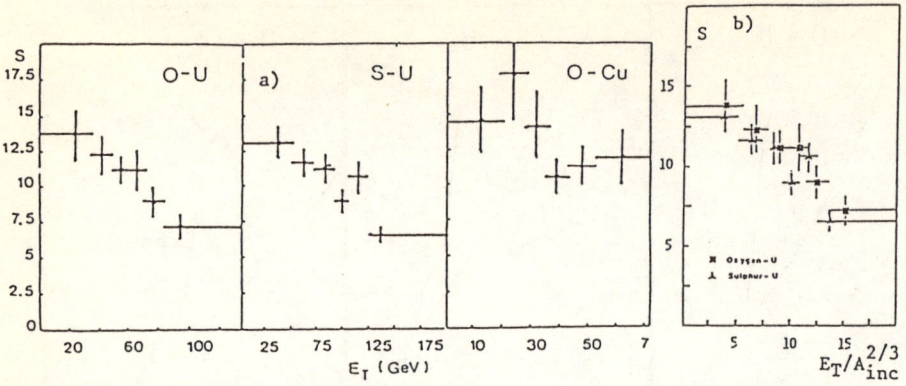

Fig. 45. a) NA38 values of the ratio $S = N_\psi/N_c$ as function of E_T, for 0-U, S-U and 0-Cu interactions : b) S as a function of $E_T/A_{inc}^{2/3}$ for 0-U and S-U collisions.

0+U, S+U and 0+Cu interactions; in b) the values of S for 0+U and S+U interactions as a function of $E_T A^{-2/3}$, a quantity which should be proportional to energy density. We see clearly that S decreases as E_T increases, or as the energy density increases. This is the main result of the NA38 experiment : <u>the production of the J/ψ relative to the continuum decreases as the energy density of the interaction increases.</u>

The ratio R of the value of S at the highest E_T band to the value of S at the lowest E_T band was found to be :

$$R = 0.52 \pm 0.09 \quad \text{for 0-U}$$
$$0.50 \pm 0.11 \quad \text{for S-U}$$
$$0.73 \pm 0.10 \quad \text{for 0-Cu}$$

The conclusion is that these ratios R represent the reduction of J/ψ production in collisions with the maximum energy density relative to the production in collisions with the minimum energy density.

7.9 DEPENDENCE OF THE J/Ψ SUPPRESSION ON THE TRANSVERSE MOMENTUM

Fig. 46 shows the dependence of the J/ψ suppression on the transverse momentum for 0+U and S+U interactions. It gives the ratio $R(P_T)$, for different P_T intervals, of the number of J/ψ's in the highest E_T band upon the number of J/ψ's in the lowest E_T band, normalized to the same number of the continuum events in the J/ψ mass region. We see that the J/ψ suppression is stronger at low values of P_T; for the 0-U events $R(P_T)$ varies by about a factor 2 as p_T goes from 0 to ~ 3 GeV/c.

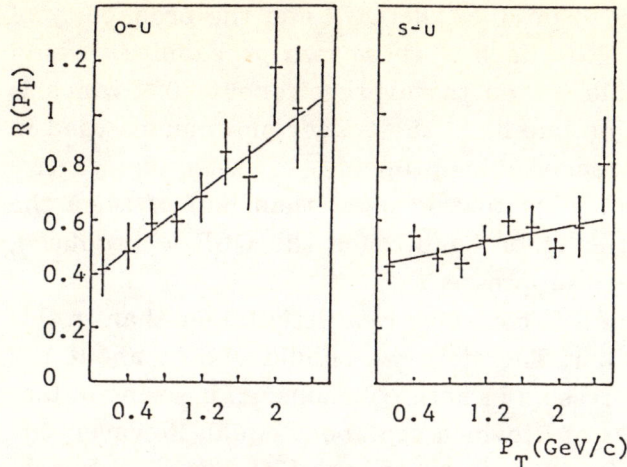

Fig. 46. Ratio $R(P_T)$ as function of P_T for J/Ψ produced in O-U and S-U collisions at 200 GeV/nucleon. NA38 experiment.

We shall see in section 8 that $R(P_T)$ is an important variable to test models of J/ψ suppression.

7.10 CONCLUSIONS FROM THE NA38 EXPERIMENT ON THE J/ψ PRODUCTION

In J/ψ production in O+U, S+U and O+Cu interactions at 200 GeV/nucleon:

1) the ratio S of the number of J/ψ's upon the number of events in the continuum mass spectrum in the J/ψ mass region <u>decreases</u> as the energy density of the interaction <u>increases</u>.

2) The ratio R of the value of S at the highest E_T band to the value of S at the lowest E_T band was found to be 0.52 ± 0.11, 0.50 ± 0.10 and 0.73 ± 0.10 for oxygen-uranium, sulphur-uranium and oxygen-copper interactions, respectively.

3) The suppression of the J/ψ production is stronger at low P_T as shown by the ratio $R(P_T)$ in **Fig. 46**.

7.11 REMARKS ON THE NA38 EXPERIMENT

I shall present some personal remarks on the NA38 experiment.

1) The ratios R ≈ 0.5, which show that there is about 50% of the J/ψ suppression in central collisions relative to peripheral collisions in O+U and S+U interactions, at first sight might seem small. If we assume the existence of the QGP these ratios are in fact rather large. Because, under this assumption, the experiment measures the product of two

probabilities: the probability to produce the QGP and the probability of J/ψ suppression due to the QGP. Each of these two probabilities would be larger than 50% and would be "on the average" about 70%, which is very high. This means that on one hand the trigger on dimuons used in the experiment has the remarkable property of selecting <u>central</u> collisions associated with J/ψ such that in more than half of them the QGP is produced; and on the other hand, once the QGP is produced, more than half of the J/ψ's are suppressed.

2) The energy density in S-U collisions is a little larger than in 0-U collisions, certainly not smaller. Therefore we should expect about the same percentage of J/ψ suppression in both collisions, as it seems in fact to be indicated by the values of R, which are about equal. However, the comparison of $R(P_T)$ versus P_T for 0-U events with S-U events, shown in **Fig. 46**, is rather disturbing because the dependence of $R(P_T)$ on P_T for S-U collisions is much less pronounced than for 0-U collisions. This is queer and is in contradiction with what we should expect. Taking the experimental data as they are, this point is difficult to be understood and should be clarified in the future.

3) As far as the future of the subject is concerned I shall make the following remarks:

i) If there is a signal of QGP in J/ψ production, we should expect a signal also in dilepton pairs with mass smaller than the J/ψ mass. We have seen that the low mass $\mu^+\mu^-$ spectrum has a large background due to the decays $\pi \rightarrow \mu\,\nu$ and $K \rightarrow \mu\,\nu$, whose probabilities are high, 100% and 64%, respectively. In order to avoid the background in the low mass region (< 2 GeV/c^2) an experiment should be done on e^+e^- pair production, because of the low rates of the π and K decays into electrons : $(\pi \rightarrow e\nu)/(\pi \rightarrow \mu\nu)$ = 1.2×10^{-4} and $(K \rightarrow e\,\nu)/(K \rightarrow \mu\,\nu) = 1.5 \times 10^{-5}$. There would be no e^+e^- background originating from these decays, and the e^+e^- low mass spectrum would be clean.

ii) In order to clarify whether the J/ψ suppression is due to the production of QGP two crucial tests should be done :

1st - In the transition from hadronic matter to QGP there must be a threshold energy density, below which there should be no deconfinement. An experiment should be repeated at lower energy densities for central collisions to see whether there is a threshold for J/ψ suppression. The existence of a threshold would be very much in favour of a GQP; the absence of a threshold would certainly be against a QGP.

2nd - An experiment should be done with higher statistics in order to study the dependence of the J/ψ suppression on the J/ψ rapidity or x_F, as well as the dependence on P_T. If the suppression is due to QGP, the dependence on the rapidity or on X_F should have the same trend as the dependence on P_T which is shown in **Fig. 46**; it should be stronger for small rapidities. Should the dependence of the J/ψ suppression on rapidity have a trend different from the dependence on P_T, this difference would be against QGP.

8. MODELS TO EXPLAIN THE J/Ψ SUPPRESSION

8.1 INTRODUCTION

In section 7.1 we have pointed out Matsui and Satz's prediction [16] that if a quark-gluon plasma is created, the production of the J/ψ should be suppressed. Their prediction was independent of and prior to the results of the NA38 experiment at CERN. After these experimental results became known, several models have been proposed to explain the J/ψ suppression. There are two classes of models : one class assumes that the quark gluon plasma is formed; the other assumes that there is no quark-gluon plasma and the J/ψ disappears because it interacts inelastically with some dense hadronic matter created in the collision. A rather detailed account of the different models and their comparison with the NA38 results can be found in Ref. [59].

8.2 TIME FOR THE PRODUCTION OF THE $c\bar{c}$ PAIR AND TIME FOR THE FORMATION OF THE J/Ψ.

All models, of both classes, assume that the $c\bar{c}$ pair, which should become a J/ψ as a bound state, is produced in a <u>nucleon-nucleon</u> interaction in a very short time, very early in the collision. If m_c is the mass of the charm quark, the time τ_P for the production of $c\bar{c}$ pair can be estimated in the $c\bar{c}$ rest frame from the uncertainty principle, $2 m_c \tau_P \lesssim \hbar$. By taking $m_c \approx 1.5$ GeV and $\hbar = 197$ MeV fm we obtain $\tau_P \lesssim 0.06$ fm. This is an estimate of the time for production of the $c\bar{c}$ pair in its rest frame. In the laboratory system if the $c\bar{c}$ pair has a momentum P this time will be $t_P = \gamma \tau_P = (P/2 m_c)\tau_P$. For $c\bar{c}$ pairs produced in 200 GeV nucleon-nucleon interaction P varies typically from ~ 20 to 60 GeV/c and, as a consequence, $t_P \approx 0.4$ to 1.2 fm, which is a very short time.

Another time which must be taken into account is the time t_F needed for the formation of the J/ψ, i.e. for the $c\bar{c}$ distance to grow and reach the resonant frequency at the J/ψ binding radius $r_\psi \approx 0.5$ fm (**Fig. 47**). If the quark intrinsec transverse momentum is $K_T \approx (0.5 - 1)$ GeV/c, the transverse velocity of c or $\bar{c}$ is $V_T \approx K_T/P$, and the formation time $t_F \approx r_\psi/V_T$, or

$$t_F \approx \frac{r_\psi P}{K_T} \approx \frac{0.5 \text{ fm} \times P \text{ GeV/c}}{(0.5-1) \text{ GeV/c}} \approx (0.5-1) \, P \text{ fm}$$

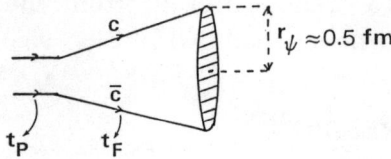

Fig. 47. The two times which must be taken into account in the formation of a resonance: t_P is the short time for the production of the $c\bar{c}$ pair in the beginning of a collision, t_F is the longer time during which the $c\bar{c}$ pair travels until the formation of the bound state J/ψ.

For $P \approx 20$ to 60 GeV/c, we obtain $t_F \approx 10$ to 60 fm which is much longer than t_P.

We must, therefore, take into account the two times: t_P, for the production of the $c\bar{c}$ pair and t_F, for the formation of J/ψ, t_F being 25 to 50 times longer than t_P in the present experiments. It is important to notice that the time t_F is long enough for the J/ψ to be formed <u>outside the nucleus.</u> Even when the target is as large as an uranium nucleus, whose radius is about 7.5 fm, most of the J/ψ's are formed outside the nucleus because the $c\bar{c}$ pair must travel ~ 10 to 60 fm before the J/ψ appears.

This brings us to the heart of the problem on the J/ψ suppression. Let us recall the fragmentation and central regions formed in the collision, **Fig. 13**, section 3.7. The $c\bar{c}$ pair is formed in an initial nucleon-nucleon interaction at the very beginning of the collision and travels through the fragmentation and central regions before the J/ψ can be formed. Therefore, if the J/ψ is not formed, something happens to the c or $\bar{c}$, or both, when crossing those regions. The heart of the problem is therefore : what are the central and fragmentation regions made of ? If they are a plasma of quarks and gluons, it is this plasma which prevents the $c\bar{c}$ to become a bound state. If those regions are not a plasma, they must be some sort of dense hadronic matter which prevents the J/ψ to be formed, presumably because either the c or the $\bar{c}$ or both interact with the quarks existing there.

The problem of the J/ψ suppression becomes the problem of knowing the composition of the central and fragmentation regions. There is, unfortunately, a lack of experimental data on this, which is the main difficulty to be faced if one wants to make models for J/ψ suppression.

Only the central region is usually taken into account in the different models, because still less is known about the fragmentation regions. We shall consider only models which try to explain J/ψ suppression under some assumption about the <u>composition</u> of the central region. We shall ignore computations which assume that the $c\bar{c}$ pair and J/ψ are produced at the same point inside the nucleus and that the J/ψ disappearance is due to J/ψ-nucleon interaction, because this is in contradiction with what we have seen above about the need to consider two distinct times, t_P and t_F. In dealing with models we shall see mainly their physical assumptions. The reader will be referred to the original papers for details, like for instance the actual computations and the parameters which must be introduced as a consequence of our ignorance on the hadronic matter.

8.3   MODELS WHICH ASSUME THE FORMATION OF THE QUARK-GLUON PLASMA

8.3.1   The prediction of Matsui and Satz

Matsui and Satz [16] have predicted the following mechanism for the suppression of the J/ψ production in the presence of a plasma. If a very high energy density is reached and a quark-gluon plasma is formed, for temperatures T greater than a critical deconfinement temperature T_c, there should be a distance r_D such that, for distances greater than r_D the strong (colour) forces that bind the $c\bar{c}$ pair together become screened, i.e., are no longer operative. This means that for distances $r > r_D$ there is no more strong interaction between c and $\bar{c}$, which therefore, cannot form bound states. This is called Debye colour screening by analogy with the electron plasma; r_D is the Debye colour screening radius.

If the J/ψ radius is $r_\psi < r_D$ the J/ψ will be formed, because the screening operates at distances outside r_ψ. If $r_\psi > r_D$ screening operates and the J/ψ will not be formed.

8.3.2 Plasma dimensions and lifetime and the formation time of resonances

In charmonium spectroscopy one computes the mean radius $<r>$, the mean square radial momentum $<p>$ and the formation (proper) time t_0 of a charmonium state. **Table 3** gives these values for the J/ψ, the ψ' and the χ_c [62, 63].

Table 3

Resonance	$<r>$ fm	$<p>$ GeV/c	τ_0 fm/c
J/ψ	0.5	0.7	0.9
ψ'	0.9	0.8	1.5
χ_c	0.7	0.5	2.0

The plasma has a duration in time and has dimensions which undergo an evolution with the time. The $c\bar{c}$ pair will stay in the plasma during certain time which depends on its velocity and on those plasma characteristics. In order to understand the connections between those quantities, let us consider a simple example. Assume that the plasma has a screening region with radius $\vec{s}$ and that the $c\bar{c}$ pair is created at the position $\vec{r}_0$ with momentum $\vec{P}$. After the time t_ψ necessary for the J/ψ formation has elapsed, the $c\bar{c}$ pair has travelled a distance $\vec{d} = t_\psi \vec{P}/M_\psi$ (Fig. 48). If $|\vec{r}_0 + \vec{d}| < |\vec{s}|$ the $c\bar{c}$ does not scape the screening region and the J/ψ will not be formed; if $|\vec{r}_0 + \vec{d}| > |\vec{s}|$ the $c\bar{c}$ scapes the screening region and the J/ψ will be formed.

From **Table 3** we see that the $c\bar{c}$ pair takes a longer time to form a ψ' or a χ_c than to form a J/ψ; therefore the screening will affect more the production of the ψ' and χ_c than that of the J/ψ. We have seen in section 7.3 that in the NA38 experiment ~ 40% of the J/ψ's come from the decay

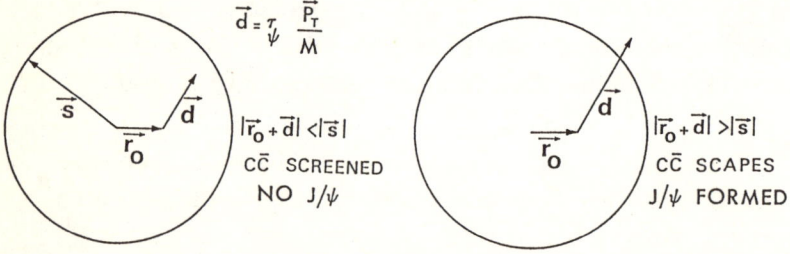

Fig. 48. Relation between plasma dimensions, $c\bar{c}$ momentum and J/Ψ formation. See text.

of χ-states. Those ~ 40% are more affected by the presence of the plasma than the ~ 60% of J/ψ's which are promptly produced.

These simple considerations show clearly that in order to compute the rate of suppression of a resonance formation due to QGP, we must know:

a) the formation time and size of the resonance;

b) the plasma lifetime;

c) the longitudinal and transverse dimensions of the plasma and their evolution with time;

d) the exact relationship between the Debye colour screening radius r_D and the plasma temperature T. Uncertainties which exist at present on this relationship should be eliminated [16].

Further difficulties appear when we want to compare a theoretical calculation of the suppression of a resonance with experimental results. The comparison has to be based on some quantities which are used in the theory and should be measured experimentally. The temperature, for instance, which could be a good variable, in the theory must be related to r_D; in the experiment it must be related to the energy density, or the entropy of the collision, but neither of these two can be directly measured. They can just be estimated under certain assumptions, from the energy released in the collision. We have seen in section 3.10 that there is not a model-free, unambiguous way of estimating experimentally the energy density.

Different models are based on different assumptions about points a) to d) mentioned above.

8.4 SPECIFIC PLASMA MODELS

Three models have been put forward to try explaining the NA38 results under the assumption that the plasma is formed. They all compute the ratio $R(p_T)$ of the J/ψ suppression as a function of the J/ψ transverse momentum, given in **Fig. 46**, section 7.9.

Chu and Matsui [64] consider only the central rapidity region of heavy ions central collisions. Transverse expansion of the plasma is neglected, only longitudinal expansion is considered and is described by Bjorken's hydrodynamical model. The evolution of the energy density with time is taken from this model and its dependence on the transverse coordinate is parametrized. The authors evaluate a minimum energy density ε_s below which the J/ψ formation is no longer suppressed; the result depends on three parameters which are introduced in the model,

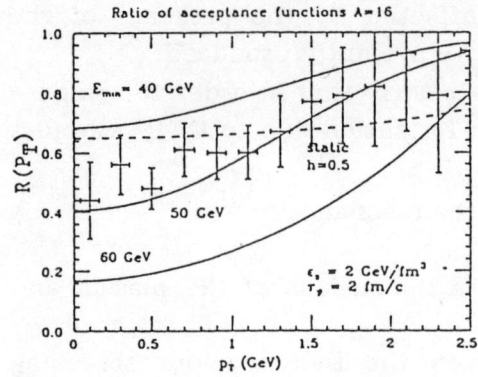

Fig. 49. $R(P_T)$ obtained by Chu and Matsui for J/ψ produced in O+U interactions, with the NA38 experimental points [64].

and with a given set of parameters ε_s is found to be 0.7 GeV/fm^3. In order to compute $R(P_T)$ and compare with the NA38 results a "theoretical" energy scale has to be made; this is very difficult and it might differ from the NA38 E_T distribution by as much as 50%. The result on $R(P_T)$ for O-U interactions is shown in **Fig. 49** under the assumptions that the energy density is 2 GeV/fm^3, the J/ψ formation time is $t_\psi = 2$ fm/c, and for three different values of the lower edge of the maximum E_T bin to be compared with NA38 : E_{min} = 40, 50 and 60 GeV. The same Figure shows the NA38 experimental points. The conclusion is that the shapes of the curves are reasonable, but the absolute normalization can easily change by factors as large as 4, depending on the values chosen for the parameters of the model.

Blaizot and Ollitrault [65] have a model similar to the one developed by Chu and Matsui, with some identical parametrizations, but their model is static; it does consider plasma expansions. The resulting $R(P_T)$ can also change by factors as large as 4, as in the previous model.

Karsch and Petronzio [63, 66] assume plasma formation and use NA38 data to estimate some of the plasma parameters, which are afterward used to compute the P_T-dependence of J/ψ suppression, $R(P_T)$. It is taken into account that $\sim$ 40% of the J/ψ's come from the decay of χ-states; critical screening conditions are computed for both, the J/ψ and the χ, and plasma suppression of prompt-produced J/ψ's as well as of the χ is considered. By using the resonances formation times given in **Table 3** and fixing the plasma dimensions by making use of the NA38 data under certain assumptions it is possible to compute the fractions of J/ψ's and of χ's which leave the screening plasma region. The result is given in **Fig. 50**, for some specific values of the parameters used in the model, together with the NA38 experimental data. The agreement is reasonable, but it should be regarded as an indication of coherence of the different

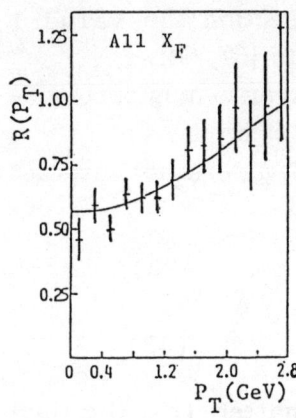

Fig. 50. $R(P_T)$ obtained by Karsch and Petronzio for J/Ψ produced in O+U interactions, with the NA38 experimental points [66].

assumptions introduced, because the result was obtained with the use of the NA38 data, and not as a proof.

8.5 MODELS OF ABSORPTION OF THE J/Ψ BY INELASTIC SCATTERING IN THE HOT HADRONIC MATTER

8.5.1 The main problem

In these models the J/ψ should disappear by interacting in the central region with a high density hadronic matter, made of normal mesons. The difficulties in computing these models can be seen from the expression which gives the number dN of interactions of the J/ψ in a volume dv during the time interval dt :

$$dN = n \, \bar{v}_{rel} \, \sigma \, dv \, dt \qquad (31)$$

where n is the particle density of the hadronic matter per unit volume, $\bar{v}_{rel}$ is the average velocity of the J/ψ relative to those particles and σ is the cross section. The particle density is usually a function of the particles momentum, position and time. The great difficulty is that the meson composition of such hadronic matter, n, $\bar{v}_{rel}$ and σ are all unknown (σ is a function of v_{rel}). In order to compute models on J/ψ inelastic scattering one has, therefore, to make assumptions about those variables, which, even if they are plausible, are not proven.

8.5.2 Basic assumptions and difficulties

These models have some common charactersitics, summarized in points a) to f):

a) - They all assume that the J/ψ is produced locally inside the nucleus.

b) - They all use for the J/ψ + nucleon cross section the value 1 to 3 mb, obtained in J/ψ photoproduction experiments [67, 68].

c) - They assume that the J/ψ has only transverse momentum P_T and that its P_L^* in the C.M.S. of a nucleon-nucleon collision is zero.

d) - They all use for the probability that the J/ψ does not interact ("survival probability") the classical expression:

$$\text{Prob.} \; \alpha \; \exp(-\lambda) \qquad (32)$$

where
$$\lambda = \int_{t_0}^{t_f} \nu\left[\vec{P}_T, \vec{r}(t), t\right] dt \qquad (33)$$

t_0 is the initial time, when the J/ψ enters the hot matter, t_f is the final time, when absorption stops, ν is the number of collisions per unit time, $\vec{P}_T$ is the J/ψ transverse momentum and $\vec{r}(t)$ is the J/ψ position at the time t. From expressions (32) and (33) we have

$$\text{Prob.} \approx \left(\frac{t_0}{t_f}\right)^\beta \qquad (34)$$

β is the <u>absorption parameter</u>, equal to the ratio ν/expansion rate. It is given by:

$$\beta = \bar{\sigma} \; \bar{v}_{rel} \; n_0 \; t_0 \qquad (35)$$

where $\bar{\sigma}$ is the average cross section and n_0 is the particle density of the hot hadronic matter at the time t_0.

e) - Bjorken's hydrodynamical model [20] gives the evolution of the product $n_0 t_0$ with time and its relationship with the distribution of rapidity dN/dy in the central region for $y \approx 0$:

$$n_0 t_0 = nt = \frac{1}{4\pi R_A^2} \left.\frac{dN_h}{dy}\right|_{y=0} \qquad (36)$$

R_A is the radius of the incident nucleus. From (35) and (36) we have:

$$\beta = \bar{\sigma} \; \bar{v}_{rel} \; \frac{1}{4\pi R_A^2} \left.\frac{dN_h}{dy}\right|_{y=0} \qquad (37)$$

f) - From expr. (36) we can have an approximate value for n_0:

$$n_0 = \frac{1}{t_0} \frac{1}{4\pi R_A^2} \left.\frac{dN_h}{dy}\right|_{y=0} \qquad (38)$$

For an oxygen nucleus $R_A \approx 3$ fm. The NA35 Collaboration measured $dN_h/dy|_{y=0} \approx 120$ [69]. For $t_0 \approx 1$ fm/c we obtain $n_0 \approx 4$ hadrons/fm³. This hadron density is too large, because as we have seen in section 3.1 the nucleon density in normal nuclear matter is ~ 0.14 nucleons/fm^3. This is a great difficulty with the J/ψ absorption models : <u>we have to understand how a mesonic hadronic matter could be in a state with a density about 25 times larger than the nucleon density in normal nuclear matter</u> and still stay as normal hadrons. Remember that the pion volume is ~ 0.8 fm^3.

8.5.3 Initial and final state parton interactions

The CERN NA10 Collaboration measured the $\langle P_T^2 \rangle$ of Drell-Yan muon pairs with mass > 4 GeV/c^2 and of J/ψ's produced in π^-+W and D+W interactions at 140 and 286 GeV/c and found the differences [70] :

$$\langle \Delta P_T^2 \rangle^{DY} = \langle P_T^2 \rangle_W^{DY} - \langle P_T^2 \rangle_D^{DY} = 0.15 \pm 0.03 \pm 0.03 (\text{GeV/c})^2$$
$$\text{at } 286 \text{ GeV/c}$$
$$= 0.16 \pm 0.03 \pm 0.03 \ (\text{GeV/c})^2 \text{ at } 140 \text{ GeV/c}$$
$$\langle \Delta P_T^2 \rangle^\psi = \langle P_T^2 \rangle_W^\psi - \langle P_T^2 \rangle_D^\psi = 0.29 \pm 0.02 \ (\text{GeV/c})^2 \text{ at } 140 \text{ GeV/c}$$

We should notice that $\langle \Delta P_T^2 \rangle^\psi \sim 2 \langle \Delta P_T^2 \rangle^{DY}$. The J/ψ result agrees with a previous measurement made by the NA3 Collaboration in 400 GeV/c proton-proton interaction in Pt and H [71]

$$\langle \Delta P_T^2 \rangle^\psi = \langle P_T^2 \rangle_{Pt}^\psi - \langle P_T^2 \rangle_H^\psi = 0.34 \pm 0.06 \ (\text{GeV/c})^2$$

It is important to analyse both the Drell-Yan and the J/ψ results. The former is interpreted as a clear evidence of an <u>initial state</u> interaction of the quark and the antiquark which undergo an elastic scattering in the target before annihilating into a virtual photon which materializes into the muon pair. The J/ψ result, compared with the Drell-Yan, is interpreted as an evidence of elastic scattering of the <u>final state</u> partons before the $c\bar{c}$ pair forms the bound state, besides the initial state interaction (it can be either the elastic scattering of c or $\bar{c}$, or both, or of the gluons which fuse to produce the $c\bar{c}$ pair). This had been predicted theoretically [72], as well as that $\langle \Delta P_T^2 \rangle^\psi \approx 2 \langle P_T^2 \rangle^{DY}$, as was experimentally shown by the NA10 Collaboration.

Some of the models for the J/ψ absorption included lately the elastic scattering of partons in the <u>initial state.</u> In this case the initial state partons are the quarks c and $\bar{c}$ or the gluons which produce c and $\bar{c}$, before the c and $\bar{c}$ form the bound state J/ψ.

8.6 SOME SPECIFIC MODELS

Four models along the lines described in section 8.5 have been proposed to interpret the NA38 results. Attention is focused particularly on the ratio $R(P_T)$, defined in section 7.9, which gives the J/ψ suppression as a function of P_T. **Table 4** gives the different hadronic gas composition which these models assume, as well as the inelastic reactions which would absorb the J/ψ. Several parameters or assumptions must be introduced in each model.

Table 4

MODEL	HADRON GAS COMPOSITION	REACTIONS
Ftačnik, Lichard and Pišut [20]	π	πψ → $D\bar{D}$
Gavin, Gyulassy and Jackson [21]	π	πψ → η_c ππ → $D\bar{D}$
	ρ	ρψ → η_c π
	ω	ωψ → η_c ππ
	η	ηψ → η_c πππ
Vogt, Prakash, Koch and Hansson [22]	π ⎤ ⎥ M,M' ρ ⎦	Mψ → $D\bar{D}$ Mψ → $\eta_{c'}$ χ M' M η_c, χ → $D\bar{D}$ M $\eta_{c'}$ χ → ψ M'
Blaizot and Ollitrault [23]	π	πψ → $D\bar{D}$

1. **Ftačnik, Lichard and Pišut** [73, 74] assume that the hadron gas is made only of π. They take into account that ~ 40% of the J/ψ's originate from χ-decays. The resulting $R(P_T)$ is given in **Fig. 51** together with the NA38 experimental data : a) for σ(ψπ) = 1 mb, σ(χπ) = 2.36 mb, t_f = 4 fm, and b) for σ(χπ) = 2 mb, σ(χπ) = 4.72 mb, t_f = 2.5 fm. The curves are intended to show the <u>shape only</u> and not the absolute normalization, which does not agree with the data.

2. **Gavin, Gyulassy and Jackson** [75] assume that the hadron gas is composed of π, ρ, ω and η in proportions that are obtained from the

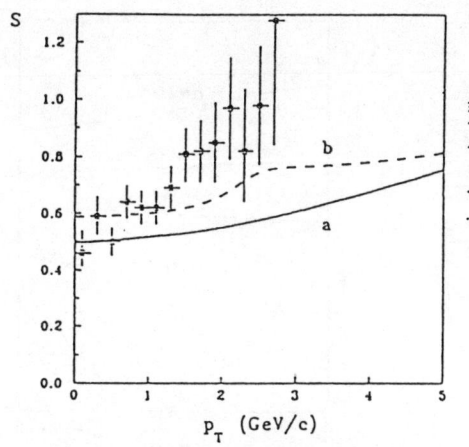
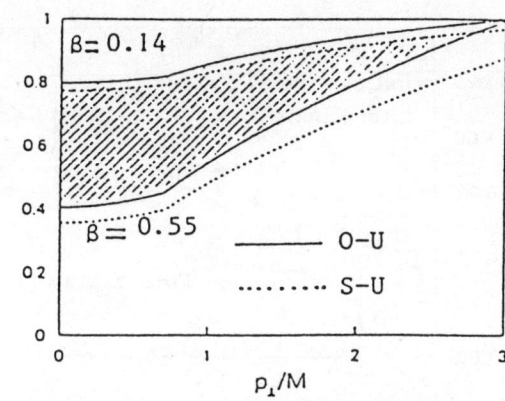

Fig. 51. R(P_T) obtained by Ftačnik et al. for J/ψ produced in O+U interactions, without initial state parton interactions, with the NA38 experimental points [73].

Fig. 52. R(P_T) obtained by Gavin et al. without initial state parton interactions [75].

FRITIOF Lund Monte Carlo. **Fig. 52** gives their results on the J/ψ survival probability as a function of the ratio P_T/M_ψ, for oxygen-uranium and sulphur-uranium interactions for two values of the absorption parameter of expr. (35), β = 0.14 and 0.55. The dashed region corresponds to all possible results for values of β in between these two values. The results can vary by a factor ~ 2 and furthermore they do not reproduce the trend of the NA38 experimental data. For the ratio P_T/M_ψ varying from 0 to ~ 1, i.e. P_T varying from 0 to ~ 3 GeV/c, the curves are rather flat whereas the experimental R(P_T) varies by a factor ~ 2 (see **Fig. 46**).

Gavin and Gyulassy later modified the model by introducing initial state gluon interactions [76]. The result is given in **Fig. 53** in which the NA38 experimental points are also plotted. We see that the shape agrees rather well with the experimental data. The problem with the absolute normalization remains.

3. **Vogt, Prakash, Koch and Hansson** [77] assume that the hadron gas is made of π and ρ, which are called indifferently M or M' in **Table 4**. This model assumes the production of the charmonium states $η_c$ and χ, besides the J/ψ. As a consequence, the reactions M $η_c$, χ → J/ψ M' would create the J/ψ, in competition with the other reactions given in **Table 4**, which would absorb the J/ψ. The result on the survival probability as a function of P_T is given in **Fig. 54** for different values of the ratio ρ = (N $η_c$ + N_χ)/N_ψ (see section 7.3). For large values of ρ(≈ 10) the survival

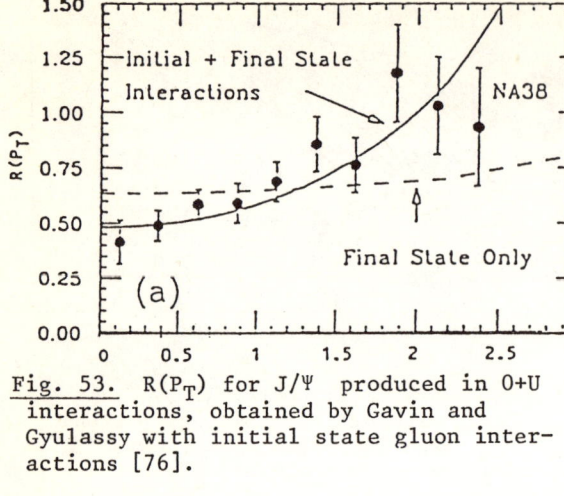

Fig. 53. $R(P_T)$ for J/Ψ produced in O+U interactions, obtained by Gavin and Gyulassy with initial state gluon interactions [76].

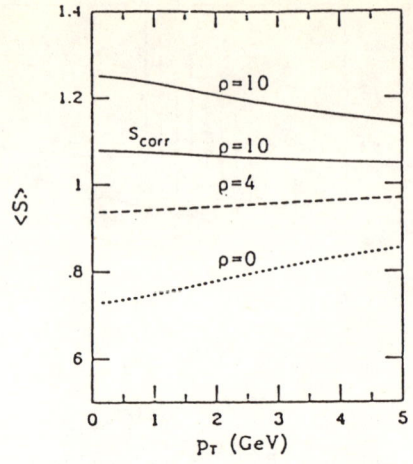

Fig. 54. $R(P_T)$ for J/Ψ produced in O+U interactions obtained by Vogt et al. for different values of the parameter ρ defined in section 7.3 [77].

probability is larger than 1, meaning that the net result from the several reactions given in **Table 4** is that there would be more J/ψ's created than absorbed. However, for $\rho = 10$ the curves have the wrong slope relative to the experimental data. For low values of ρ the curves do not reproduce the increase of $R(P_T)$ as a function of P_T given in **Fig. 46**.

4. Blaizot and Ollitrault [78] assume that the hadron gas is made of massless π only and make a detailed analysis of the relationships among the different parameters under three assumptions about the pion gas: a static uniform gas, a one-dimensional uniform expanding gas and a gas with a longitudinal and a transverse expansion. The survival probability of the J/ψ as a function of P_T is given in **Fig. 55** for two values of the impact parameter, b = 0 and = b = 6 fm. We see that in the range of P_T in which there are experimental data, $P_T < 3$ GeV/c, the increase of $R(P_T)$ is much smaller than in the experimental data shown in **Fig. 46**.

Blaizot and Ollitrault later introduced in the model initial gluon interaction [79]. The result is given in **Fig. 56** together with the NA38 experimental points only for a comparison of the shapes (the theoretical curve was normalized to the experimental one at $P_T = 1$ GeV/c). The inclusion of initial state interactions improve the shape of the curve but does not improve the discrepancy with the absolute normalization.

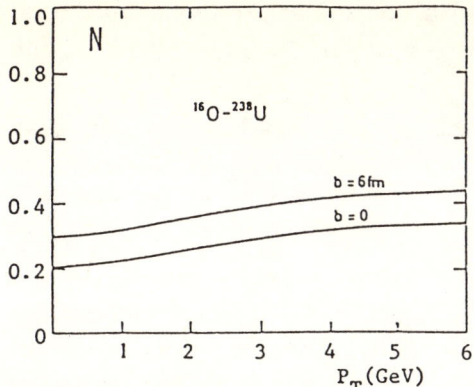

Fig. 55. $R(P_T)$ for J/Ψ produced in O+U interactions obtained by Blaizot and Ollitrault for two values of the impact parameter, b=0 and 6 fm, without initial state interactions [78].

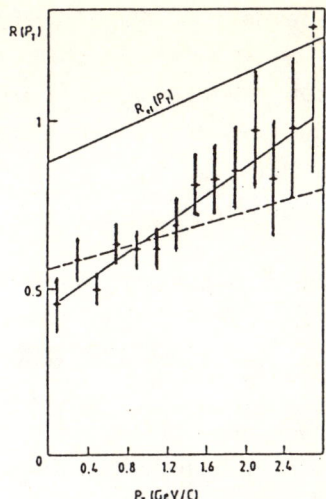

Fig. 56. $R(P_T)$ for J/Ψ produced in O+U interactions, obtained by Blaizot and Ollitrault, with initial state gluons interactions [79].

8.7 CONCLUSIONS FROM THE MODELS ON J/ψ ABSORPTION BY INELASTIC SCATTERING

1 - These models have to face enormous difficulties due to our ignorance on hadronic matter in the central and fragmentation regions; they must use as input several physical assumptions which may be plausible but are not proven.

2 - These models use for the J/ψ + nucleon cross section σ, which appears in expr. (31), (35) and (37), the value 1 to 3 mb which was obtained in J/ψ photoproduction experiments [67, 68]. These experiments did not measure directly this cross section, but indirectly by using the Vector Dominance Model and assuming implicitly that the $c\bar{c}$ pair and the J/ψ are produced at the same point inside the nucleus. Brodsky and Mueller [123] made a critical analysis of these experiments and concluded that their interpretations are not correct, because the experiments did not take into account the existence of the two different times, t_P and t_F, for the production of the $c\bar{c}$ pair and for the formation of the J/ψ, respectively. Because the J/ψ is formed outside the nucleus the experiments did not measure the J/ψ + nucleon cross section. Therefore, this cross section is not known.

3 - The parametrization of the variables is rather loose and final results may vary within factors about 2 to 4, even if we assume that the J/ψ + nucleon cross section is 1 to 3 mb.

4 - The present models assume that the J/ψ has P_T only and that $P_L^* = 0$. We have seen in section 7 that this not true; in the NA38 experiment the P_L^* of the J/ψ extends up to ~ 4 GeV/c, as shown in **Fig. 41**.

5 - The introduction of initial state parton interactions gives to the shape of the curve $R(P_T)$ versus P_T a trend which approaches the experimental one, but the discrepancies with absolute normalization remain.

6 - These models require the existence of a state of normal hadrons with a very high density, about 4 hadrons/fm^3. This is difficult to understand.

8.8 QCD AND OTHER THEORETICAL APPROCHES

In view of the great success of QCD as a theory of strong interactions we should expect that the ultimate explanation of J/ψ suppression should come from QCD. The QGP of course fits well into a QCD picture, but another approach, without the QGP assumptions, was shown by Close in his lectures at this school [80] : the EMC effect. Unfortunately, it is not possible to do an experimental test on this approach because there are no data on the EMC effect for gluons. But the idea is very appealing because it is based on an effect which is known to exist with quarks and is intellectually very elegant.

8.9 GENERAL CONCLUSIONS ON THE MODELS TO EXPLAIN THE J/ψ SUPPRESSION

We have seen that in the two classes of models, one assuming the formation of the quark-gluon plasma, the other assuming inelastic scattering of the J/ψ in some dense hadronic matter, the theoretical $R(P_T)$ differs from the experimental values by factors between 2 and 4. Taking into account on the one hand the uncertainties due to our ignorance on the plasma and resonances characteristics in the first class of models, and on the composition of the hadronic matter and on J/ψ - hadron cross section in the second class, and on the other hand the rather large NA38 experimental errors, there are no objective criteria to prefer one of these class of models to the other. Both are acceptable. However, if the quark-gluon plasma is not formed we have to understand the high hadron density per unit volume in the central region of the collision, which is obtained from a combination of the Bjorken model with the NA35 results.

It seems that within these two classes of models whatever explanation we choose for the J/ψ suppression, we need the $c\bar{c}$ pair going through a medium of high density, either of quark-gluon (plasma) or of hadronic matter. Finally, due to the success of QCD as a theory of strong interactions we should expect that the ultimate explanation for J/ψ suppression should come from QCD.

9. BOSE-EINSTEIN INTERFERENCE. BOSON-INTERFEROMETRY

9.1 HISTORY

The Bose-Einstein interference was discovered accidentally by Goldhaber et al. in 1959, in an experiment on antiproton-proton annihilation at 1.05 GeV/c in a propane bubble chamber [81]. Those authors were studying the reactions $\bar{p}p \to p^+p^+\pi^-\pi^-$ and $\bar{p}p \to \pi^+\pi^+\pi^+\pi^-\pi^-\pi^-$ in a search for a meson resonance which should decay into $\pi^+\pi^-$, whose existence had been predicted by Glashow. This resonance did not show up in the $\pi^+\pi^-$ effective mass spectrum, probably due to the small statistics of their experiment, and was later discovered by the Alvarez group at Berkeley; it is the ρ-meson. Since the resonance peak did not appear in the mass spectrum, Goldhaber et al. investigated whether some effect would appear in the angular distributions. They found out that pion pairs of the same charge have a higher probability for emission under small opening angles than pairs of opposite charges. The results are sketched in **Fig. 57**, in which $\theta^*_{\pi\pi}$ is the angle between two pions in the $\bar{p}p$ center of mass system. Just as another historical curiosity, in their data the unlike-charge pion pairs obeyed rather well the angular distribution predicted by the Fermi statistical model of particle production, which was fashionable at that time, but the like-charge pion pairs did not. This led the authors to conclude that perhaps that model should be improved (which was true, for several reasons).

Later, Goldhaber, Goldhaber, Lee and Pais interpreted that result as a consequence of pions obeying Bose-Einstein statistics : there is the interference of the wave functions of the two pions [82]. The asymmetry is then due to general principles of quantum mechanics. This became known as "GGLP effect".

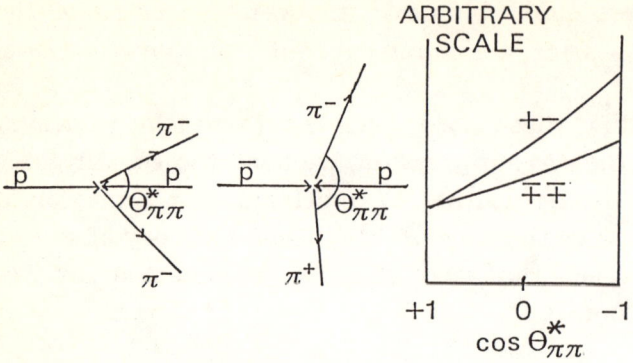

Fig. 57. The "GGLP effect". Bosons with the same-sign electric charges are emitted at smaller angles than bosons with opposit-sign charges.

Many experiments on the GGLP effect have been performed in the last 30 years, with different incident beams : pions, kaons, protons, antiproton, α-particles, neutrinos, heavy ions, as well as in e^+e^- annihilations. Good reviews on the subject were made by Goldhaber [83] and Lörstad [84]. In the latter paper an extensive list of references on different experiments is given.

9.2 BOSE-EINSTEIN INTERFEROMETRY AND THE DIMENSIONS OF THE BOSON SOURCE

Cocconi pointed out that the Bose-Einstein interference allows a measurement of the dimensions of the interaction region from which the bosons are emitted, as well as a measurement of the interaction time under certain assumptions [85]. He proposed a method analogous to the method used in radio-astronomy to measure the angular dimensions of radio sources, the Hanbury-Brown Twiss effect [86]. In radio-astronomy it consists in detecting two photons emitted by a distance source arriving simultaneously into two telescopes. In particle physics it consists in detecting two identical bosons emitted from an interaction. There is, however, a difference between radio-astronomy and particle production : in radio-astronomy the interference occurs near the photon detectors (telescopes), while in particle production the interference occurs near the source of the particles. For the Bose-Einstein interference of the wave functions of two bosons see some book on quantum mechanics, for instance Feynman's book [87].

9.3 CONDITIONS FOR THE MEASUREMENT OF THE INTERFERENCE

Let us assume two <u>identical</u> bosons produced simultaneously, within a short time interval Δt, with equal momenta, within a small difference Δp, and with random relative phases. The geometry is shown in **Fig. 58**, in which α and β are the points of emission of the two bosons, R the distance $\alpha\beta$, A and B two detectors placed at the distance r from the source, D the distance AB, $\Theta = D/r$ and $\rho = R/r$, a_1, a_2, b_1 and b_2 are the distances αA, βA, βB and αB respectively.

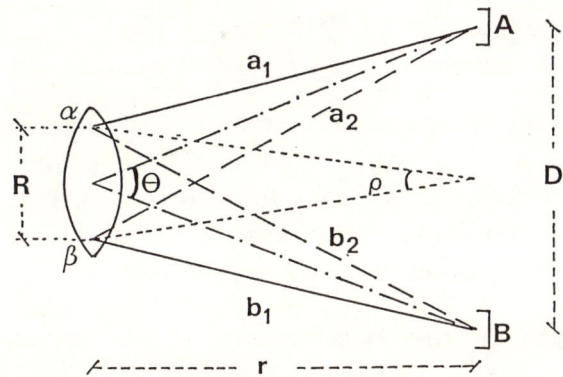

Fig. 58. Detection of the interference of two bosons wave functions. α and β are the points of emission of the bosons, A and B are detectors.

Let us call $K = 1/\lambdabar = p/\hbar$.

There are two indistinguishable coincidences:

1) A detects a particle issued from α and B detects a particle issued from β; the amplitude for this is $e^{ika_1} e^{ikb_1}$.

2) A detects a particle issued from β and B detects a particle issued from α; the amplitude for this process is $e^{ika_2} e^{ikb_2}$.

For two <u>bosons</u> the total amplitude is the <u>sum</u> of the two partial amplitudes. Therefore the <u>amplitude</u> of the coincidence is <u>proportional</u> to

$$e^{ika_1} e^{ikb_1} + e^{ika_2} e^{ikb_2} = e^{ik(a_1+b_1)} + e^{ik(a_2+b_2)}$$

The <u>rate of the coincidences</u> is proportional to the <u>square</u> of the amplitude:

$$\{e^{ik(a_1+b_1)} + e^{ik(a_2+b_2)}\} \{e^{-ik(a_1+b_1)} + e^{-ik(a_2+b_2)}\} =$$
$$\{2 + e^{ik[(a_2-a_1) + (b_2-b_1)]} + e^{-ik[(a_2-a_1) + (b_2-b_1)]}\}$$

We see that $(a_2 - a_1) + (b_2 - b_1) = \rho D$ and, therefore, the rate of coincidences is proportional to $2(1 + \cos KR\Theta)$.

The coincidence rates as a function of Θ has a modulation with maxima at $KR\Theta/2\pi = 0, 1, 2,...,n$. These maxima allow a measurement of R, the dimension of the source.

9.4 CONDITION ON THE MOMENTUM

The condition for the observation of the interference of the wave functions is that there must be the path ambiguity; this means that there must be simultaneity, within Δt, and the momenta of the two bosons must be equal, within Δp. The uncertainty relation must be satisfied for the effect of the Bose-Einstein statistics to operate:

$$\Delta p . \Delta x \leq \hbar = 197 \text{ MeV.fm} \qquad (39)$$

From here we conclude that for Δx of the scale of about 1 fm, Δp is about 200 MeV/c. The conclusion is that the momenta of both bosons must be nearly equal, they should differ by just few hundred MeVs. This condition must be satisfied experimentally when we choose the two identical bosons to study the interference of their wave functions.

9.5 HOW TO SELECT EXPERIMENTALLY THE TWO BOSONS WHICH SHOULD SATISFY THE CONDITION Δp SMALL

Several criteria are used in different experiments:

1) If $\vec{p}_1$, and $\vec{p}_2$ are the momenta of the two bosons, we can define the difference

$$\delta = ||\vec{p}_1| - |\vec{p}_2|| \qquad (40)$$

and study the interference as a function of δ. The interference should be greater for the smallest values of δ.

2) If $\vec{p}_{1T}, \vec{p}_{2T}$ and $\vec{p}_{1L}, \vec{p}_{2L}$ are the transverse and longitudinal momenta of the bosons, respectively, we can define the differences:

$$Q_T = ||\vec{p}_{1T}| - |\vec{p}_{2T}|| \text{ and } Q_L = ||\vec{p}_{1L}| - |\vec{p}_{2L}|| \qquad (41)$$

and study the interference as a function of Q_T and Q_L. The interference should be greater for the smallest values of Q_T and Q_L.

3) Let us call the four-momenta of the two bosons $q_1 = (E_1, \vec{p}_1)$ and $q_2 = (E_2, \vec{p}_2)$, respectively. The invariant mass squared of the two bosons is $M_{12}^2 = (E_1+E_2)^2 - (\vec{p}_1+\vec{p}_2)^2$ and the four-momenta transfer squared is

$$Q^2 = -(q_1 - q_2)^2 = M_{12}^2 - (m_1 + m_2)^2 \qquad (42)$$

We can study the interference as a function of Q^2, and it should increase as Q^2 decreases.

9.6 CORRELATION COEFFICIENTS

The use of the angular variable $\cos\theta_{\pi\pi}$ which was first introduced by Goldhaber et al. [81, 82] is rather cumbersome. It is more convenient to use some variables called <u>correlation coefficients</u>, which will be defined below.

Let us call σ the total cross section of the reactions in which the two bosons are produced, $d\sigma/dp_1$ and $d\sigma/dp_2$ the differential cross sections as functions of the two momenta. We define the <u>one-particle density</u> as:

$$\rho(p_1) = \frac{1}{\sigma}\frac{d\sigma}{dp_1} \quad \text{and} \quad \rho(p_2) = \frac{1}{\sigma}\frac{d\sigma}{dp_2}$$

and the <u>two-particle density</u> as

$$\rho(p_1, p_2) = \frac{1}{\sigma}\frac{d^2\sigma}{dp_1\,dp_2}$$

The <u>two-body correlation coefficient</u> C_2 is defined as the ratio:

$$C_2 = \frac{\rho(p_1, p_2)}{\rho(p_1)\rho(p_2)} \qquad (43)$$

The idea is to have the ratio of a quantity which would be affected by the interference to a quantity which would not. However, $\rho(p_1)$ and $\rho(p_2)$ are difficult to be determined for a given reaction and a given experimental detector geometry. Instead of this we usually take as a reference some other two-body density of a sample of events which do not have Bose-Einstein interference; very frequently we use the unlike-charge pairs of bosons. Calling $\rho_0(p_1,p_2)$ this reference, the correlation coefficient is defined experimentally as:

$$C_2 = \frac{\rho(p_1, p_2)}{\rho_0(p_1, p_2)} \qquad (44)$$

In practice, $\rho(p_1, p_2)$ is the number of bosons with charges -- or ++, and $\rho_0(p_1, p_2)$ is the number of bosons with charges +- (which do not have Bose-Einstein interference).

9.7 EXPERIMENTAL C_2 AND ANALYTICAL EXPRESSION FOR C_2

The way to work is the following:

1 - First, we measure C_2 experimentally, as a function of either δ, or Q_T and Q_L, or Q^2. The quantities $\rho(p_1, p_2)$ and $\rho_0(p_1, p_2)$ are just numbers of events.

2 - Second, we choose an analytical expression for C_2 which contains the radius R of the source, or its transverse and longitudinal dimensions, R_T and R_L, and sometimes contains also the proper-time τ for the boson freeze-out.

3 - Third, we fit the experimental values of C_2 to the analytical expression and from the fit we obtain R, or R_T and R_L, and τ if it is used in the equation for C_2.

9.8 ANALYTICAL EXPRESSIONS FOR C_2

There are several models which produce analytical expressions for C_2, for instance : a gaussian distribution of a source [83, 84], a model by Kolehmainen and Gyulassy [88], a model by Kapylov and Podgoretskii [89]. We shall describe the first two.

Gaussian distribution - The gaussian distribution of a source in the boson-pair center of mass system is :

$$\rho(r) = \frac{1}{4\pi^2 R^4} \exp(-r^2/2\ R^2) \qquad (45)$$

where R is the width of the source. The corresponding correlation function is :

$$C_2(Q^2) = A\left[1 + \exp(-Q^2\ R^2)\right] \qquad (46)$$

where Q^2 is the four-momentum transfer squared and A is a normalization constant.

We can see that there are two extreme conditions:

a) $Q^2 \to \infty$, in which case $C_2(Q^2) = 1$, therefore constant. In this case there is <u>no interference at all</u>. The production of the bosons is <u>completely coherent</u>.

b) $Q^2 = 0$, in which case $C_2(Q^2) = 2$. In this case there is <u>maximum interference</u>. The production of the bosons is <u>incoherent</u>; we say that there is the <u>maximum of chaoticity</u>.

The real experimental situation is in between these two extremes. We define a <u>chaoticity parameter</u> λ. For the gaussian source distribution the correlation function becomes

$$C_2(Q^2) = A\left[1 + \lambda \exp(-Q^2\ R^2)\right] \qquad (47)$$

$\lambda = 0$ corresponds to totally coherent boson emission from the source; $\lambda = 1$ corresponds to totally chaotic boson emission; $\lambda \neq 0,1$ may be due to a physical process or to experimental conditions.

We can also introduce in the gaussian distribution a transverse radius parameter R_T of the source and a longitudinal radius parameter R_L. The correlation coefficient becomes:

$$C_2(Q_T, Q_L) = A\left[1 + \lambda \exp(-Q_T^2 R_T^2/2) \exp(-Q_L^2 R_L^2/2)\right] \qquad (48)$$

Kolehmainen and Gyulassy model [88]. This model incorporates appropriately the relativistic collision dynamics of a high energy reaction. The correlation coefficient is given by:

$$C(Q_T, \Delta y, m_{T1}, m_{T2}) = A\left[1 + \lambda \frac{|G(p_1, p_2)|^2}{G(p_1, p_1) G(p_2, p_2)}\right] \qquad (49)$$

where : $G(p_1, p_2) = a\, K_0(\sqrt{u}) \exp(-Q_T^2 R_T^2/4)$

$K_0(\sqrt{u})$ = modified Bessel function of the complex argument u

$$u = 2m_{T1} m_{T2} \left(\tfrac{1}{4} T^2 + \tau_0^2\right)\cosh \Delta y$$
$$+ (m_{T1}^2 + m_{T2}^2)\left(\tfrac{1}{4} T^2 - \tau_0^2\right)$$
$$+ i\left(\tfrac{\tau_0}{T}\right)(m_{T1}^2 - m_{T2}^2)$$

λ = chaoticity parameter
R_T = transverse radius of the source
$Q_T = ||\vec{p}_{1T}| - |\vec{p}_{2T}||$
t_0 = proper time for pion freeze-out
T = temperature of the source
Δy = rapidity difference between the bosons
m_{T1}, m_{T2} = bosons transverse masses.

The experimental values of C_2 are fitted to the expr. (49) and the three free parameters λ, R_T and τ_0 are obtained from the fit.

9.9 EXPERIMENTAL RESULTS OBTAINED IN ULTRARELATIVISTIC HEAVY ION COLLISIONS

Two experiments with ultrarelativistic heavy ion collisions at CERN did pion interferometry : WA80 and NA35.

9.9.1 The WA80 experiment

The WA80 Collaboration did π° interferometry in the reactions 0 + Au and 0 + C at 200 GeV/nucleon [90, 91]. The π°'s were determined from the $\gamma\gamma$ invariant mass as described in section 5.2 and shown in **Fig. 25**. The experimental correlation coefficient was obtained by using

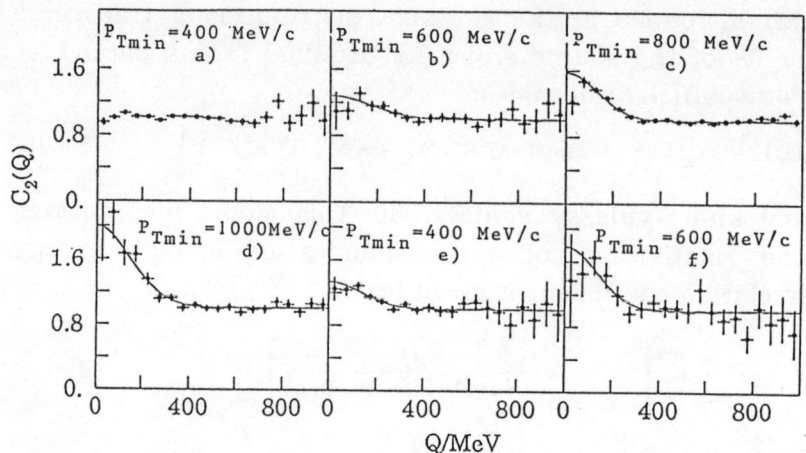

Fig. 59. WA80 correlation functions for two π°'s produced in 200 GeV/nucleon collisions : a, b, c and d are for O+Au, e and f for O+C [90, 91].

expr.(44) in which the reference $\rho_0(p_1,p_2)$ was obtained with two π°'s coming from two different events. The correlation coefficient was expressed as a function of the four-momentum transfer given by expr. (42). The source was assumed to have a gaussian distribution and the analytical correlation coefficient used was that given by expr. (47). Because there is a large background under the π° mass peak (see **Fig. 25**) several analyses were made with different cuts on the π° transverse momentum.

The experimental $C_2(Q)$ values and the fitted curves, expr. (47), are given in **Fig. 59**, as functions of Q; a) to d) correspond to O + Au, e) and f) to O + C interactions. The values of $P_{T\,min}$ in MeV/c in each of them are the p_T minimum cuts which were applied. We see that in the O + Au events there is no interference for P_{Tmin} = 400 and 600 MeV/c and there is interference for the cuts at 800 and 1000 MeV/c, for Q < 200 MeV. In O + C events the interference only appears for $p_{T\,min}$ = 600 MeV/c, also for Q < 200 MeV. The results of the fit are given in **Table 5** as a function of $P_{T\,min}$ and of the ratio π°-peak/background, which increases as P_T minimum increases. The conclusion from this analysis is that in the cases in which interference occurs the source radius is about 1.3 to 1.5 fm. This is the range of values for R which are usually found in different experiments, without any special feature. The chaoticity parameter increases as $P_{T\,min}$ increases.

Table 5 : *Results of the WA80 Collaboration on two π^{o}'s boson interferometry in $0 + C$ and $0 + Au$ interactions at 200 GeV/nucleon, as a function of $P_{T\ min}$ and of the ratio π^{o} mass peak/background. R is the source radius and λ is the chaoticity parameter.*

Cut-off $P_{T\ min}$ MeV/c	π^{o} mass peak / background	R fm	λ
400	O + C 0.23	1.41 ± 0.23	0.33 ± 0.06
600	0.39	1.47 ± 0.29	0.79 ± 0.23
400	O + Au 0.05		
600	0.11	1.16 ± 0.22	0.28 ± 0.07
800	0.13	1.50 ± 0.07	0.57 ± 0.05
1000	0.15	1.32 ± 0.12	1.0 ± 0.3

9.9.2 The NA35 experiment

The NA35 Collaboration did π^- interferometry in the reaction $0 + Au$ at 200 GeV/nucleon, with a streamer chamber in a magnetic field (**Fig. 21**) [92-94]. The correlation coefficient was measured as a function of Q_T and Q_L as given by expr. (41); the Q_L-dependence was set by selecting the events with $Q_L < 100$ MeV. The experimental correlation coefficients were obtained by using expr. (44), in which the reference $\rho_0(p_1, p_2)$ was obtained by taking two π^-'s coming from two different events. The events were analysed in different rapidity intervals of the π^-'s and for each interval two analyses were made : one assuming a gaussian distribution of the source expr. (48), the other assuming the Kolehmainen-Gyulassy's expr. (49). **Fig. 60** shows the experimental correlation coefficients $C(Q_T)$ as functions of Q_T for four rapidity intervals. The continuous lines in this Figure are the fit to the gaussian distribution. The results of the fits are given in **Table 6**. We should notice the high statistics of this experiment.

Fig.60 shows that interference occurs for $Q_T < 100$ MeV in all rapidity intervals, but it is stronger in the interval $2 < y < 3$. **Table 6** shows that :

a) there is agreement between the values of R_T and λ obtained with gaussian and Kolehmainen-Gyulassy source distributions in all rapidity intervals;

b) In the first, second and fourth intervals, $R_T \approx 4$ fm, $\lambda \approx 0.3$ and $\tau_0 \approx 3$ fm/c. However, in the interval $2 < y < 3$ the values are much higher : $R_T \approx 8$ fm, $\tau_0 \approx 6$ fm/c and $\lambda \approx 0.8$. If we consider a coherent central collision of an ^{16}O with an Au nucleus, the 16 nucleons of the oxygen will interact with 50 nucleons of the gold nucleus. The center of mass of the system composed of 16 nucleons at 200 GeV/nucleon + 50 nucleons at rest has rapidity $y = 2.5$ in the laboratory system (see exercise 4 in section 2). This rapidity is inside the interval $2 < y < 3$ in which the radius R_T of the source, λ and τ_0 are about twice as large in the other rapidity intervals.

Table 6 : *Results of the NA35 Collaboration on two π^-'s boson interferometry in 0+Au interactions at 200 GeV/nucleon*

y intervals	Number of π^- pairs	Gaussian distribution R_T fm R_L fm λ	Kolehmainen Gyulassy distribution R_T fm τ_0 fm/c λ
$1 < y < 4$	203612	4.1 ± 0.4 3.1+0.7-0.4 0.31+0.07-0.03	3.6 ± 0.3 2.9 ± 0.7 0.29 ± 0.05
$1 < y < 2$	24633	4.3 ± 0.6 2.6 ± 0.6 0.34+0.09-0.06	4.0 ± 0.7 2.5 ± 1.0 0.30 ± 0.12
$2 < y < 3$	39310	8.1 ± 1.6 5.6+1.2-0.8 0.77 ± 0.19	7.3 ± 1.6 6.4 ± 1.0 0.84 ± 0.15
$3 < y < 4.5$	20282	4.3+1.2-0.8 5.8 ± 2.2 0.55 ± 0.20	

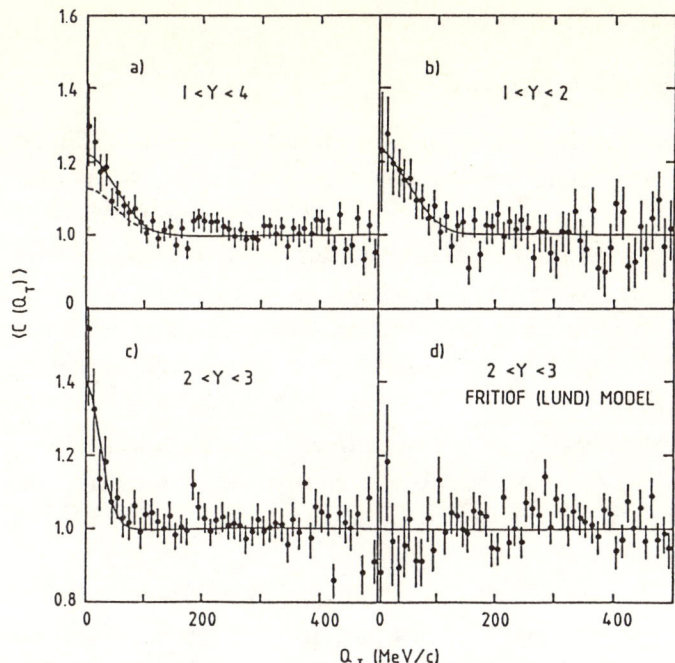

Fig. 60. NA35 correlation functions for two π^-'s produced in O+Au collisions at 200 GeV/nucleon, for different rapidity intervals [92-94].

9.10 CONCLUSIONS FROM BOSON INTERFEROMETRY

The WA80 experiment results on two π°'s interferometry give pion source dimensions about 1.3 to 1.5 fm, which is within the range usually found in experiments with hadrons and leptons beams. We should notice that in this experiment the π°'s are selected with a large background (see **Fig. 25**).

The NA35 experiment results on two π^-'s interferometry give pion source dimensions about 8 fm in the rapidity interval $2 < y < 3$, which is much higher than what is found in experiments with hadrons and lepton beams, and about twice as large as the values found in the other rapidity intervals in this experiment. In this same rapidity interval, also the values of λ and τ_0 are about twice as large as in the other three rapidity intervals. This is a very interesting result, and perhaps it is an indication of something new. We should recall again that the rapidity interval $2 < y < 3$ contains the y_{cm} of a coherent collision of the oxygen nucleons with 50 nucleons of gold nucleus.

10. FLUCTUATIONS AND INTERMITTENCY

10.1 STATISTICAL AND PHYSICAL FLUCTUATIONS

The experimental distribution of a variable has fluctuations. There is always a statistical fluctuation due to the finite number of events. However, fluctuations in a distribution might be due to some physical process. For instance, just as a conceptual consideration, let us assume that the quark-gluon plasma is created in some collision and that in some events many pions coming from the plasma are concentrated in a particular solid angle; the rapidity distribution of pions will have a fluctuation in that solid angle.

It is important to know whether a fluctuation is in fact statistical or is due to some physical process. We usually consider that a fluctuation is not just due to statistics when it is larger than 5 or 6 standard deviations.

<u>Question</u> : Is it possible to decide if the fluctuations observed in a distribution of a variable have a physical component? Let us consider as an example the pseudorapidity distribution of charged particles of the cosmic ray events found by the JACEE Collaboration, shown in **Figs. 6 and 7**, section 2.5.3; are the fluctuations due to the limited number of particles, even if this number is large for a single event, or are they due to some physics, for example, the production of quark-gluon plasma ?

A new method of analysis of fluctuations has recently been proposed by Bialas and Peschanski [95-97], based on <u>scaled factorial moments</u>. The method allows the detection of large non-statistical fluctuations as well as investigating the pattern of the fluctuation (eventually leading to their physical origin). Those authors call <u>*intermittency*</u> the physical fluctuations, by analogy with the phenomenon of intermittency in fluid turbulence. The method was originally applied to the pseudorapidity distribution of the charged particles detected in the JACEE cosmic ray events just mentioned.

Large fluctuations have been observed in different types of collisions, for example $\bar{p}p$ [98], $\pi^{\pm}p$ [99], e^+e^- annihilations [100] and μp [101].

10.2 THE SCALED FACTORIAL MOMENTS

The scaled factorial moments can be applied to the distribution of any variable. To fix the idea we shall consider the distribution of rapidity.

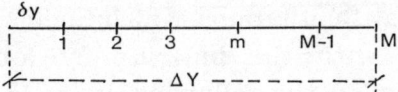

Fig. 61. The overall rapidity interval Δy is divided into M bins of equal size δy.

Let us call : Δy an overall rapidity interval in which a measurement was made and N the total number of particles in the interval Δy. Let us divide Δy into M bins of equal size δy each (**Fig. 61**):

$$\delta y = \frac{\Delta y}{M}$$

and call K_m the number of particles in the m th bin. We have

$$N = \sum_{m=1}^{M} K_m$$

We shall consider two cases :

case 1 - there is one single event with a given multiplicity N in the interval Δy;

case 2 - there is a sample of many events, each one with a different multiplicity.

In case 1, the <u>scaled factorial moment of the ith order</u> is defined as :

$$F_i = \frac{1}{M} \sum_{m=1}^{M} M^i \frac{K_m (K_m-1)\ldots(K_m-i+1)}{N.(N-1)\ldots(N-i+1)} \tag{50}$$

In case 2, an average F_i is defined as :

$$\langle F_i \rangle = \frac{1}{M} \sum_{m=1}^{M} \frac{M^i}{\langle N \rangle^i} K_m(K_m-1)\ldots(K_m-i+1) \tag{51}$$

where $\langle N \rangle$ is the average multiplicity of all events in the overall interval Δy.

10.3 IMPORTANT PROPERTIES OF $\langle F_i \rangle$

Bialas and Peschanski have shown that:

1 - The scaled factorial moments F_i averaged over many events are equal to the moments of a true probability distribution. The statistical fluctuations are present, but their effect does not hide a true probability distribution.

2 - Different patterns of non-statistical fluctuations will give different absolute values of $\langle F_i \rangle$.

3 - Different patterns of non-statistical fluctuations will influence differently the variation of $< F_i >$ with the bin size δy (or equivalently, with M). This can be seen in the following example. Let us assume a fluctuation which occurs in a fixe range S, i. e. in a fixed scale S of the variable. The scale S can be of the order of, or greater, or smaller than the small bin size δy in the overall interval Δy, in which the fluctuation occurs. Then, if δy < S the fluctuations occur in intervals which are larger than δy, and as a consequence the values of $< F_i >$ are indepen of the size of δy; we can make δy of smaller sizes and this will not affect the fluctuations, because they occur in intervals which contain δy. If δy ⪆ S the fluctuations occur in intervals which are smaller than δy; in this case the values of $< F_i >$ depend on the size of δy.

4 - For an intermittent pattern, the values of $< F_i >$ obey a power law :

$$< F_i > = \left(\frac{\Delta y}{\delta y}\right)^{\phi_i} = M^{\phi_i} \qquad (52)$$

From this expression we have :

$$\ln < F_i > = \phi_i \ln \Delta y - \phi_i \ln \delta y \qquad (53)$$

By making $\phi_i \ln \Delta y = a_i$, we have

$$\ln < F_i > = a_i - \phi_i \ln \delta y \qquad (54)$$

The conclusion from expr. (52) is that there must be a linear relationship between $\ln < F_i >$ and $-\ln \delta y$.

5 - The following relationship should hold :

$$\Phi_i = \binom{i}{2} \phi_2 \qquad \text{for } i \geq 2 \qquad (55)$$

where we use the notation $\binom{i}{2}$ for the combination of i objects 2 by 2.

10.4 EXPERIMENTAL RESULTS WITH ULTRARELATIVISTIC HEAVY ION COLLISIONS

Three experiments on intermittency have been performed with heavy ion beams; two of them with nuclear emulsions, EMU07 or KLM Collaboration and EMU01 Collaboration, the third was done by the WA80 Collaboration.

10.4.1 Nuclear emulsion experiment by the KLM Collaboration

The KLM Collaboration [102] studied intermittency in the following interactions : proton-emulsions at 200 and 800 GeV/c, with emulsions exposed at FERMILAB; ^{16}O-emulsions at 60 and 200 GeV/nucleon, with emulsions exposed at CERN.

Table 7 gives the number of events, the pseudorapidity intervals covered and the average multiplicity of minimum ionizing particles, $< N_s >$ inside the correponding pseudorapidity intervals. We should notice that the numbers of events are relatively small, specially in the cases of the ^{16}O beams, the η intervals are large in all cases and $< N_s >$ is also large (in the proton interactions, only events with $N_s > 10$ were kept for the analysis).

Table 7 - *Data of the KLM Collaboration*

Beam	Number of events	Number of events with $N_s > 10$	η interval	$< N_s >$
p 200 GeV/c	2595	1542	0.5 - 5.5	17.3 ± 0.4
p 800 GeV/c	1749	1336	0.5 - 6.5	22.2 ± 0.6
^{16}O 60 GeV/nucleon	226	226	0.5 - 4.5	93.1 ± 6.0
^{16}O 200 GeV/nucleon	146	146	0.5 - 5.5	154 ± 13

Results - **Fig. 62** shows $\ln < F_i >$ as a function of $-\ln \delta\eta$ from the second order to the fifth order scaled factorial moments for proton-emulsion interactions at 200 GeV/c in a) and 800 GeV/c in b). **Fig. 63** shows the same plots from the second to the sixth order scaled factorial moments for ^{16}O-emulsion interactions at 60 GeV/nucleon in a) and 200 GeV/nucleon in b). In both Figures the solid lines are linear fits to the data. The conclusion is that in the four cases the experimental data follow the linear relationship (54) and shows clearly an intermittent pattern.

In order to check whether the linear dependence of $\ln < F_i >$ on $-\ln \delta y$ is not due to the method itself, a comparison was made with Monte Carlo events; these have been generated with random multiplicities and pseudorapidity distributions. Results on $\ln < F_2 >$ and

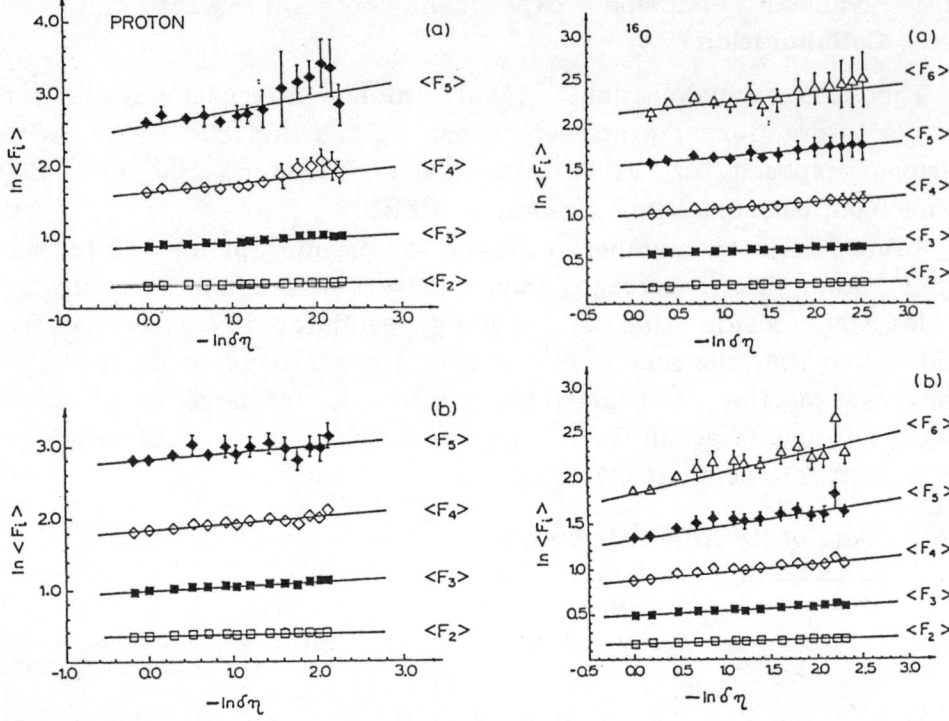

Fig. 62. KLM Collaboration ln $<F_i>$ versus bin size $\delta\eta$, for proton-emulsion interactions : a) 200 GeV/c, b) 800 GeV/c [102].

Fig. 63. KLM Collaboration ln $<F_i>$ versus bin size $\delta\eta$ for oxygen-emulsion interactions :
a) 60 GeV/nucleon, b) 200 GeV/c nucleon [102].

ln$< F_4 >$ are given as examples in **Fig. 64**, in which the full squares correspond to experimental data and the open ones to Monte Carlo events : a) for protons at 200 GeV/c, b) for ^{16}O at 200 GeV/nucleon. We see clearly that the Monte Carlo events have a flat distribution and, therefore, show no intermittency pattern.

Fig. 65 gives the slopes ϕ_i of eq. (54) as a function of the order i, obtained from the straight lines of **Figs. 62** and **63**; in a) for proton events, in b) for ^{16}O events.

Conclusions - The conclusions from this experiment are :
   a) the pseudorapidity distributions of particles produced in proton-emulsion and the ^{16}O-emulsion collision show an intermittent pattern.
   b) Monte-Carlo generated events do not show intermittent pattern. For the proton events **Fig. 65a** shows that :

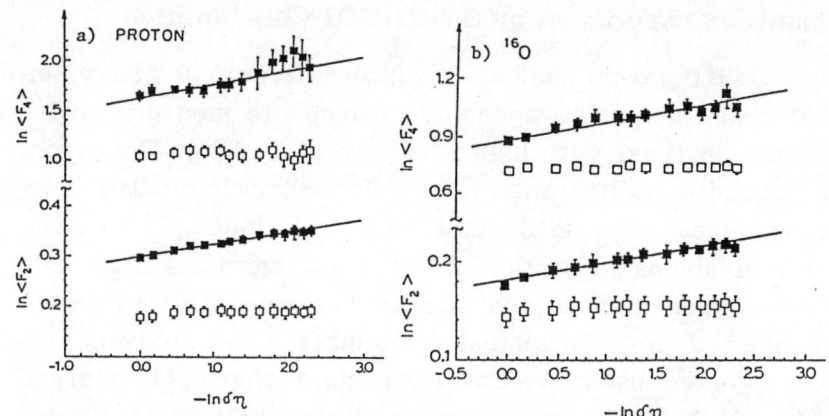

Fig. 64. KLM Collaboration comparison of $\ln <F_2>$ and $\ln <F_4>$ versus bin size $\delta\eta$ from data (full squares) with Monte Carlo events (open square) : a) proton emulsion interactions at 200 GeV/c, b) oxygen emulsion interactions at 200 GeV/nucleon [102].

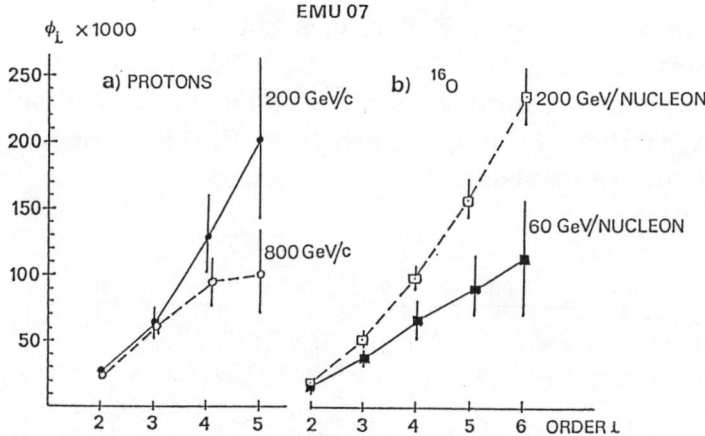

Fig. 65. KLM Collaboration results on the slopes Φ_i : a) for proton emulsion interactions; b) for oxygen emulsion interactions. From Ref. [102]. See text.

c) the slopes ϕ_i increase with the order i : $\phi_5 > \phi_4 > \phi_3 > \phi_2$;

d) for the same order i there is <u>no correlation</u> between ϕ_i at 200 GeV/c and ϕ_i at 800 GeV/c.

For the ^{16}O events **Fig. 65b** shows that :

e) again $\phi_6 > \phi_5 > \phi_4 > \phi_3 > \phi_2$;

f) for the same order i there is a correlation, ϕ_i at 200 GeV/nucleon is larger than ϕ_i at 60 GeV/nucleon; this seems to show that for the heavy ion interactions the slope is larger for larger interacting systems.

10.4.2 Nuclear emulsion experiment by the EMU01 Collaboration

The EMU01 Collaboration looked for intermittency in heavy ion collisions with emulsion chambers specially designed to measure angles of produced charged particles with high accuracy [103, 104]. The errors were kept in the pseudorapidity range $1 < \eta < 7$ smaller than 0.013 units of pseudorapidity (see eq. (22) and the relationship between pseudorapidities and angles in section 2.5.3). Data were taken at 200 GeV/nucleon with oxygen and sulphur beams. For the latter thin gold foils have been inserted into the emulsion chambers. The analysis was made with 80 oxygen-emulsion events with more than 150 charged particles and 40 sulphur-gold events with more than 300 charged particles. This multiplicity cuts automatically exclude the contribution from lighter target components (H, C, N and O). Analysis was made also with FRITIOF Monte Carlo events having the same cuts as the data.

The results for the second scaled factorial moment F_2 are shown in **Fig. 66** as a function of the bin size. Full circles correspond to O + emulsion and open circles to S + Au events, the solid line to result from FRITIOF. Identical plots are obtained for the moments F_i of other orders i. We can see that there is a departure from FRITIOF for the S + Au events, but not for O + emulsion.

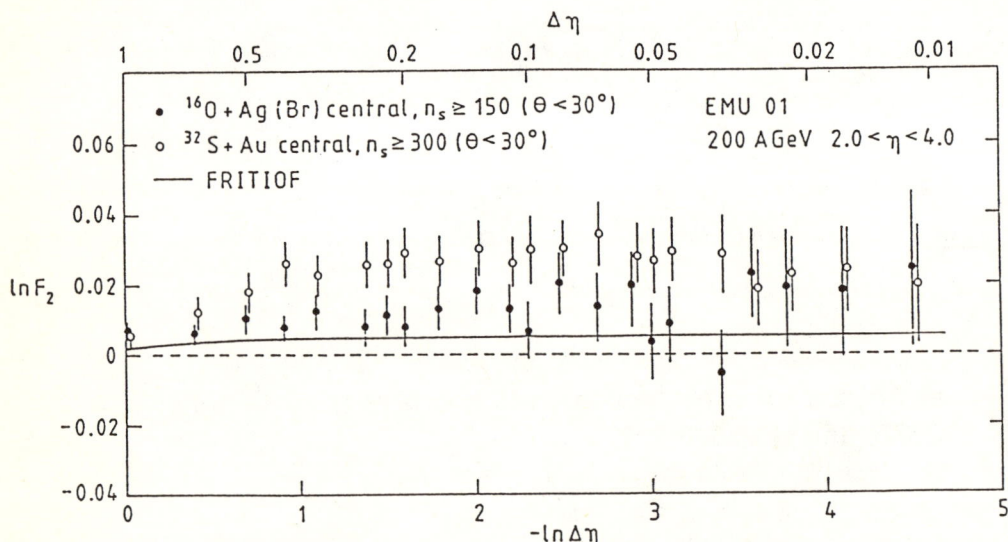

Fig. 66. EMU01 dependence of $\ln F_2$ on the bin width : full circles for oxygen emulsion central interactions, open circles for sulphur gold central interactions, both at 200 GeV/nucleon. Full line is FRITIOF result [104].

The authors also looked for effects in the transverse direction by studying the formation of clusters of particles as a function of the azimuthal angle. A small effect was detected in S + Au events and no effect in O + emulsion events.

The conclusions from this experiment is that a small effect of non-statistical fluctuations in the longitudinal direction and a small effect for clusters in the transverse direction are observed in S + Au interactions, but not in O + emulsions interactions, at 200 GeV/nucleon.

10.4.3 The WA80 experiment

The WA80 experiment measures multiplicities with Iarocci type streamer tubes in the rapidity range $2.4 \leq \eta \leq 4.0$ (see **Fig. 18** in section 4.3). This Collaboration looked for intermittency in $^{16}O+C$ and $^{16}O+Au$ interactions at 200 GeV/nucleon [105]. The energy measured by the zero-degree calorimeter, E_{ZDC}, is correlated with the number of nucleon-participants (or with the centrality of the collision). From Monte Carlo generated events the collisions were classified into <u>medium</u> and <u>central</u>, when the numbers of projectile participants are 7 to 11 and 15 to 16, respectively. **Table 8** gives the energy measured by the zero-degree calorimeter, the number of events and the mean multiplicity in each case. We should notice the good statistics of this experiment and the high mean multiplicities.

Table 8 - *Main characteristics of the WA80 Collaboration data*

Collision (200 GeV/nucleon)	E_{ZDC} TeV	Number of events	Mean multiplicity
O + C medium	2.2 - 2.4	38 K	23.0 ± 0.1
O + C central	$\leq$ 1.7	114 K	45.5 ± 0.1
O + Au medium	1.0 - 1.8	31 K	66.1 ± 0.4
O + Au central	$\leq$ 0.4	92 K	117.9 ± 0.7

<u>Results</u> - The analysis was done both with the data and with FRITIOF Monte Carlo generated events. The $\ln \langle F_i \rangle$ as a function of $-\ln \delta\eta$ is given in **Fig. 67** for O + Au interactions at 200 GeV/nucleon, from the second to the sixth order scaled factorial moments : in a) for medium, in b) for central collisions. The corresponding results for O + C

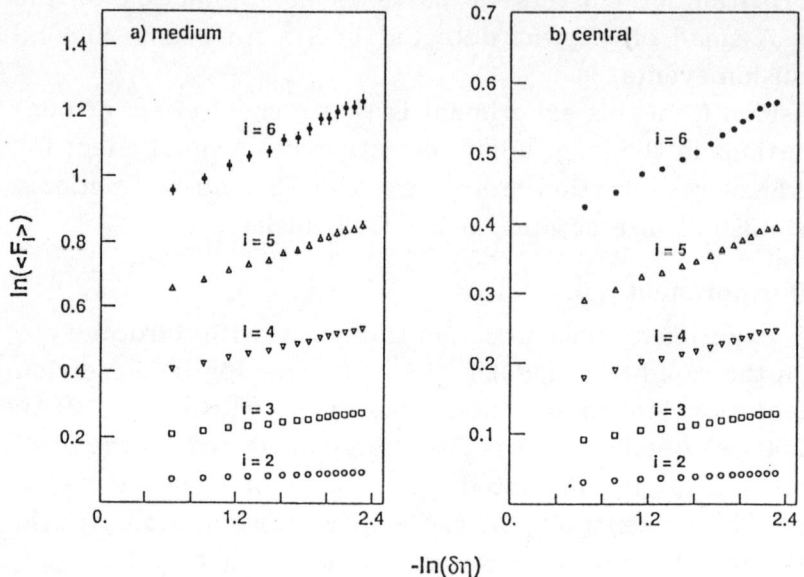

Fig. 67. WA80 result on the dependence of $\ln <F_i>$ on the bin size for i from 2 to 6 for a) medium and b) central O+Au interactions at 200 GeV/nucleon [105].

interactions at 200 GeV/nucleon are shown in **Fig. 68**. We see that in all cases there is intermittent pattern in the fluctuations of the pseudorapidity distributions. **Fig. 69** shows the results obtained with the Fritiof Monte Carlo events also for i = 2 to i = 6; there is <u>no</u> intermittent pattern.

The slopes ϕ_i of eq. (54), obtained from the straight lines of **Figs. 67 and 68**, are given in **Fig. 70**, for O+Au an O+C interactions. Finally the validity of eq. (55) was checked by making the plot given in **Fig. 71** in which one sees that $2\phi_i/i(i-1)$ is practically constant as a function of i.

Conclusions -The conclusions from the WA80 Collaboration are :

a) The pseudorapidity distributions of particles produced in O+Au and O+C collisions at 200 GeV/nucleon show intermittent pattern.

b) Fritiof Monte Carlo generated events do <u>not</u> show intermittent pattern.

c) Fig. 70 shows that both for O+C and O+Au collisions the slope ϕ_i are larger in medium than in central collisions, and for each type of collision they are larger for O+C than for O+Au.

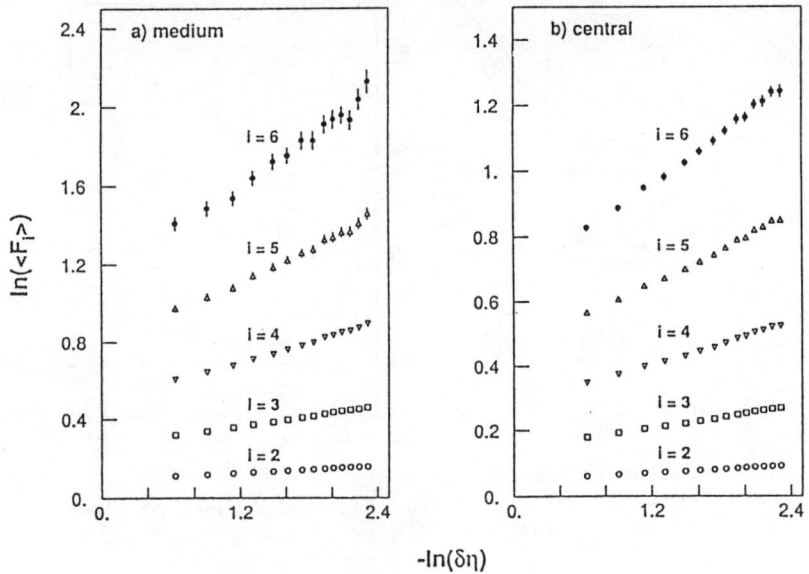

Fig. 68. The same as Fig. 67 for O+C interactions at 200 GeV/nucleon [105].

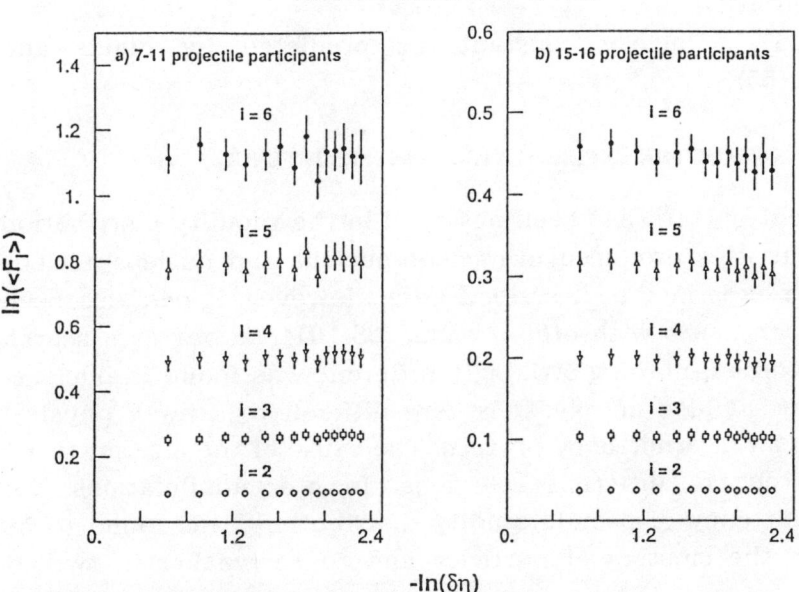

Fig. 69. WA80 dependence of $\ln <F_i>$ of FRITIOF events on the bin size for a) medium and b) central O+Au interactions at 200 GeV/nucleon [105].

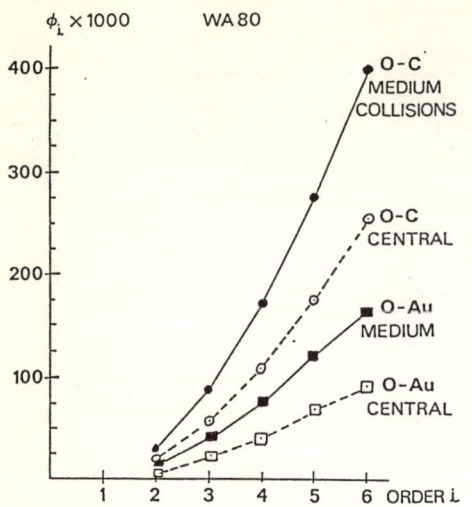

Fig. 70. WA80 results on the slopes ϕ_i for medium and central O+C and O+Au interactions at 200 GeV/nucleon. From ref. [105].

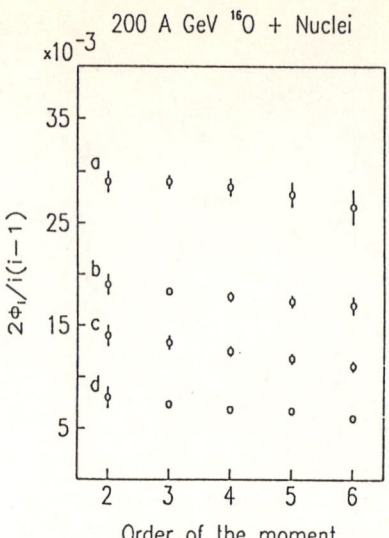

Fig. 71. WA80 check of the validity of eq. (55) [105]. See text.

d) Conclusion c) seems to show that the slopes ϕ_i are smaller for larger interacting system. This in contradiction with conclusion f) from the KLM Collaboration.

e) $2\phi_i/i(i-1)$ is nearly constant, as predicted by Bialas and Peschanski, eq. (55).

10.5 CONCLUSIONS FROM THE EXPERIMENTS ON INTERMITTENCY

Intermittent pattern has been observed in the rapidity distribution of secondary particles produced in nucleon-nucleus and nucleus-nucleus collisions in the experiments described here. However, it has also been observed in experiments with other beams [98-101]. As far as a search for QGP is concerned, nothing strikingly different was found in nucleus-nucleus collisions at high energy. It is very difficult to draw a physical conclusion from an intermittency pattern. The cause of the intermittency might be different in different reactions. In e^+e^- annihilations, for example, intermittency of pseudorapidity distributions was found to be associated with the clusters of particles and to be weaker in two-jet events [100].

The subject is needing new ideas and new experiments. We should watch its evolution.

11. STRANGE PARTICLES PRODUCTION

11.1 STRANGE PARTICLES AS A POSSIBLE SIGNATURE OF THE QGP

It has been suggested that one of the possible signatures of the QGP would be an enhancement of the production of strange particles relative to the production observed in nucleon-nucleon collisions. For reviews see Koch, Müller and Rafelski [106], and Rafelski [107]. We shall give some of the arguments which have been suggested.

In a QGP gluons are abundantly produced and many quark-antiquark pairs are created via gluon-gluon fusion. In this process, strange quark-antiquark pairs would be produced more frequently than in the usual nucleon-nucleon collisions because the coupling of gluons to strange quarks and to light quarks is the same and the higher mass of the strange quarks would not be important due to the high available energy. Another argument comes from the fact that the transverse energy spectrum measured in heavy ion collisions at CERN at 200 GeV/nucleon extends only to about half of the kinematical limit [108], implying that the central rapidity region should be rich in baryons. In a baryon-rich environment which also contains high density of u and d quarks and antiquarks, further production of u and d quarks from $u\bar{u}$ and $d\bar{d}$ pairs should be suppressed by the Pauli principle : this should favour a higher relative abundance of strange quarks. Because of the abundance of strange quarks and antiquarks in the QGP, Jacob and Rafelski [109] have pointed out that the ratio of antihyperons to antiprotons might increase. As examples, $\bar{\Lambda}/\bar{p}$ or $\bar{\Xi}/\Xi^-$ ratios could be better indicators of the presence of QGP than Λ/π or K/π ratios, because both Λ and K can be produced by other mechanisms in secondary interactions, like $\pi N \rightarrow \Lambda K+X$, $\bar{K} N \rightarrow \Lambda +X$, etc. These mechanisms would produce Λ and K as serious backgrounds to a possible increase in their numbers due to QGP. In contrast, <u>strange antibaryons</u> are not abundantly produced in secondary interactions; an enhancement in their production rates should be explained by some new process, like QGP for example. It has been proposed by Rafelski and Danos [110] that one should look for strange particles with relatively large transverse momentum (say $P_T \gtrsim 1$ GeV/c), because they should originate from the early conditions of the QGP; this should be more favourable to the observation of the QGP than in the entire range of P_T, because of the large number of low P_T particles produced during the hadronization phase.

Results on strange particle production in heavy ion collisions have been obtained by the E802, NA34, NA35 and WA85 Collaborations. The NA38 Collaboration tried to obtain information on K production via an indirect method by using the muon from the K-decay. The NA36 Collaboration [111] should have results soon. An interesting experimental review on strangeness production in nucleus-nucleus collisions in all available energies, from the Bevalac at 2.1 GeV/nucleon up to CERN at 200 GeV/nucleon was made by Odyniec [112].

11.2 THE E802 EXPERIMENT

This experiment, described in section 4.2 and **Fig. 17**, has a powerful identification system for charged particles. The E802 Collaboration studied the production of $K^\pm$ and $\pi^\pm$, and their ratios, in Si+Au <u>central</u> collisions at 14.5 GeV/nucleon [113,114]. The ratios they found for all events integrated on P_T and the ratios known in proton-proton collisions at equivalent energies are:

	K^+/π^+ (%)	K^-/π^- (%)
Si + Au	24 ± 5	4 ± 2
p + p	7 ± 2	4 ± 2

showing a net enhancement of K^+/π^+ in Si+Au relative to p+p. The most interesting results are those ratios measured as a function of P_T. This is shown in **Fig. 72**, where those ratios are plotted with ratios obtained in p+p and p+Pb interactions at comparable energies. The enhancement is more pronounced in K^+/π^+ than in K^-/π^- and increases with P_T. The results are compared with the ones obtained with FRITIOF Monte Carlo and shown in **Fig. 73** where is plotted the "double ratio" K/π, i.e. the ratio (K/π) experimental / (K/π) FRITIOF. Since FRITIOF results agree with their π-production, the authors conclude that the enhancement in the K/π ratio is due to an enhancement in K-production.

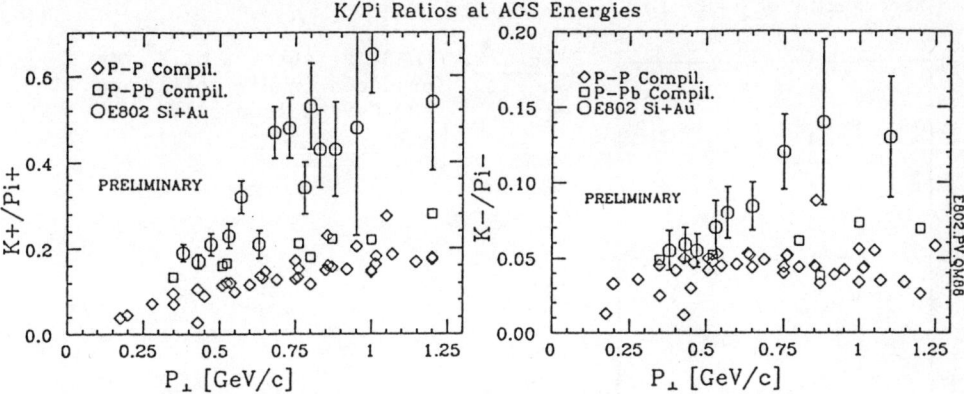

Fig. 72. E802 compilation of K/π ratios versus P_T in p-p and p-A collisions at the AGS energies compared to Si+Au [113].

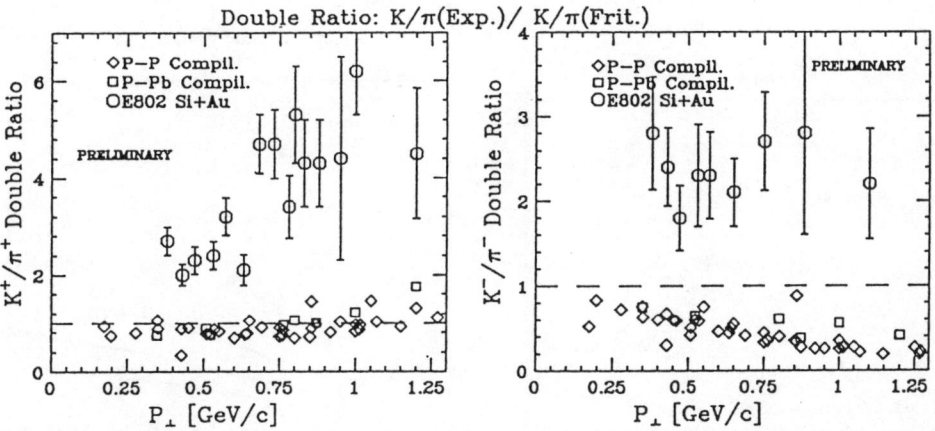

Fig. 73. E802 "double ratio" of experimental K/π yields to that of FRITIOF, versus P_T [114].

11.3 THE NA34 EXPERIMENT

The NA34 Collaboration measured in the External Spectrometer of their detector (section 4.5, **Fig. 20**) the $K^{\pm}$ and $\pi^{\pm}$ transverse momentum and transverse mass spectra in S+W collisions at 200 GeV/nucleon, in the rapidity range 1−1.3 and P_T range 0.15−0.45 GeV/c [115]. **Fig.74** shows the K^+ and K^- P_T spectra. The K^+ spectrum is flatter than the K^-; this is in contrast to what is observed in proton-proton collisions, where the two spectra have approximately the same slope. It could be due to an enhancement of the K^+/π^+ ratio in the ions collisions relative to proton-proton collision.

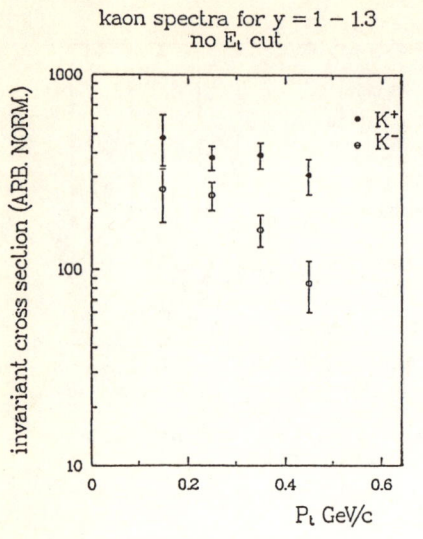

Fig. 74. NA34 P_T spectra of K^+ and K^- produced in S+W collisions at 200 GeV/nucleon [115].

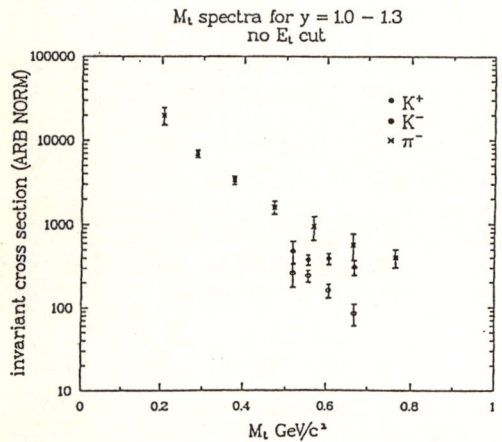

Fig. 75. NA34 transverse mass spectra for K^+, K^- and π^- produced in S+W collisions at 200 GeV/nucleon [115].

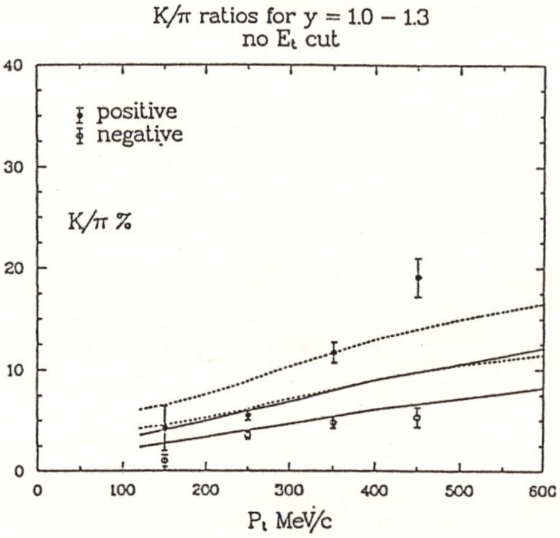

Fig. 76. NA34 K^+/π^+ and K^-/π^- ratios versus P_T, compared with results from the ISR [115]. See text.

Fig. 75 shows the transverse mass spectra for K^+, K^- and π^-. It is not surprising that the K^+ spectrum of M_T is flatter than the K^- spectrum, because it should reflect the behaviour of P_T, which is shown in **Fig. 74**. However, it is interesting that the K^+ spectrum is also flatter than the π^-. **Fig. 76** shows the K^+/π^+ and K^-/π^- ratios. The dashed and solid lines show the K/π ratios observed in proton-proton interactions at the CERN ISR for $\sqrt{s}$ = 23 GeV for positive and negative particles, respectively, normalized to the rapidity interval 1—1.3. Only the K^+/π^+ point at 450 MeV/c is higher than the proton curves, the other points of K^+/π^+ and K^-/π^- are not.

We should emphasize that this analysis was made without any transverse energy cuts, i.e., the selected events are a mixture of central and peripheral collisions.

Experiments NA34 and E802 both measured K/π ratios, but there are two important differences between their events. First, in the NA34 there is no selection of central collisions, whereas the E802 events are all selected as central collisions. Second, the NA34 events have rather low P_T, in the range $\sim$ 0.15 to 0.45 GeV/c; whereas in the E802 the P_T range is higher about 0.4 to 1.2 GeV/c. The excess of K/π relative to proton-proton in E802 appears at $P_T \sim$ 0.6 GeV/c, outside the range of NA34.

11.4 THE NA38 EXPERIMENT

The NA38 apparatus detects neither pions nor kaons (section 4.7, **Fig. 22**), but detects a very large number of muon pairs. From these, the like-sign pairs, $\mu^+\mu^+$ and $\mu^-\mu^-$, come from π – and K – decays. The Collaboration studied the single muons from the like-sign pairs produced in oxygen-uranium, sulphur-uranium and proton-uranium collisions at 200 GeV/nucleon in order to obtain, indirectly, information about the P_T distribution and the strangeness content of the sample of the parent pions and kaons [116, 117]. The analysis is based on large statistics, about 437 000 muons from oxygen-uranium collisions, a comparable number for sulphur-uranium and 4800 from proton-uranium; about 60% of them originate from pion decays and about 40% from kaons. The trigger accepts muons in the pseudorapidity range 2.8 – 4 in the L.S. with $P_T > 0.6$ GeV/c. The P_T distributions of the single muons were fitted to $(1/P_T)\, dN/dP_T \propto \exp(-P_T/P_{T0})$ in the ranges $0.7 < P_T < 2.8$ GeV/c for O-U and S-U, and $0.7 < P_T < 2.1$ GeV/c for p-U. The inverse slope parameter P_{T0} was found to be 199 ± 3 MeV/c for O-U, 203 ± 2 MeV/c for S-U and 206 ± 4 MeV/c for p-U. The O-U and S-U data fitted for different transverse energy bins show a slight increase of P_{T0} with E_T (**Fig. 77**).

The K/π ratio is obtained from a model which uses previous measurements made in proton-tungsten interactions at 200 GeV/c [50]; no appreciable difference was found between O-U, S-U and p-U collisions.

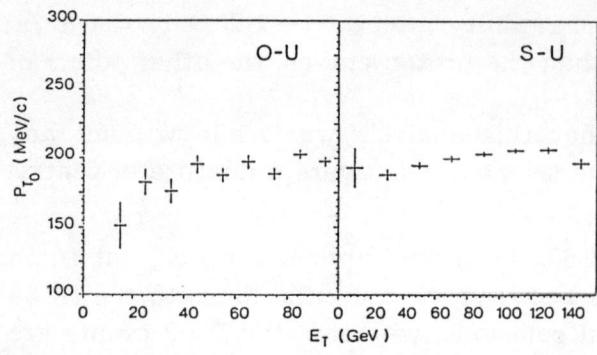

Fig. 77 NA38 inverse slope P_{T_0} as a function of E_T of O-U ans S-U reactions [116, 117].

This method of studying indirectly the π and K production by using the decay muons did not show any significant differences between p-U, O-U and S-U collisions regarding both the K/π ratio and the inverse slope P_{T0}, apart the slight increase of P_{T0} with E_T.

In spite of the large statistics, nothing striking was found in this analysis.

11.5 THE NA35 EXPERIMENT

The NA35 Collaboration has interesting results on strange particles production in two experiments : a first on production of K_S^o, Λ and $\overline{\Lambda}$ in oxygen-gold and proton-gold interactions at 60 and 200 GeV/nucleon [118-120], and a second, on production of K_S^o and Λ in sulphur-sulphur interactions at 200 GeV/nucleon [118, 121]. The strange particles decays were identified in the streamer chamber of the detector (section 4.6, **Fig. 21**).

In the first experiment the numbers of K_S^o's and of Λ's were roughly between 300 and 500 for each beam, and those of $\overline{\Lambda}$'s were smaller, roughly between 30 and 80. In order to see whether there is an enhancement of strange particles production we must have a reference. In this experiment the numbers of strange particles were compared with the numbers of negative particles, which are predominantly pions. Obviously, any particle should be more abundantly produced in 0 + Au than in p + Au interactions and the rate of production should increase with the beam energy. **Table 9** gives the ratios of the average multiplicities of K_S^o, Λ and $\overline{\Lambda}$ to the average multiplicities of the negative particles in p + Au and 0 + Au collisions, at 60 and 200 GeV/c. In each case, the events in p + Au and 0 + Au interactions were taken in a same region of phase space, i.e. same P_T and rapidity intervals. This table shows a first interesting result: within the errors the ratios for K_S^o,

Table 9 - *Ratios of the average multiplicities of produced strange particles to the average multiplicities of negatively charged particles in p+Au and 0+Au collisions at 60 and 200 GeV/nucleon (see text). Predictions from FRITIOF Monte Carlo are also given.*

	60 GEV / NUCLEON			
	p + Au		0 + Au	
	Data	Fritiof	Data	Fritiof
$\langle K_s^o \rangle / \langle n_- \rangle$	0.254±0.108	0.126	0.150±0.039	0.145
$\langle \Lambda \rangle / \langle n_- \rangle$	0.267±0.070	0.075	0.244±0.038	0.100
$\langle \bar{\Lambda} \rangle / \langle n_- \rangle$	0.024±0.016	0.015	0.014±0.007	0.020
	200 GeV / NUCLEON			
	p + Au		0 + Au	
	Data	Fritiof	Data	Fritiof
$\langle K_s^o \rangle / \langle n_- \rangle$	0.150±0.029	0.149	0.142±0.022	0.152
$\langle \Lambda \rangle / \langle n_- \rangle$	0.153±0.019	0.059	0.179±0.017	0.072
$\langle \bar{\Lambda} \rangle / \langle n_- \rangle$	0.028±0.010	0.024	0.025±0.007	0.028

Λ and $\bar{\Lambda}$ are the same in p+Au and 0+Au, at both energies. Another interesting analysis was made by selecting only <u>central</u> collision events in 0+Au and comparing with p+Au events. **Table 10** gives the ratios R of the multiplicities of K_s^o, Λ, $\bar{\Lambda}$ and negative particles in <u>central</u> 0+Au collisions to the multiplicities in p+Au interactions, at 60 and 200 GeV/nucleon. In each case the intervals of P_T and of rapidity are the same as in **Table 9**. Predictions from FRITIOF are also shown. **Table 10** shows a second interesting result : within the errors the ratios R are compatible with the value 16.

The main conclusions from the NA35 experiment on K_s^o, Λ and $\bar{\Lambda}$ production in 0+Au and p+Au collisions at 60 and 200 GeV/nucleons are :

1) the ratios of the average multiplicities of K_s^o, Λ and $\bar{\Lambda}$ to the average multiplicities of the negative particles are the same in 0+Au and p+Au collisions, at 60 and at 200 GeV/nucleon.

2) The ratios of the multiplicities of K_s^o, Λ, $\bar{\Lambda}$ and negative particles in <u>central</u> 0+Au collisions to the multiplicities in p+Au collisions are compatible with the value 16, at 60 and 200 GeV/nucleon. However, one should be careful in trying to draw firm physical conclusions from this, because the errors on the

Table 10 - *Ratios R of K_S°, Λ, $\bar{\Lambda}$ and negative particles multiplicites in* central *O+Au collisions to the multiplicities in p +Au interactions, at 60 and 200 GeV/nucleon. Predictions from FRITIOF Monte Carlo are also given.*

R=central OAu/pAu	60 GeV/nucleon		200 GeV/nucleon	
	Data	Fritiof	Data	Fritiof
R_{K°	12.8 ± 4.8	18.8 ± 1.0	15.1 ± 3.4	16.5 ± 0.8
R_Λ	14.6 ± 3.0	21.2 ± 1.0	18.6 ± 2.5	20.1 ± 2.1
$R_{\bar{\Lambda}}$	13.7 ± 7.0	19.6 ± 3.3	16.3 ± 7.0	18.4 ± 3.8
R_-	23.0 ± 3.6	16.4 ± 0.2	15.9 ± 1.2	16.2 ± 0.3

ratios R are large (a consequence of the rather small numbers of K_S°, Λ and $\bar{\Lambda}$ mentioned above).

3) A final conclusion, as a consequence of the first two conclusions, is that this experiment did not find evidence for an enhancement of K_S°, Λ, or $\bar{\Lambda}$ production in 0+Au collisions at 60 or 200 GeV/nucleon.

In a second experiment the NA35 Collaboration studied the average Λ and K_S° multiplicities in S-S interactions at 200 GeV/nucleon [118, 121]. The interesting point is that the two colliding nuclei are identical. Two types of trigger have been used : a minimum bias trigger which selected any inelastic interaction, and a forward energy veto trigger which selected central interactions by imposing that the energy measured by the zero-degree calorimeter should be small (see **Fig. 21**). Let $\langle N_{+-} \rangle$, $\langle N_\Lambda \rangle$ and $\langle N_{K^\circ} \rangle$ be the average multiplicities of the charged particles, the Λ and the K_S°, respectively. The events were divided into three categories : the minimum bias trigger events with $N_{+-} < 100$ were defined as "peripheral", the ones with $N_{+-} > 100$ were defined as "intermediate" and the third category were the "central" events, defined by the energy measured by the zero-degree calorimeter.

Fig. **78a)** and **b)** shows $\langle N_\Lambda \rangle$ and $\langle N_{K^\circ} \rangle$, respectively, as a function of $\langle N_{+-} \rangle$ for the three categories of events. The dashed lines are predictions from the FRITIOF Monte Carlo and the dotted lines are predictions from a model made by the authors which they called NN model.

The NN model uses an extrapolation of Λ and K_S° production in p-p interactions to S-S collisions. The Λ multiplicity, for example, is

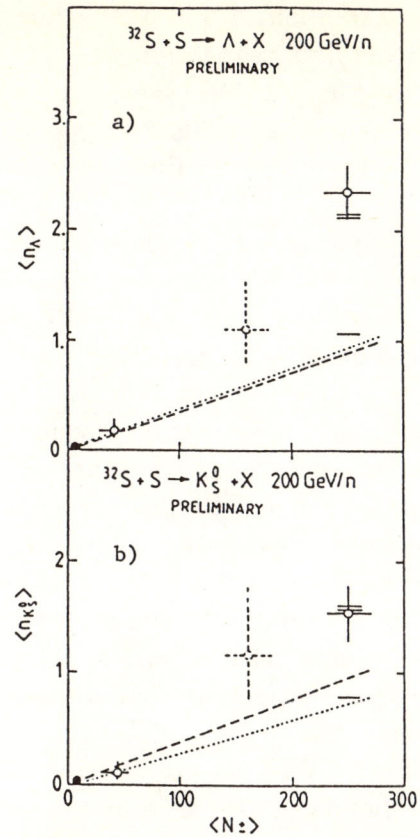
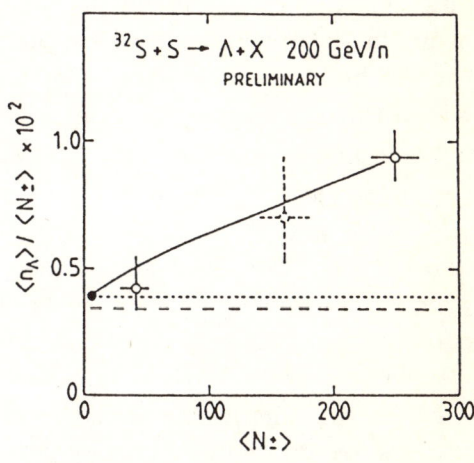

Fig. 78. NA35 average numbers $<N_\Lambda>$ and $<N_{K^0_S}>$ as a function of $<N_{+-}>$ for S+S collisions at 200 GeV/nucleon [121].

Fig. 79. NA35 ratio $<N_\Lambda>/<N_{+-}>$ versus $<N_{+-}>$ for S+S collisions at 200 GeV/nucleon [121].

calculated by:

$$<N_\Lambda> = acc. <N_\Lambda>^{pp} N_{NN}$$

where acc is the probability to detect a Λ in the apparatus, $<N_\Lambda>^{pp}$ is the average multiplicity of Λ in p-p interactions and N_{NN} is the effective number of nucleon-nucleon interactions in the S-S collisions, calculated as $N_{NN} = <N_{+-}>^{SS}/<N_{+-}>^{NN}$. We see that this phenomenological model assumes independent nucleon-nucleon collisions, as does FRITIOF. Therefore, it is not suprising that the two models give similar results, as is seen in Figs. **78 and 79**.

Fig. 78 shows that there is an enhancement of Λ and K^0_S production from peripheral to central collisions, relative to both models. For central collisions the $<N_\Lambda>$ is a factor ~ 2.3 greater than the predictions from both models, and the $\langle N_{K^\circ} \rangle$ is about 1.5 and 2 higher than the predictions from FRITIOF and the NN model, respectively.

The ratio $\langle N_\Lambda \rangle / \langle N_{+-} \rangle$ is given in **Fig. 79**. We see that it increases approximately linearly with $\langle N_{+-} \rangle$. Whereas in **Fig. 78** $\langle N_\Lambda \rangle$ increases faster than linearly with $\langle N_{+-} \rangle$. The authors investigated whether this enhancement of strange particles production could be explained by some secondary interactions mechanism, without any appeal of QGP formation. It was assumed that in the S-S collisions there is formation of matter in two fireballs whose energy and temperature were calculated from data on transverse momentum distribution and multiplicities of produced particles [118]. Secondary interactions would occur in the fireballs. Two simple models were considered : one in which the fireballs are a hadron gas, formed of normal hadrons, the other in which they are a quark-gluon gas. Non-strange hadrons in the former and light quarks and gluons in the latter were assumed to be in thermodynamical equilibrium. The initial numbers of strange particles and of strange quarks were obtained from the NN model. In the case of hadron gas, the only secondary interaction which was considered to produce Λ and K_S° was $\pi N \rightarrow \Lambda K_S^\circ + X$. Another important reaction to produce Λ is $\bar{K}N \rightarrow \Lambda + X$ but it was not considered.

The results of these models for central S-S collisions are shown in **Fig. 78**; the single solid horizontal line corresponds to the a hadron gas and the double solid horizontal line to the a parton gas assumption. We see that the hadron gas assumption does not explain the Λ and K_S° enhancements, but the parton gas would explain.

The conclusion from this experiment is that the observed enhancements of Λ and K_S° in central S-S collisions relative to FRITIOF and to the NN model could be related to QGP but could also be due to final state interactions of partons in a parton gas.

11.6 THE WA85 EXPERIMENT

The WA85 detector was described in section 4.4, **Fig. 19**. This experiment has preliminary results on the production rates of negative particles, Λ and $\bar{\Lambda}$ as a function of the multiplicity of charged particles, as well as on the ratios of $\bar{\Lambda}$ to Λ and of $\bar{\Xi}^-$ to Ξ^- production, in sulphur-tungstene central collisions at 200 GeV/nucleon [42, 43, 122]. The trigger selected central collision events, i.e events with small energy measured by the zero-degree hadron calorimeter. Due to the acceptance of the apparatus the strange particles and the negative particles were selected

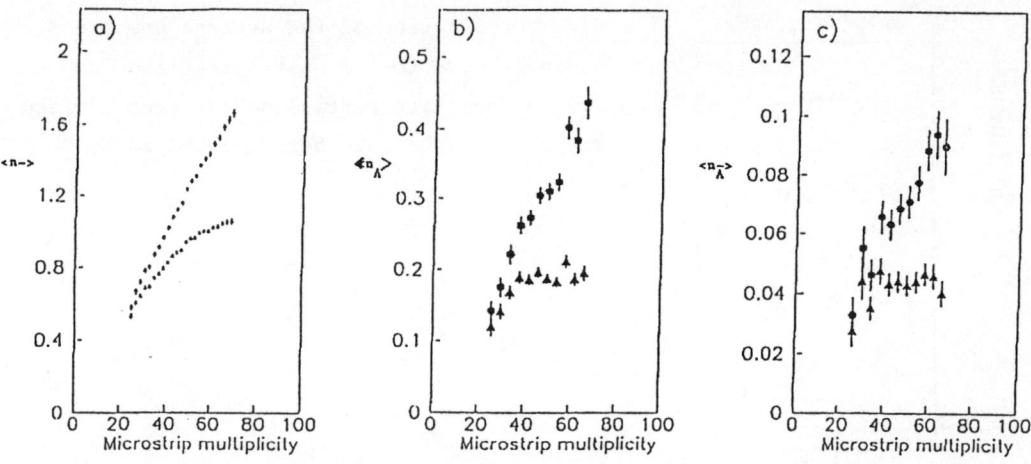

Fig. 80. WA85 average numbers $<n_->$, $<n_\Lambda>$ and $<n_{\bar\Lambda}>$ versus multiplicity [43].

with $P_T \geq 1$ GeV/c in the rapidity interval $2.2 < y_{lab} < 3.0$, yielding $\sim$ 8700 Λ's, $\sim$ 2100 $\bar\Lambda$'s, $\sim$ 120 Ξ^- and $\sim$ 50 $\bar\Xi^-$.

Fig. 80 a, b, c gives the average numbers, per event, of negative hadrons, $<n_->$, of Λ's, $<n_\Lambda>$ and of $\bar\Lambda$'s, $<n_{\bar\Lambda}>$, respectively as a function of the multiplicity. The triangles correspond to events corrected for geometrical acceptance and the circles to events with both geometrical acceptance and reconstruction efficiency corrections. The multiplicities were measured by the silicon microstrips (**Fig. 19**). We see that the three average numbers increase very approximately linearly with the multiplicity, except perhaps $<n_\Lambda>$ for multiplicities less than $\sim$ 40. This is confirmed by **Fig. 81**, which shows the ratios of $<n_->$, $<n_\Lambda>$ and $<n_{\bar\Lambda}>$ to the multiplicities. In this figure the points for the negative hadrons have been divided by 2 and those for the $\bar\Lambda$ have been multiplied by 2. We see that the three plots show no increase of the three ratios with multiplicity; for the Λ there is perhaps a slight increase for multiplicities smaller than $\sim$ 40, but none of the three ratios increase at high multiplicities. This result is consistent with models of particle production based on a superposition of independent nucleon-nucleon collisions; such models, including FRITIOF, do not predict any increase of the raio $<n>$/multiplicity as the multiplicity increases.

A second topic studied in this experiment was the ratio of $\bar\Lambda$ to Λ yields. It was found [122]:

$$\frac{\bar\Lambda}{\Lambda} = 0.24 \pm 0.02$$

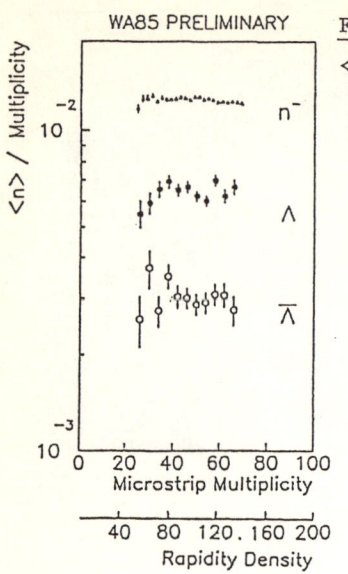

Fig. 81. WA85 ratios of the average numbers $<n_->$, $<n_\Lambda>$ and $<n_{\bar\Lambda}>$ to the multiplicities. The points for negative particles have been divided by 2, points for $\bar\Lambda$ have been multiplied by 2 [43].

An analysis of this ratio as a function of the multiplicity gives:

for multiplicity ≤ 48, $\bar\Lambda/\Lambda = 0.26 \pm 0.02$
for multiplicity > 48, $= 0.22 \pm 0.02$

Therefore, no variation with multiplicity is observed. Since the multiplicity is directly related to the centrality of the collision, this means that no variation of the ratio $\bar\Lambda/\Lambda$ was observed with the centrality of the collision, i.e. with the energy density.

A third topic in this experiment was the production of $\bar\Xi^-$ and Ξ^-. These two hyperons are well identified. The following ratio was found

$$\frac{\bar\Xi^-}{\Xi^-} = 0.44 \pm 0.10$$

which is very interesting because it is higher than in normal hadronic production.

The conclusions from these preliminary results from the WA85 Collaboration on S-W collisions at 200 GeV/c are : 1) the study of Λ and $\bar\Lambda$ production has not so far given any abnormal new results; 2) the ratio of $\bar\Xi^-/\Xi^-$ is new and very interesting.

11.7 CONCLUSIONS FROM THE PRODUCTION OF STRANGE PARTICLES

All collaborations have presented their results as preliminary; analysis is still going on in all of them, statistics will be improved with ion as well as with proton beams for comparisons, and with better statistics results might improve or change.

We should stress, however:

1) that the results from E802 on the ratio K/π are very important and even more so because they are given as a function of P_T in a rather wide range of P_T for the beam energy involved.

2) The absence of an enhancement of K_s^o, Λ and $\bar{\Lambda}$ found so far by the NA35 Collaboration in 0+Au collisions is based on rather small statistics, as can be seen from the errors quoted in **Tables 9 and 10**. We should wait for more events to be analysed. The different behaviour of Λ production in S-S collisions (two identical ions) relative to 0+Au collisions may become interesting and should be followed. We should wait for a direct, experimental comparison of Λ production in heavy ions with Λ production in proton collisions.

3) The ratios $\bar{\Lambda}/\Lambda$ and $\bar{\Xi}/\Xi^-$ obtained by the WA85 Collaboration are important and it will be interesting to see the evolution of this, specially if the collaboration can measure those ratios as a function of P_T.

12. CONCLUSIONS

We would like to emphasize that the experiments which we described in this series of lecture have obtained many other interesting results which are important for our understanding of high energy nucleus-nucleus collisions as such, besides the results which we have presented. However, as is written in the Introduction, we have deliberately concentrated the lectures on specific topics concerning the search for the quark-gluon plasma.

Conclusions from each topic have been presented at the end of each section. We could see that the most encouraging results on possible signatures of the quark-gluon plasma have been so far the production of strange particles and of the J/ψ.

The first conclusion is that there is an enormous lack of experimental data both in nucleus-nucleus and nucleon-nucleus interactions. More information on particle production in nucleon-nucleus interactions is fundamental for the interpretation of nucleus-nucleus collisions. One of the most important lacking information is the A-dependence of differential cross sections of inclusive particle production as a function of transverse momentum and rapidity. This is particularly important whenever we measure the ratios of particle production rates, as has been done with the J/ψ and with strange particles. Different A-dependences for different particles should affect the ratios of their production rates.

A second conclusion is that some of the experiments, like the J/ψ or strange particles production, for instance, must be done at different energies, to see whether there are threshold effects. There must be a threshold in energy density for the transition from hadronic matter to a quark-gluon plasma.

The final conclusion is that due to the uncertainties in our knowledge of hadronic matter, none of the experiments can clame to have an unambiguous evidence for the quark-gluon plasma.

ACKNOWLEDGEMENTS

I would like to express my gratitude to Professors SANTANU PAL, SIBAJI RAHA and BIKASH SINHA for their kind invitation which gave me the opportunity of delivering this series of lectures at the "Winter School on Quark-Gluon Plasma" in Puri. It was a great pleasure to participate in the extremely well organized school, in a stimulating intellectual atmosphere maintained by an audience eager to discuss physics.

Finally, I am very grateful Mrs. Françoise ESCHENBRENNER for her dedication in the difficult task of typing the manuscript and for her patience in understanding the manias of the author for many changes in presentation in the course of the typing.

REFERENCES

[1]   C. ALCOCK, "The Astrophysics and Cosmology of Quark-Gluon Plasma", lectures at this school, in these Proceedings.

[2]   R. BROCKMANN, "Instrumentation in Ultra-Relativistic Heavy-Ion Experiments", lectures at this school, in these Proceedings.

[3]   J.C. PARIKH, "Physics of the Quark-Gluon Plasma", lectures at this school, in these Proceedings.

[4]   M.J. TANNENBAUM, "Hadron Production in Relativistic Heavy Ion Interactions and the Search for the Quark-Gluon Plasma", lectures at this school, in these Proceedings.

[5]   H. SATZ, H.J. SPECHT and R. STOCK, eds. "Quark Matter'87", Proc. of the Sixth Int. Conf. on Ultra-Relativistic Nucleus-Nucleus Collisions, Nordkirchen, (F.R.G.), 24-28 August 1987, Z. Phys. C38 (1988) N°S 1/2.

G. BAYM, P. BRAUN-MUNZINGER and S. NAGAMIYA,eds. "Quark Matter '88", Proc. of the Seventh Int. Conf. on Ultra-Relativistic Nucleus-Nucleus Collisions, Lenox, Massachusetts, U.S.A., September 26-30, 1988, Nucl. Phys. A498 (1989).

[6] P. CARRUTHERS and J. RAFELSKI, eds., Hadronic Matter in Collision 1988, Tucson, Arizona, U.S.A. 6-12 October 1988 (World Scientific, Singapore 1989).

[7] B. SINHA and S. RAHA, eds. Proc. of the Int. Conf. on Physics and Astrophysics of Quark-Gluon Plasma, Bombay, India, 8-12 February 1988 (World Scientific,Singapore).

[8] J. TRAN THANH VAN, ed. Proc. of the XXIVth Rencontre de Moriond, Les Arcs, Savoie, France, 12-18 March 1989 (Editions Frontières).

[9] M.J. TANNENBAUM, Int. J. Modern Physics A4(1989)3377.

[10] W.M. GEIST, Int. J. Modern Physics A4 (1989) 3717.

[11] H.J. SPECHT, Proc. of the Int. Europhysics Conf. on High Energy Physics, Madrid, Spain, 6-13 September 1989 (North Holland, Amsterdam, in printing).

[12] H.R. SCHMIDT and H. GUTBROD, "Highly Relativistic Heavy-Ion Experiments at CERN", lectures delivered at the Int. Advanced Course on the Nuclear Equation of State, Peniscola, Spain, May 21-June 3, 1989.

[13] P. CAPILUPPI et al., Nucl. Phys. B79 (1974) 189.

[14] K. GUETTLER et al., Phys. Lett. B64 (1976) 111.

[15] T. BURNETT et al., Phys. Rev. Lett. 50 (1983) 2062.

[16] T. MATSUI and H. SATZ, Phys. Lett. B178 (1986) 416.

[17] P. BRAUN-MUNZINGER et al., E814 Collaboration, Z. Phys. C 38(1988) 45.

[18] T.J. HUMANIC et al., NA35 Collaboration, Z. Phys. C 38(1988) 79, Proc. of the Int. Conf. on Ultra-Relativistic Nucleus-Nucleus Collisions. Quark Matter 1987-Nordkirchen, FRG, 24-28 August 1987, eds. H. Satz, H.J. Specht and R. Stock.

[19] J. BENECKE, T.T. CHOU, C.N. YANG and E. YEN, Phys. Rev. 118 (1969) 2159.

[20] J.D. BJORKEN, Phys. Rev. D27 (1983) 140.

[21] S.Z. BELENSKI and L.D. LANDAU, Suppl. Il Nuovo Cimento 3 (1956) 15.

[22] L.E. ROBERTS, Il Nuovo Cimento 102A (1989) 1519.

[23] B. ANDERSSON, G. GUSTAFSON and B. NILSSON-ALMSQUIST, Nucl. Phys. B281 (1987) 289.

[24] A. CAPELLA and J. TRAN THANH VAN, Phy. Lett. 93B (1980) 146; Z. Phys. C10 (1981) 249; Z. Phys. C38(1988) 177.

[25] A. CAPELLA et al., Z. Phys. C33 (1987) 541; Phys. Rev. D35 (1987) 2921.

[26] A. BREAKSTONE et al., Phys. Lett. B132 (1983) 458.

[27] A.L.S. ANGELIS et al., Phys. Lett. B168 (1986) 158.

[28] B. NILSSON-ALMSQUIST and E. STENLUND, Computer Phys. Comm. 43(1987) 387.

[29] J.P. PANSART, Saclay report D.Ph.P.E. 86-06 (1986).

[30]   K. WERNER et al., Z. Phys. C37 (1987) 57.

[31]   J. RANFT, Phys. Rev. D37(1988) 1842.

[32]   A. SANDOVAL et al., NA35 Collaboration, Nucl. Physics A461 (1987) 465 c.

A. BAMBERGER et al., NA35 Collaboration, Phys. Lett. B184 (1987) 271.

[33]   C.R. GRUHN et al., NA36 Collaboration, Hadronic Matter in Collision 1988, Tucson, Arizona, U.S.A., October 6-12 1988, eds. P. Carruthers and J. Rafelski (World Scientifc, Singapore 1989) p. 656.

J.M. NELSON et al., NA36 Collaboration, Nucl. Phys. A498 (1989) 515c, Proc. of the Seventh Int. Conf. on Ultra-Relativistic Nucleus-Nucleus Collisions, Lenox, Massachusetts, U.S.A., September 26-30, 1988, eds. G. Baym, P. Braun-Munzinger and S. Nagamiya.

[34]   S.J. LINDENBAUM, Hadronic Matter in Collision 1988, Tucson, Arizona, U.SA., October 6-12, 1988, eds. P. Carruthers and J. Rafelski (World Scientific, Singapore, 1989), p. 673.

W.A. LOVE et al., E810 Collaboration, Nucl. Phys. A498 (1989) 523c, Proc. of the Seventh Int. Conf. on Ultra-Relativistic Nucleus-Nucleus Collisions, Lenox, Massachusetts, U.S.A., September 26-30, 1988, eds. G. Baym, P. Braun-Munzinger and S. Nagamiya.

[35]   Y. TAKAHASHI et al., EMU05 Collaboration, Nucl. Phys. A498 (1989) 529 c, Proc. of the Seventh Int. Conf. on Ultra-Relativistic Nucleus-Nucleus Collisions, Lenox, Massachusetts, U.S.A., September 26-30, 1988, eds. G. Baym, P. Braun-Munzinger and S. Nagamiya.

[36]   L.M. BARBIER et al., KLM Collaboration, Nucl. Phys. A498(1989) 535c, Proc. of the Seventh Int. Conf. on Ultra-Relativistic Nucleus-Nucleus Collisions, Lenox, Massachusetts, U.S.A., September 26-30, 1988, eds. G. Baym, P. Braun-Munzinger and S. Nagamiya.

[37]   E. STENDLUND et al., EMU01 Collaboration, Nucl. Phys. A498 (1989) 541c, Proc. of the Seventh Int. Conf. on Ultra-Relativistic Nucleus-Nucleus Collisions, Lenox, Massachusetts, U.S.A., September 26-30, 1988, eds. G. Baym, P. Braun-Munzinger and S. Nagamiya.

[38]   P.L. JAIN, K. SENGUPTA and G. SINGH, EMU08 Collaboration, Nucl. Phys. A498 (1989) 547c, Proc. of the Seventh Int. Conf. on Ultra-Relativistic Nucleus-Nucleus Collisions, Lenox, Massachusetts, U.S.A., September 26-30, 1988, eds. G.Baym, P. Braun-Munzinger and S. Nagamiya.

[39]   H. GUTBROD et al., Preprint GSI-85-32, August 1985.
R. ALBRECHT et al., Nucl. Instr. and Meth. A276 (1989) 131.
T.C. AWES et al., Nucl. Instr. and Meth. A279 (1989) 479.

[40]   S.G. STEADMAN et al., E802 Collaboration, Hadronic Matter in Collision 1988, Tucson, Arizona, U.S.A., October 6-12, 1988, eds. P. Carruthers and J. Rafelski (World Scientific, Singapore, 1989)p.607.

[41]   H. GORDON et al., NA34 COllaboration, Proposal CERN-SPSC 84-43.
T. AKESSON et al., NA34 Collaboration, Nucl. Instr. and Meth. A262 (1987) 243.
C. LEROY et al., Nucl. instr. and Meth. A252 (1986) 4.
G. LONDON et al., NA34 Collaboration, paper presented at the Int. Europhysics

Conf. on High Energy Physics, Madrid, Spain 6-13 September 1989 (North Holland, Amsterdam, in the press).

[42] F. NAVACH et al., WA85 Collaboration, Hadronic Matter in Collision 1988, Tucson, Arizona, U.S.A., October 6-12, 1988, eds. P. Carruthers and J. Rafelski (World Scientific, Singapore, 1989) p. 617.

[43] J.B. KINSON et al., WA85 Collaboration, Int. Europhysics Conf. on High Energy Physics, Madrid, Spain, 6-13 September 1989, ed. F. Barreiro (North Holland, Amsterdam, 1990) in the press.

[44] A. BUSSIERE et al., NA38 Collaboration, Z. Phys. C38 (1988) 117, Proc. of the Sixth Int. Conf. on Ultra Relativistic Nucleus-Nucleus Collisions - Quark Matter'87 Nordkirchen, FRG, 24-28 August 1987, eds. H. Satz, H.J. Specht and R. Stock.

A. DEVAUX et al., NA38 Collaboration, Nucl. Phys. A498 (1989) 509c, Proc. of the Seventh int. Conf. on Ultra-Relativistic Nucleus-Nucleus Collisions, Lenox, Massachusetts, U.S.A., September 26-30, 1988, eds. G. Baym, P. Braun-Munzinger and S. Nagamiya.

C. BAGLIN et al., NA38 Collaboration, to be submitted to Nucl. Instr. Methods.

[45] T.H. BURNETT et al., Phys. Rev. Lett. 57 (1986) 3249.

[46] L. VAN HOVE, Phys. Lett. B118 (1982) 138.

[47] E.V. SHURYAK and O. ZHIROV, Phys. Lett. B171 (1986) 99.

[48] R. SANTO et al., WA80 Collaboration, Nucl. Phys. A498 (1989) 391c, Proc. of the Seventh Int. Conf. on Ultra-Relativistic Nucleus-Nucleus Collisions, Lenox, Massachusetts, U.S.A., September 26-30, 1988, eds. G. Baym, P. Braun-Munzinger and S. Nagamiya.

H.H. GUTBROD et al., WA80 Collaboration, Hadronic Matter in Collision 1988, Tucson, Arizona, U.S.A., October 6-12, 1988, eds. P. Carruthers and J. Rafelski (World Scientific, Singapore, 1989) p. 342.

R. ALBRECHT et al., WA80 Collaboration, Preprint GSI-87-82, December 1987.

R. ALBRECHT et al., WA80 Collaboration, Preprint GSI-89-03, January 1989.

[49] M.A. FAESSLER, Phys. Reports 115 (1984) 1.

[50] J.W. CRONIN et al., Phys. Rev. D11 (1975) 3105.
D. ANTREASYAN et al., Phys. Rev. D19 (1979) 764.

[51] T. AKESSON et al., HELIOS Collaboration : a) Preprint CERN-EP/89-111, August 1989; b) submitted to Z. Phys. C, in the press.

[52] J.Y. GROSSIORD et al., NA38 Collaboration, Nucl. Physics A498 (1989) 249c, Proc. of the Seventh Int. Conf. on Ultra-Relativistic Nucleus-Nucleus Collisions, Lenox, Massachusetts, U.S.A., September 26-30, 1988, eds. G. Baym, P. Braun-Munzinger and S. Nagamiya - and references therein.

[53] R.C. HWA and K. KAJANTIE, Phys. Rev. D32 (1985) 1109.

[54] S. RAHA and B. SINHA, Phys. Rev. Lett. 58 (1987) 101.

[55] T. AKESSON et al., HELIOS Collaboration : a) Preprint CERN-EP 89/113, September 1989; b) submitted to Z. Phys. C, in the press.

[56] A. CHILINGAROV et al., Nucl. Phys. B151 (1979) 29.
R. SINGER et al., Phys. Lett. B60 (1976) 385.
C. KOURKOUMELIS et al., Phys. Lett. B84 (1979) 277.
M.G. ALBROW et al., Nucl. Phys. B155 (1979) 39.
M. DIAKONOU et al., Phys. Lett. B89 (1980) 432.
P. PEREZ et al., Phys. Lett. B112 (1982) 260.
M. BASILE et al., Il Nuovo Cimento A65 (1981) 421
T. AKESSON et al., Z. Phys. C 18(1983) 5.
T. AKESSON et al., Phys. Lett. B178 (1986) 447.

[57] C. BAGLIN et al., NA38 Collaboration, Phys. Lett. B220(1989) 471 and references therein.

[58] C. RACCA et al., NA38 Collaboration, Hadronic Matter in Collision 1988, Tucson, Arizona, U.S.A., October 6-12, 1988, eds. P. Carruthers and J. Rafelski (World Scientific, Singapore 1989), p. 552.

[59] R.A. SALMERON, New Results in Hadronic Interactions, Proc. of the 24th Rencontre de Moriond, Les Arcs, Savoie, France, 12-18 March 1989, ed. J. Tran Thanh Van (Editions Frontières) p. 413.

[60] Y. LEMOINE et al., Phys. Lett. B113 (1982) 509.

[61] C. KOURKOUMELIS et al., Phys. Lett. B81 (1979) 405.

[62] F. KARSCH, M.T. MEHR and H. SATZ, Z. Phys. C 37(1988) 617.

[63] F. KARSCH and R. PETRONZIO, Z. Phys. C37 (1988) 627.

[64] M.C. CHU and T. MATSUI, Phys. Rev. D37 (1988) 1851.

[65] J.P. BLAIZOT and J.Y. OLLITRAULT, Phys. Lett. B199 (1987) 499.

[66] F. KARSCH and R. PETRONZIO, Phys. Lett. B193 (1987) 105.

[67] R. ANDERSON et al., Phys. Rev. Lett. 38 (1977) 263.

[68] U. CAMERINI et al., Phys. Rev. Lett. 35 (1975) 483.
B. KNAPP et al., Phys. Rev. Lett. 34 (1975) 1040.
J. BRANSON et al., Phys. Rev. Lett. 38 (1977) 1334.
M.D. SOKOLOFF et al., Phys. Rev. Lett. 57 (1986) 3003.

[69] H. STROBELE et al., NA35 Collaboration, Z. Phys. C38 (1988) 89.

[70] P. BORDALO et al., Phys. Lett. B193 (1987) 373.

[71] J. BADIER et al., Z. Phys. C20 (1983) 101.

[72] G.T. BODWIN, S.J. BRODSKY and G.P. LEPAGE, Phys. Rev. Lett. 47 (1981) 1799.
C. MICHEL and G. WILK, Z. Phys. C10 (1981) 169.

[73] J. FTAČNIK, P. LICHARD and J. PIŠUT, Phys. Lett. B207 (1988) 194.

[74] J. FTAČNIK, P. LICHARD, N. PIŠUTOVÁ and J. PIŠUT, Z. Phys. C42 (1989) 139.

[75] S. GAVIN, M. GYULASSY and A. JACKSON, Phys. Lett. B207 (1988) 257.

[76] S. GAVIN, M. GYULASSY, Phys. Lett. B214 (1988) 241.

[77] R. VOGT, M. PRAKASH, P. KOCH and T.H. HANSSON, Phys. Lett. B207 (1988) 263.

[78] J.P. BLAIZOT and J.Y. OLLITRAULT, Phys. Rev. D39 (1989) 232.

[79] J.P. BLAIZOT and J.Y. OLLITRAULT, Phys. Lett. B217 (1989) 392.

[80] F. CLOSE, lectures at this school, in these Proceedings.

[81] G. GOLDHABER, W.B. FOWLER, S. GOLDHABER, T.F. HOANG, T.E. KALOGEROPOULOS and W. POWELL, Phys. Rev. Lett. 3(1959) 181.

[82] G. GOLDHABER, S. GOLDHABER, W. LEE and A. PAIS, Phys. Rev. 120 (1960) 300.

[83] G. GOLDHABER, Proc. of the First Int. Workshop on Local Equilibrium in Strong Interaction Physics, eds. D.K. Scott and R.M. Weiner (World Scientific, Singapore, 1985) p. 115.

G. GOLDHABER, Proc. of the Int. Conf. on High Energy Physics, Lisbon, Portugal, 9-15 July 1981, ed. J. Dias de Deus and J. Soffer, p. 767.

[84] B. LÖRSTAD, Int. J. Mod. Phys. A4 (1989) 2861.

[85] G. COCCONI, Phys. Lett. B49 (1974) 459.

[86] R. HANBURY-BROWN and R.Q. TWISS, Phil. Mag. 45(1954) 663; Nature 178 (1956) 1046.

[87] R.P. FEYNMAN, Lectures in Physics (Addison-Wesley, Reading, Mass., 1965), Volume 3, chapters 3 and 4.

[88] K. KOLEHMAINEN and M. GYULASSY, Phys. Lett. B180 (1986) 203.

[89] G. KOPYLOV and M. PODGORETSKI, Sov. J. Nucl. Phys. 18 (1974) 336 and 19(1974) 215; G. KOPYLOV, Phys. Lett. B50 (1974) 472.

[90] H. GUTBROD et al., WA80 Collaboration, Hadronic Matter in Collision 1988, Tucson, Arizona, U.S.A., October 6-12, 1988, eds. P. Carruthers and J. Rafelski, (World Scientifc, 1989) p. 342.

[91] T. PEITZMANN et al., WA80 Collaboration, Preprint GSI-89-05, January 1989; Nucl. Phys. A498 (1989) 397c, Proc. of the Seventh Int. Conf. on Ultra-Relativisytic Nucleus-Nucleus Collisions, Lenox, Massachusetts, U.S.A., September 26-30, 1988, eds. G. Baym, P. Braun-Munzinger and S. Nagamiya.

[92] T.J. HUMANIC et al., NA35 Collaboration, Z. Phys. C38 (1988) 79, Proc. of the Sixth Int. Conf. on Ultra-Relativistic Nucleus-Nucleus Collisions - Quark Matter 1987- Nordkirchen, F.R.G., 24-28 August 1987, eds. H. Satz, H.J. Specht and R. Stock.

[93] J.W. HARRIS et al., NA35 Collaboration, Nucl. Phys. A498 (1989) 133c, Proc. of the Seventh Int. Conf. on Ultra-Relativistic Nucleus-Nucleus Collisions, Lenox, Massachusetts, U.S.A., September 26-30, 1988, eds. G. Baym, P. Braun-Munzinger and S. Nagamiya.

[94] A. BAMBERGER et al., NA35 Collaboration, Phys. Lett. B203 (1988) 320.

[95] A. BIALAS and R. PESCHANSKI, Nucl. Phys. B273 (1986) 703; Nucl. B308(1988) 857; Phys. Lett. B207 (1988) 59.

[96] A. BIALAS, K. FIALKOWSKI and R. PESCHANSKI, Europhys. Lett. 7 (1988) 125.

[97] R. PESCHANSKI, New Results in Hadronic Interactions, Proc. of the 24th Rencontre de Moriond, Les Arcs, Savoie, France, March 12-18, 1989, ed. J. Tran Thanh Van (Editions Frontières) p. 247, and references therein.

[98] G.J. ALNER et al., UA5 Collaboration, Phys. Rep. 154 (1987) 247.

[99] a) M. ADAMUS et al., NA22 Collaboration, Phys. Lett. B185(1987) 200.
b) I.V. AJINENKO et al., NA22 Collaboration, Phys. Lett. B222 (1989) 306.

[100] W. BRAUNSCHWEIG et al., TASSO Collaboration, Phys. Lett. B231 (1989) 548.

[101] I. DERADO, G. JANCSO, N. SCHMITZ and P. STOPA, Preprint Max Planck Institut, 1989, München, F.R.G.

[102] R. HOLYNSKI et al., KLM Collaboration, Phys. Rev. Lett. 62 (1989) 733.

[103] M.I. ADAMOVICH et al., EMU01 Collaboration, Phys. Lett. B201 (1988) 397.

[104] E. STENLUND et al., EMU01 Collaboration, Nucl. Phys. A498 (1989) 541c, Proc. of the Seventh Int. Conf. on Ultra-Relativistic Nucleus-Nucleus Collisions, Lenox, Massachusetts, U.S.A., September 26-30, 1988, eds. G. Baym, P. Braun-Munzinger and S. Nagamiya.

[105] R. ALBRECHT et al., WA80 Collaboration, Phys. Lett. B221 (1989). 427.
C.R. YOUNG et al., WA80 Collaboration, Nucl. Phys. A498 (1989) 53c, Proc. of the Seventh Int. Conf. on Ultra-Relativistic Nuclus-Nucleus Collisions, Lenox, Massachusetts, U.S.A., September 26-30, 1988, eds. G. Baym, P. Braun-Munzinger and S. Nagamiya.

[106] For a review see : P. KOCH, B. MÜLLER and J. RAFELSKI, Strangeness in Relativistic Heavy Ion Collisions, Phys. Rep. 142 (1986) 167.

[107] For a recent review on specific topics see : J. RAFELSKI, Hadronic Matter in Collision 1988, Tucson, Arizona, U.S.A., October 6-12, 1988, eds. P. Carruthers and J. Rafelski (World Scientific, Singapore, 1989) p. 776.

[108] See for example C. FABJAN, Current Issues in Hadron Physics, Proceeding of the 23rd Rencontre de Moriond, Les Arcs, Savoie, France, March 13-19, 1988, ed. J. Tran Thanh Van (Editions Frontières) p. 617.

[109] M. JACOB and J. RAFELSKI, Phys. Lett. B190 (1987) 173.

[110] J. RAFELSKI and M. DANOS, Phys. Lett. B192 (1987) 432.

[111] C.R. GRUHN et al., NA36 Collaboration, Hadronic Matter in Collision 1988, Tucson, Arizona, U.S.A., October 6-12, 1988, eds. P. Carruthers and J. Rafelski (World Scientific, 1989) p. 656;
D. E.GREINER et al., NA36 Collaboration, ib. p. 665.

[112] G. ODYNIEC, Hadronic Matter in Collision 1988, Tucson, Arizona, U.S.A., October 6-12, 1988, eds. P. Carruthers and J. Rafelski (World Scientific, Singapore 1989) p. 721.

[113]   P. VINCENT et al., E802 Collaboration, Nucl. Physics A498 (1989) 67, Proc. of the Seventh Int. Conf. on Ultra-Relativistic Nucleus-Nucleus Collisions, Lenox, Massachusetts, U.S.A., September 26-30, 1988, eds. G. Baym, P. Braun-Munzinger and S. Nagamiya.

[114]   S.G. STEADMAN et al., E802 Collaboration, Hadronic Matter in Collision 1988, Tucson, Arizona, U.S.A., October 6-12, 1988, eds. P. Carruthers and J. Rafelski (World Scientific, 1989) p. 607.

[115]   G. LONDON et al., NA34 Collaboration, Proceedings of the Int. Europhysics Conf. on High Energy Physics, Madrid, Spain, September 6-13, 1989 (North Holland, Amsterdam, in the press).

[116]   P. SONDEREGGER et al., NA38 Collaboration, Z. Physik C38 (1988) 129, Proc. of the Sixth Int. Conf. on Ultra-Relativistic Nucleus-Nucleus Collisions -Quark Matter'87- Nordkirchen, FRG, 24-28 August 1987, eds. H. Satz, H.J. Specht and R. Stock.

[117]   J.Y. GROSSIORD et al., NA38 Collaboration, Nucl. Physics A498 (1989) 249c, Proc. of the Seventh Int. Conf. on Ultra-Relativistic Nucleus-Nucleus Collisions, Lenox, Massachusetts, U.S.A., September 26-30, 1988, eds. G. Baym, R. Braun-Munzinger and S. Nagamiya.

[118]   J.W. HARRIS et al., NA35 Collaboration, Nucl. Phys. A498 (1989), 133c, Proc. of the Seventh Int. Conf. on Ultra-Relativistic Nuclus-Nucleus Collisions, Lenox, Massachusetts, U.S.A., September 26-30, 1988, eds. G. Baym, P. Braun-Munzinger and S. Nagamiya.

[119]   I. DERADO, Hadronic Matter in Collision 1988, Tucson, Arizona, U.S.A, October 6-12, 1988, eds. P. Carruthers and J. Rafelski (World Scientific, Singapore, 1989) p. 636.

[120]   A. BAMBERGER et al., NA35 Collaboration, Z. Phys. C43 (1989) 25.

[121]   M. GAZDZICKI et al., NA35 Collaboration, Hadronic Matter in Collision 1988, Tucson, Arizona, U.S.A., October 6-12, 1988, eds. P. Carruthers and J. Rafelski (World Scientific Singapore, 1989) p. 647.

[122]   E. QUERCIGH et al., WA85 Collaboration, Hadronic Matter in Collision 1988, Tucson, Arizona, U.S.A., October 6-12, 1988, eds. P. Carruthers and J. Rafelski (World Scientific, Singapore, 1989) p. 625.

[123]   S.J. BRODSKY and A.H. MUELLER, Phys. Lett. B206 (1988) 685.

Hadron Production in Relativistic Heavy Ion Interactions and the Search for the Quark–Gluon Plasma*

M.J. Tannenbaum

Brookhaven National Laboratory, Upton, NY 11973, USA

1. Introduction

High energy collisions of nuclei provide the means of creating nuclear matter in conditions of extreme temperature and density. At large energy density, or baryon density, a phase transition is expected from a state of nucleons containing confined quarks and gluons to a state of "deconfined" (from their individual nucleons) quarks and gluons covering the entire volume of nuclear matter, or a volume that is many units of the characteristic length scale. This state is expected to be in thermal and chemical equilibrium. In the terminology of high energy physics, this is called a "soft" process, related to the QCD confinement scale

$$\Lambda_{QCD}^{-1} \sim (0.1 GeV)^{-1} \sim 2\, fm$$

This state is called the Quark-Gluon-Plasma (QGP)[1,2].

A schematic drawing of a relativistic heavy ion collision is shown in Figure 1. Two energy regimes are discussed for the QGP[3]. At lower energies, typical of the AGS, the colliding nuclei are expected to stop each other, leading to a Baryon-Rich system. This will be the region of maximum baryon density. At very high energy, 100 to 200 GeV per nucleon pair in the center of mass, the nuclear fragments will be well separated from a central region of particle production. This is the region of the Baryon-Free or Gluon plasma.

There has been considerable work over the past few years in making quantitative predictions for the QGP. A recent calculation of a phase diagram for "isentropic expansion trajectories for a hadronizing QGP"[4] is shown in figure 2. The transition temperature from a state of hadrons to the QGP varies, from $T_c = 140$ MeV at zero baryon density, to zero temperature at a critical baryon density ~ 6.5 times the normal nuclear density:

$$\rho_0 = 0.15\, \text{nucleons}/fm^3\ .$$

Predictions for the transition temperature are constrained to a relatively narrow range $140 < T_c < 250$ MeV, while the critical baryon or energy density is prediected to be 5 to 20 times the normal density[5].

From the point of view of an experimentalist there are two major questions in this field. The first is how to relate the thermodynamical properties (temperature, energy density, entropy...) of the QGP or hot nuclear matter to properties that can be measured in the laboratory. The second question is how the QGP can be detected.

---

* This research has been supported in part by the U.S. Department of Energy under Contract DE-AC02-76CH00016

INITIAL STATE BEFORE COLLISION

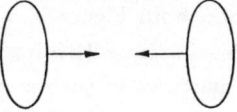

$\sqrt{S}/A \lesssim 5$ GeV : BARYONS STOPPED IN OVER-ALL CM

AT HIGHER ENERGY, NUCLEI ARE TRANSPARENT TO EACH OTHER

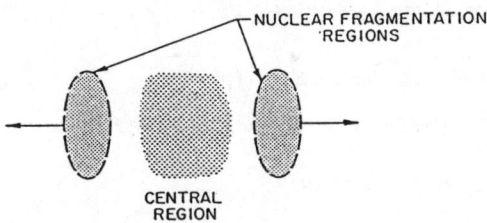

Fig. 1: Schematic of Relativistic Heavy Ion Collision from RHIC Conceptual Design Report BNL 51932(1986)

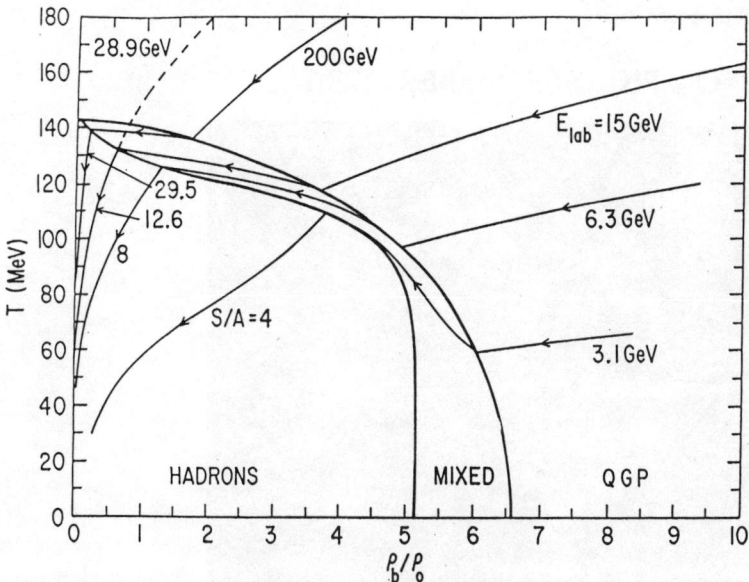

Fig. 2: Phase Diagram of nuclear matter from U. Heinz et al.[5]

2. Observables

The challenge of RHI collisions can be understood from Figure 3, which is a streamer chamber photograph of a 200 GeV/u oxygen projectile colliding with a lead nucleus[6]. It would appear to be a daunting task to reconstruct all the particles in such events. Consequently, it is more common to use single-particle or multi-particle inclusive variables to analyze these reactions.

A single particle inclusive reaction involves the measurement of just one particle coming out of a reaction. For any particle, the momentum can be resolved into transverse (p_T) and longitudinal (p_L) components; and in many cases the mass (m) of the particle can be determined. The longitudinal momentum is conveniently expressed in terms of the rapidity (y):

$$y = \ln\left(\frac{E + p_L}{m_T}\right) \tag{2.1}$$

$$\cosh y = E/m_T \qquad \sinh y = p_L/m_T \tag{2.2}$$

where

$$m_T = \sqrt{m^2 + p_T^2} \quad \text{and} \quad E = \sqrt{p_L^2 + m_T^2} \tag{2.3}$$

In the limit when ($m \ll E$) the rapidity reduces to the pseudorapidity(η)

$$\eta = -\ln\tan\theta/2 \tag{2.4}$$

$$\cosh\eta = \csc\theta \qquad \sinh\eta = \cot\theta \tag{2.5}$$

where θ is the polar angle of emission. The rapidity variable has the useful property that it transforms linearly under a Lorentz transformation so that the invariant differential single particle inclusive cross section becomes:

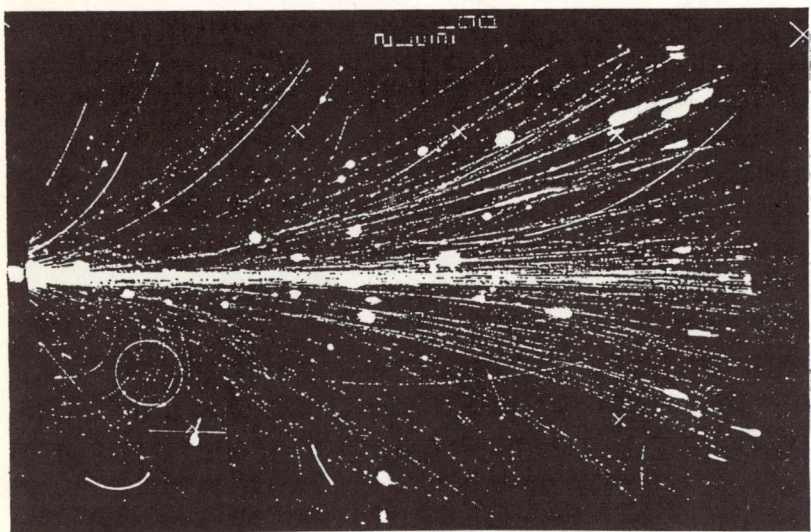

Fig. 3: Streamer chamber photograph[6] of a $^{16}O + Pb$ collision at 200 GeV/nucleon.

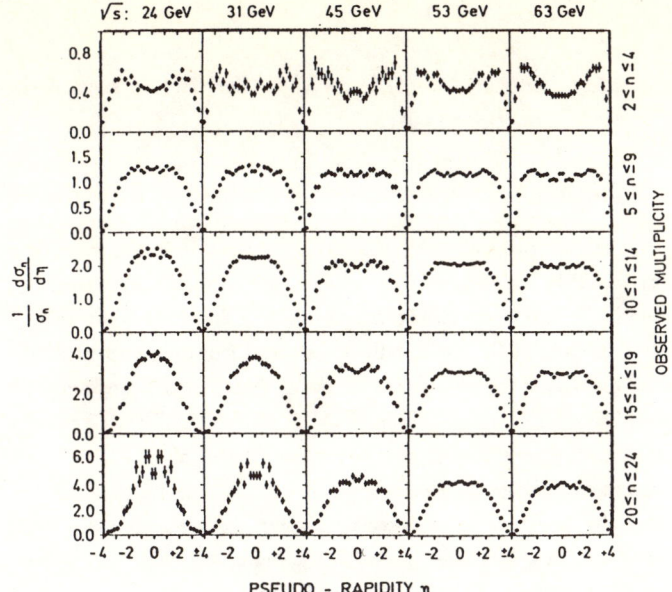

Fig. 4: Measurements in a streamer chamber at the CERN ISR of the normalized charged particle densities (corrected for acceptance up to $|\eta| \simeq 4$) in various intervals of the total observed multiplicity, as a function of the c.m. energy, $\sqrt{s}$, of the $p-p$ collisions [7].

$$\frac{E d^3\sigma}{dp^3} = \frac{d^3\sigma}{p_T dp_T dy d\phi} \quad (2.6)$$

where

$$dy = \frac{dp_L}{E}. \quad (2.7)$$

In the region near the projectile or target rapidity, the Feynman x fragmentation variable is also used:

$$x_F = 2p_L/\sqrt{s} \quad (2.8)$$

where $\sqrt{s}$ is the center-of-mass energy of the collision.

The transverse momentum distributions can be determined for the different particles, and typically the average transverse momentum, $<p_T>$ is taken as a measure of the temperature, T. The charged particle multiplicity, either over all space, or in resricted intervals of rapidity, is taken as a measure of entropy.

A convenient description of high energy collisions is provided by the charged particle density in rapidity, dn/dy. A classical measurement in a streamer chamber from $p-p$ collisions at the CERN ISR[7] is shown in figure 4. Regions of nuclear fragmentation take up the first 1-2 units around the projectile and target rapidity and if the center-of-mass energy is sufficiently high, a central plateau is exhibited. Another, similar variable is the transverse energy density in rapidity or $dE_T/dy \sim <p_T> \times dn/dy$. This is thought to be related to the co-moving energy density in a longitudinal expansion, and according to Bjorken[7] is proportional to the energy density in space ϵ:

$$\epsilon_{Bj} = \frac{d<E>}{dy} \frac{1}{2\tau_0 \pi R^2} \qquad (2.9)$$

where τ_0, the formation time, is usually taken as 1 (or $\frac{1}{2}$!) fm, πR^2 is the effective area of the collision, and $d<E>/dy$ is the co-moving energy density.

3. Signatures of the Quark-Gluon Plasma

One of the more interesting signatures proposed for the QGP is that it could trigger a catastrophic transition from the metastable vacuum of the present universe to a lower energy state[9], " a possibility naturally occurring in many spontaneously broken quantum field theories ". A more likely outcome is that the existence of the QGP will be inferred from a comprehensive and systematic set of experimental data exhibiting several striking features or "anomalies", "which can be interpreted in a unified way as manifestations of QGP production" [10]. Examples of the features expected for the QGP and signatures to find them are given below:

a) Characteristic Temperature Entropy Curve[11]:
Note that this curve (Figure 5) has the features of a phase transition with which we are all familiar. The $<p_T>$, acting as temperature, increases with increasing entropy (dn/dy); then as the phase transition takes place (e.g. water changing to steam) the temperature remains constant and begins rising again when the transition to the new phase is complete.

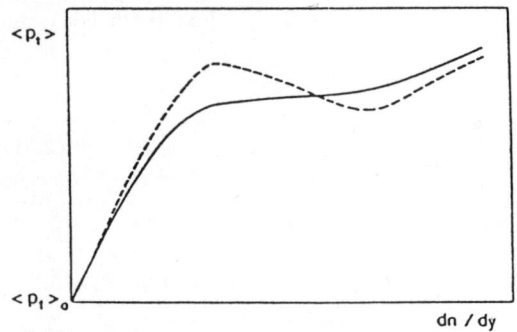

Fig. 5: Characteristic Temperature-Entropy Curve for phase transition[11]

b) Plasma Droplets Caused by Deflagration[12]:
These would be manifested by large fluctuations in dn/dy or dE_T/dy covering a range of ~1 unit on an event by event basis. The hope would be to observe the other plasma signatures only in the region of the fluctuation and not in the other regions.

c) Thermal Equilibrium :
One of the best probes of thermal equilibrium is lepton pair production[13]. There are two characteristic features of thermal production of lepton pairs. The number of lepton pairs per unit of rapidity is proportional to the square of charged particle density, and furthermore this ratio is proportional to the transition temperature T_c :

$$\frac{dN_{e^+e^-}(m_{ee} > 200 MeV)/dy}{[dn/dy]^2 \times 10^{-7}} \simeq 15 \, T_c \, (GeV)$$

Also, the p_T and mass dependence of the cross section are not independent but depend only on the transverse mass m_T. This means that at any fixed value of m_{ee}, the $< p_T(m_{ee}) >$ is linearly proportional to m_{ee}.

d) Volume of Thermal Matter :

The size of the thermal source is thought to be measured by identical particle interferometry using the GGLP effect[14]. When two identical pions occupy nearly the same coordinates in phase space, the amplitudes interfere constructively due to the symmetry of the wave-function imposed by Bose-Einstein statistics. The characteristic momentum difference leading to decorrelation in momentum space can be measured, and is taken to be the fourier transform of the size of an extended source in position space. It should be noted that dynamical effects due to final state interactions can be large, and make the interpretation of such measurements a very specialized subject[15].

e) Chemical Equilibrium[16,17]:

In the QGP there will be gluons, quarks and anti-quarks. They will continuously react with each other via the QCD subprocesses:

$$gg \to q\bar{q} \qquad q\bar{q} \to gg \qquad q\bar{q} \to q'\bar{q}'$$

where q' represents a different flavor quark $(u, d\,or\,s)$. After several interactions have taken place, the reaction rates and the abundances of the gluons and the different flavor quarks (and anti-quarks) will become equilibrated, so that they no longer change with time. This is called chemical equilibrium. Since the transition temperature T_c is comparable to the strange quark mass $\sim$150 MeV, the strange quarks $s, \bar{s}$ should have the same abundance as the $u, \bar{u}$ and $d, \bar{d}$ in the gluon plasma. In the baryon-rich plasma, the $s, \bar{s}$ will be enhanced compared to u and d, since $u, \bar{u}$ and $d, \bar{d}$ are "Pauli" blocked by valence u and d quarks.

The principal probe of chemical equilibrium is the particle composition. For instance, the abundance of strange mesons and baryons as well as anti-baryons should be quite different in a QGP than in a hadron gas or in an ordinary nuclear collision

f) Deconfinement :

It has been proposed[18] that J/Ψ production in A+A collisions will be suppressed by Debye screening of the quark color charge in the QGP. The J/Ψ is produced when two gluons interact to produce a $c, \bar{c}$ pair which then resonates to form the J/Ψ. In the plasma the $c, \bar{c}$ interaction is screened so that the $c, \bar{c}$ go their separate ways and eventually pick up other quarks at the periphery to become *open charm*. This would be quite a spectacular effect since the naive expectation is that J/Ψ production, being a pointlike process, should go like $A \times A$ in an A+A collision, and thus would be enhanced relative to the total interaction cross section, which increases only as $A^{2/3}$.

Another signature of deconfinement has recently been proposed[19]. If one could start off with a correlated two-quark object ("diquark") which then becomes deconfined due to the the formation of a QGP, the predictions of the effect of Debye screening should contain fewer uncertainties than in the case of J/Ψ suppression. It is suggested that the proton is composed of a quark-diquark bound system and that proton production at large transverse momenta in $p-p$ collisions is the result of hard scattering of diquarks. Hence, a **decrease** in the yield of protons and other baryons at large transverse momenta is predicted, in the case of QGP production, which should be directly related to the deconfinement of diquarks due to Debye screening. It is argued that such an effect is not likely to occur in dense hadronic matter. However, there is also the possibility[20] that a system of diquarks may be the preferred configuration of quark matter at temperatures and densities just above the deconfinement transition. This may lead to a diquark enhancement and an *increase* in large p_T baryon production... Clearly this subject quite interesting and requires further analysis.

4. Relation to Experimental Techniques

Each of the probes of the QGP tends to have a different experimental technique associated with it. In all cases the multiplicities in nuclear collisions are so large that all the detectors used are very highly segmented. For measuring the charged multiplicity or dn/dy, a segmented multiplicity detector is used, usually an array of proportional tubes with pad readout, or a silicon pad array. For measuring transverse energy flow, dE_T/dy, a hadron calorimeter is used. Some groups use an electromagnetic shower counter for this purpose. This has the advantage of being smaller, cheaper and higher in resolution than a full hadron calorimeter; but has the disadvantage of being biased, since only π^0 and η^0 mesons are detected (via their two photon decay). Nuclear fragmentation products are detected by calorimeters in the projectile direction and by E, dE/dx scintillator arrays in the target fragmentation region. The particle composition and transverse momentum distributions are measured using magnetic spectrometers with particle identification. Typically, time-of-flight, gas and aerogel Cerenkov counters, and dE/dx are used to separate pions from kaons, protons, deuterons, etc. Drift chambers are generally utilized for charged particle tracking, although streamer chambers and time projection chambers (TPC) are also in use. Lepton pair detectors are very specialized, and usually combine magnetic spectrometers with lepton identification (muons by penetration, and electrons by "gas" and "glass").

One of the specific problems in this field is how to detect, with minimum bias, when a nucleus-nucleus collision has taken place. Two techniques are used. The first is to put a calorimeter at zero degrees to determine whether the projectile has the full beam energy or has lost some energy. The second uses a so-called bullseye counter downstream of the target, sized just large enough to detect all the beam particles. The bullseye also measures the charge of the beam particles since the pulse height is proportional to Z^2. If a particle misses the bullseye, or the charge changes, this is taken as an indication of a nuclear interaction (see Fig. 6).

With that quick overview of the experimental techniques, the following "photo album" of the first round of experiments at CERN and Brookhaven should be easier to comprehend. The CERN heavy ion program has provided ^{16}O and ^{32}S beams at 60 and 200 GeV/u and will possibly improve the source to provide ^{207}Pb beams. There are 5 major experiments: WA80 (Figure 7), NA34 (Figure 8), NA35 (Figure 9), NA36 (Figure 10), and NA38 (Figure 11). The BNL heavy ion program has provided ^{16}O and ^{28}Si beams at 14.6 GeV/c per nucleon and is scheduled to accelerate ^{197}Au beams to 11.7 GeV/c per nucleon in 1992. A major improvement is planned for 1996 when the Relativistic Heavy Ion Collider (RHIC) is scheduled to begin operation. RHIC will provide colliding beams, covering the full mass number spectrum, with center-of-mass energies from 5 to 200 GeV per nucleon pair. At present, there are 3 major RHI experiments at the BNL-Tandem-AGS: E802 (Figure 12), E810 (Figure 13) and E814 (Figure 14).

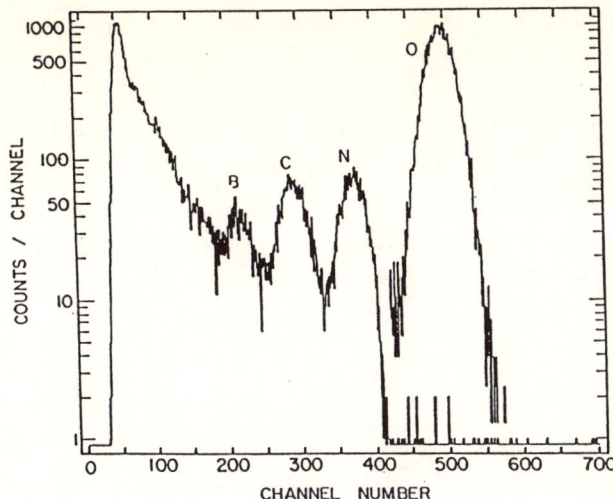

Fig. 6: Spectrum in Bullseye Counter from AGS-E802. The $^{16}_{8}O$ peak is obtained from a sample of non-interacting beam particles, and superimposed onto the spectrum for a detected interaction.

CERN PROGRAM

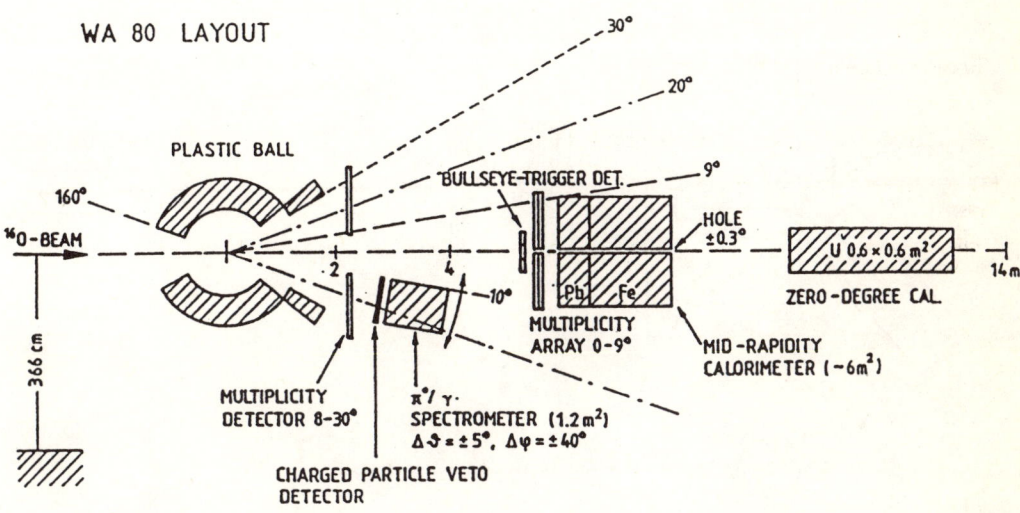

Experiment WA80: Study of Relativistic Nucleus-Nucleus Collisions at the CERN SPS

Fig. 7: WA80

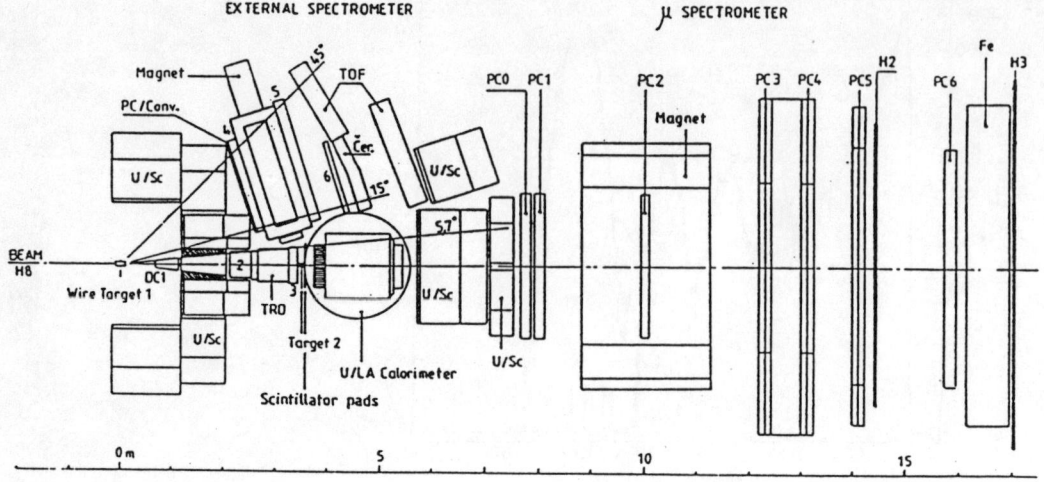

Experiment NA34/2: Study of High Energy Densities over Extended Nuclear Volumes via Nucleus-Nucleus Collisions at the SPS

Fig. 8: NA34

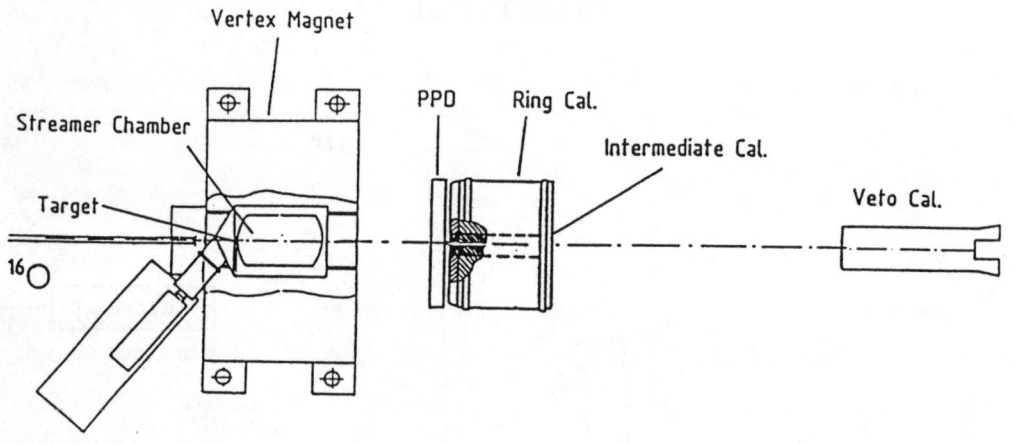

Experiment NA35: Study of Relativistic Nucleus-nucleus Collisions

Fig. 9: NA35

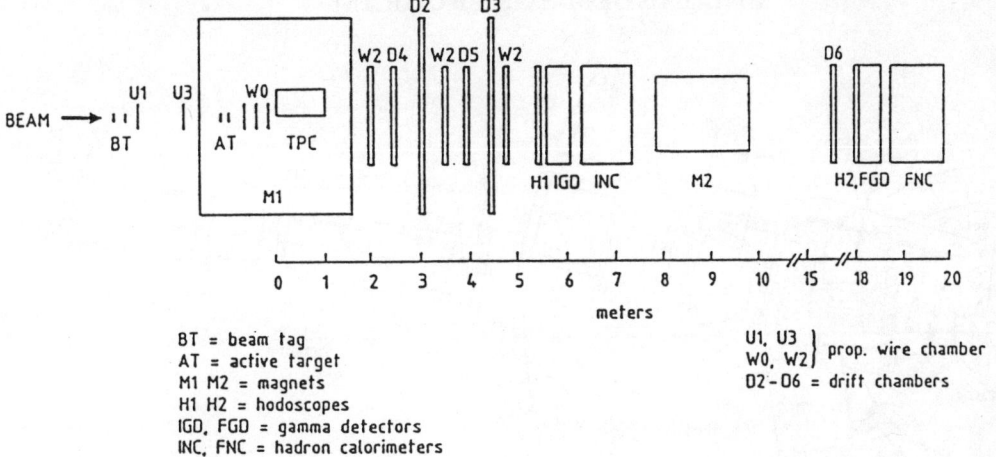

Experiment NA36: Production of Strange Baryons and Antibaryons in Relativistic Ion Collisions

Fig. 10: NA36

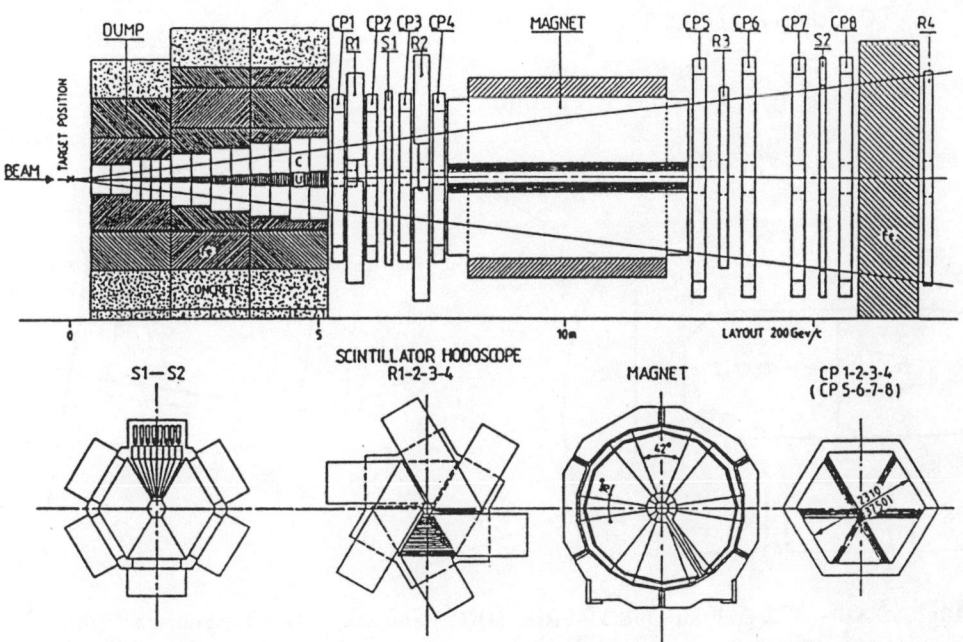

Fig. 11: NA38—Study of High-Energy Nucleus-Nucleus Interactions with the Enlarged NA10 Dimuon Spectrometer

BNL-TANDEM-AGS PROGRAM

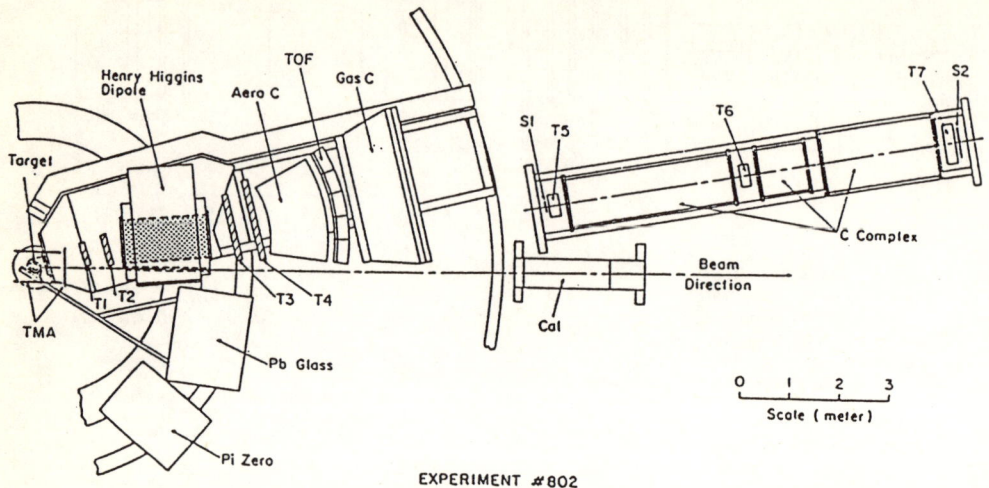

Fig. 12: E802—Studies of Particle Production at Extreme Baryon Densities in Nuclear Collisions at the AGS

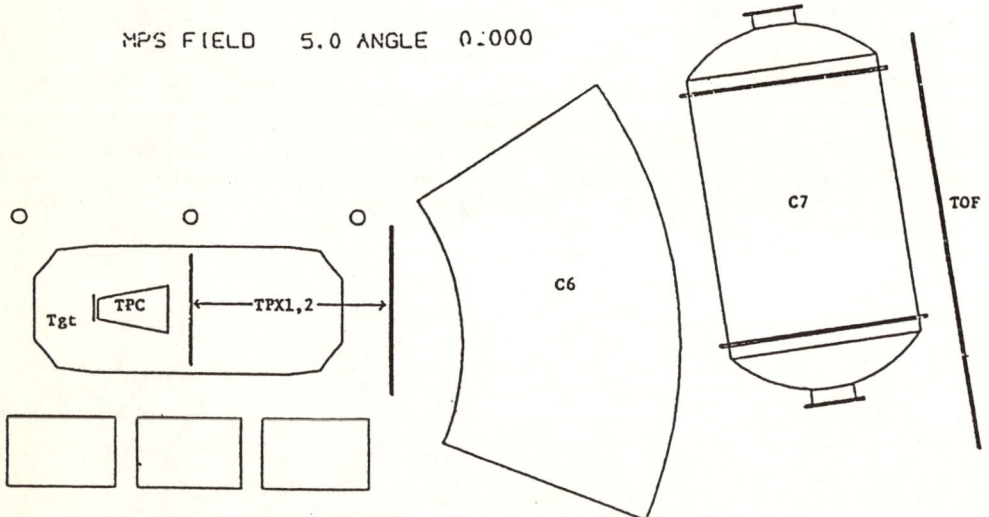

Fig. 13: E810—A Search for Quark Matter (QGP) and other New Phenomena Utilizing Heavy Ion Collisions at the AGS

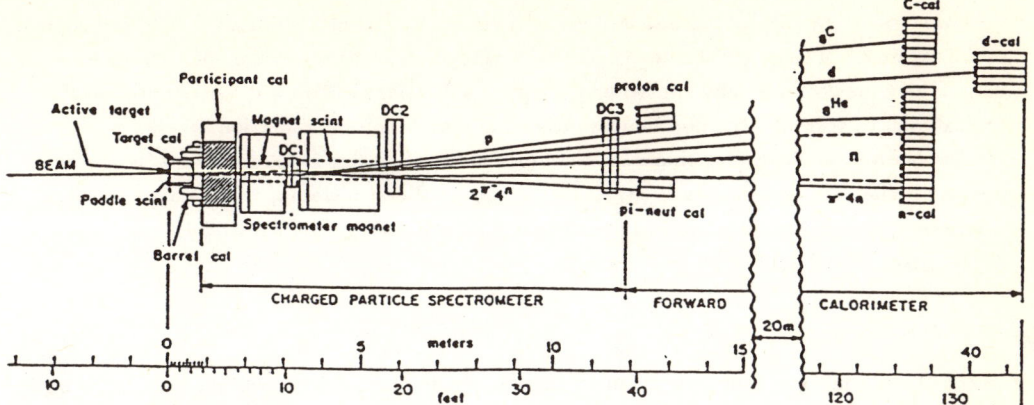

Fig. 14: E814—Study of Exotic Nuclear States Via Coulomb or Diffractive Projectile Excitation

5. Some Details of Experimental Technique and Analysis

Although it is beyond the scope of these lectures to give a full course in experimental physics, a few topics will be mentioned here which will be of use for the detailed discussions of the experimental results to follow.

5.1 Passage of charged particles through matter

The electric charge of a particle passing through matter[21] causes an electromagnetic force to be exerted on the atomic electrons of the material and causes them to be ejected from the atom, or **ionized**. The energy gained by the ions is lost by the incident particle, which slows down as it passes through matter, leaving a wake of ion-pairs in its path. Detection of these ion-pairs forms the basis of most charged particle detectors.

This is true for all particles except high energy electrons, which are so light that they essentially lose all their energy by radiation (Bremsstrahlung) due to interaction with the highly charged target nuclei. The probability for an electron of energy E to emit a photon of energy k, in range dk in passing through a thickness of material x is:

$$d\mathcal{P} = \frac{x}{X_o} \frac{dk}{k} \times F(k/E) \tag{5.1}$$

For thin radiators,

$$F(k/E) = \left[1 + (1 - k/E)^2 - \frac{2}{3}(1 - k/E)\right] \simeq 1,$$

so that the energy lost by electrons per unit length of material to photon emission can be written

$$-\frac{dE}{dx} \equiv k\frac{d\mathcal{P}}{dx} \simeq \frac{\int_0^E dk}{X_o} = \frac{E}{X_0}. \tag{5.2}$$

From this equation, it is clear that the **radiation length,** X_o is the length over which the energy of an incident electron is degraded to a fraction $1/e$ of its original value. The radiation length is inversely proportional to Z^2, where Z is the atomic number (nuclear charge) of the medium, and X_o is usually tabulated in units of $gm\ cm^{-2}$ (strictly $\rho \times X_o$ where ρ is the density of the medium). The **critical energy** of a medium is defined as the energy of electrons for which dE/dx by ionization loss and radiation are equal. Thus, for electrons above the critical energy, typically 10 to 40 MeV, the dominant source of energy loss is by radiation.

For heavier particles (starting with the muon) ionization loss is predominant (until the TeV regime). The rate of this loss is given by the Bethe-Bloch formula, which takes the form of a kinetic factor which depends only on the velocity of the incident particle, $\beta = v/c$, times another factor with a slight β and $\log Z$ dependence:

$$\frac{-dE}{dx} = \frac{D\,n_e\,Z_p^2}{\beta^2}\mathcal{F}(\beta, \ln Z)\ ,$$

where D is a constant. The important features of the ionization loss are that it depends on the square of the charge of the incident particle Z_p, and linearly on n_e the number of electrons per unit volume in the target, because it is the result of coulomb scattering of the incident particle by the atomic electrons in the target. Apart from the dominant $1/\beta^2$ dependence, the ionization loss is slowly varying, with a broad minimum; and it is usually sufficiently accurate to represent the ionization loss of an incident particle of charge Z_p, invariant mass m and momentum P as:

$$\frac{-dE}{dx} = \left[1 + \left(\frac{mc}{P}\right)^2\right] \times Z_p^2 \times \left.\frac{dE}{dx}\right|_{\min} \quad (5.3)$$

Thus, for fixed momentum, heavy particles have a larger ionization loss. As the momentum becomes much larger than the mass of the particle, the ionization loss reduces to a constant value for all particles, represented as the charge of the particle squared times $dE/dx|_{\min}$, the **minimum ionization** for singly charged particles, which is typically 1 to 2 MeV/$gm\ cm^{-2}$.

In contrast to the elastic scattering of the incident particle by the atomic electrons in a medium, which results in significant energy transfer due to the light mass of the atomic electrons, elastic scattering of the incident particle by the highly charged atomic nuclei results in negligible energy transfer, since the nuclei are heavy. However the incident particle suffers a very large number of very small angle Coulomb elastic scatterings with the target nuclei, in passing through a medium, resulting in a smearing of the angles of the incident particles by a random walk process, **Multiple Scattering**. The angular distribution for small angle scattering is given by the famous Rutherford formula, from which the mean squared scattering angle can be derived. After passing through a thickness x of material, the root mean squared multiple scattering angle in space (polar angular deflection) of a particle of momentum P, velocity β is:

$$\theta|_{\rm rms} = \sqrt{<\theta^2>} = \frac{21.1\,MeV/c}{(P\beta)} Z_p \sqrt{\frac{x}{X_o}} \quad (5.4)$$

Note that the multiple scattering increases as the square root of the number of radiation lengths (X_o) traversed. Although multiple scattering has nothing to do with radiation, the fact that the medium dependence of both processes scales as the number of radiation lengths is because both are Coulomb interactions with the target nuclei.

5.2 Strongly Interacting Particles—hadrons

In addition to the electromagnetic interactions which predominate in the target, an incident hadron may also suffer an occasional nuclear interaction. This is usually represented by a

nuclear interaction length λ_I. For an interaction cross section σ_I, the probably of suffering a nuclear interaction in a thin slab of material is just the interaction cross section (cm^2) times the number of target nuclei per cm^2 presented by the slab of material. If the thickness is x, measured in $gm\ cm^{-2}$, where $x = t \times \rho$, t is the thickness of material in cm, and ρ is the density of the material in $gm\ cm^{-3}$, then

$$\mathcal{P} = \sigma_I \times \frac{N_o \rho t}{A} \equiv \frac{x}{\lambda_I}$$

so that

$$\frac{1}{\lambda_I} = \frac{N_o \sigma_I}{A} \tag{5.5}$$

and λ_I is in $gm\ cm^{-2}$. Here N_o is Avogadro's number, the number of atoms per A grams of material of atomic weight A, and $\frac{N_o \rho t}{A}$ is the number of atoms per cm^2 in the slab of material. Note that nuclear interaction lengths are usually tabulated for incident nucleons. When an incident nucleus is involved, experimenters should usually check the computation themselves.

5.3 Passage of photons through matter

Photons are uncharged, and so do not have any long range electromagnetic forces. Thus, photons do not suffer ionization loss or multiple scattering when passing through a material. However, photons couple directly to electric charge with a "pointlike" coupling. Thus, when a photon does interact, it is usually catastrophic, resulting in the absorption or loss of the photon. The three major electromagnetic processes via which photons interact in matter are: photoelectric effect and Compton scattering with the atomic electrons; and pair production from the highly charged nuclei. For energies below $\sim 1\,MeV$, the atomic phenomena are dominant and photons are strongly absorbed in material, with the absorbtion being inversely proportional to the energy: the lower the energy, the stronger the absorption. Above a few MeV, photopair production becomes dominant, with a cross section that increases logarithmically with the photon energy until **complete screening** by the atomic electrons sets in and the cross section saturates. The interplay of these phenomena has the experimentally important consequence that photons of energies near the critical energy, typically 5 to 30 MeV, have the minimum absorption in matter, and thus can travel relatively long distances in a solid medium. Many of the troublesome mysterious backgrounds (or *albedo*) in experiments are caused by these photons near the critical energy, for which detectors are relatively transparent.

For high energy photons, the pair production process is very strongly related to the Bremsstrahlung process for high energy electrons. The probablilty of a high energy photon to pass through a thickness x of material without undergoing pair production[22] is

$$\mathcal{P}_{NC} = \exp\left(-\frac{7}{9}\frac{x}{X_o}[1-\zeta]\right) \tag{5.6}$$

where $\mathcal{P}_{NC}$ is the nonconversion probability, X_o is the radiation length, and $\zeta \lesssim 0.05$. Here the radiation length comes in because radiation and pair production are really just two aspects of the same process.

5.4 Electromagnetic Showers, Electromagnetic Calorimeters

In actual fact the concept of single Bremsstrahlung or pair production is only valid for very thin radiators or converters, $x \ll X_o$. For thick radiators, $x \gtrsim 10 X_o$, an electromagnetic cascade shower develops. An incident photon converts into an e^+e^- pair, then each member of the pair radiates photons, these photons in turn convert, making more electrons, etc. At first the number of electrons in the shower increases with the depth, and then decreases roughly exponentially. The total depth of the shower increases logarithmically with the energy. It is as if the incident electron or photon gets converted by radiative processes into electrons and positrons at the critical energy, which then stop radiating and lose the rest of their energy by ionization. The measurement of electromagnetic showers forms the basis of high resolution electron and photon detectors, or in today's jargon, **Electromagnetic Calorimeters**.

There are two classes of shower counters: sampling or total absorption. In a sampling counter, layers of high Z plates such as Pb or U create the showers, and the ionization is detected or sampled in layers of active material such as liquid argon or plastic scintillator. In a total absorption shower counter (TASC) there is only one high Z medium which both creates the showers and detects them. The showers are detected either by scintillation or Cerenkov light. The most popular scintillating TASC are made of NaI or CsI. These counters have the best energy resolution and are used over the range from keV to TeV photons. A much cheaper, and hence very popular, TASC is the Lead Glass (PbGl) Cerenkov counter. Lead Glass for particle detection contains $\sim 55\%$ PbO by weight, or about twice that of high quality crystal used in glassware and chandeliers. The radiation length of Lead Glass (SF-5) is about 2.4 cm, the critical energy is 15.8 MeV and the index of refraction is 1.67. The electrons in the shower are detected by Cerenkov radiation. The number of Cerenkov photons per unit length saturates very quickly for particles above the threshold $\beta_t = 1/n = 0.6$ so that electrons at the critical energy travel to the end of their range while emitting a constant amount of Cerenkov radiation per unit length.

An extimate of the energy resolution of a PbGl counter can be made by assuming that the incident electron or photon of energy E gets converted into electrons at the critical energy E_c, which then each emit a constant amount of Cerenkov light, represented by n_C detected Cerenkov photons, in coming to the end of their range. In this simple model, all fluctuations are ignored except in the statistics of the total number N of detected Cerenkov photons:

$$N = n_C \times \frac{E}{E_c}$$

Then,

$$\frac{\sigma_E}{E} = \frac{1}{\sqrt{N}} = \sqrt{\frac{E_c/n_C}{E}} = \frac{\sqrt{0.0158 GeV/n_C}}{\sqrt{E(GeV)}} = \frac{13\%/\sqrt{n_C}}{\sqrt{E(GeV)}}$$

The actual energy resolution of clear Lead Glass, where absorption of the Cerenkov photons is not important is (in r.m.s)

$$\frac{\sigma_E}{E} = \frac{4\%}{\sqrt{E(GeV)}}$$

implying a few detected Cerenkov photons per critical energy electron. In real life, lead glass gets radiation damaged and turns brown...

5.5 Hadron Calorimeters

At sufficiently high energies, $\gtrsim 10$ GeV, hadron interactions in thick absorbers $x \gg \lambda_I$ create cascade showers. At these high energies, multiparticle production dominates inelastic hadron reactions. Thus, a shower develops as an incident hadron suffers an inelastic interaction, many new hadrons are produced, they in turn interact, etc. (It is interesting to point out here that multiple hadron interactions inside a nucleus do not in general produce cascade showers and this is an important difference in the microscopic and macroscopic physics of hadron interactions with nuclei). Hadron showers are not nearly as elegant as electromagnetic showers and in general the fluctuations are much worse. The typical resolution for a hadron calorimeter[23] is:

$$\frac{\sigma_E}{E} = \frac{64\%}{\sqrt{E(GeV)}}$$

Also, since the nuclear interaction length of most materials is much longer that the radiation length, hadron calorimeters tend to be very large objects.

6. Measurement of the Momentum of a Charged Particle

The previous discussion on the intertactions of charged particles in matter led to the method of **energy** measurements of electrons, photons, and hadrons by total absorption counters or calorimeters. The **momentum** of charged particles is more conventionally measured by a magnetic spectrometer.

A particle with charge e momentum P in a uniform magnetic field B undergoes uniform circular motion with radius:

$$r = \frac{Pc}{eB}.$$

In practical units, r=3 meters for P=1 GeV/c and B=1.111... Tesla, for unit charged particles like those encountered by particle physicists. This is the principle of particle accelerators and solenoidal magnetic spectrometers: once the B field is given, the momentum determines the radius (or vice versa). In fixed target experiments, a magnetic spectrometer usually takes the form of a magnet with a rectangular shaped field region (see figure 15). The magnetic field deflects the charged particle, and the momentum is determined from the measured deflection. Taking the z direction, normal to the face of the magnet, and the magnetic field in the y direction, the particle deflection is in the $x - z$ plane, the plane of the drawing of figure 15. The vector equation:

$$\mathbf{F} = \frac{d\mathbf{P}}{dt} = \frac{e}{c}\mathbf{v} \times \mathbf{B}$$

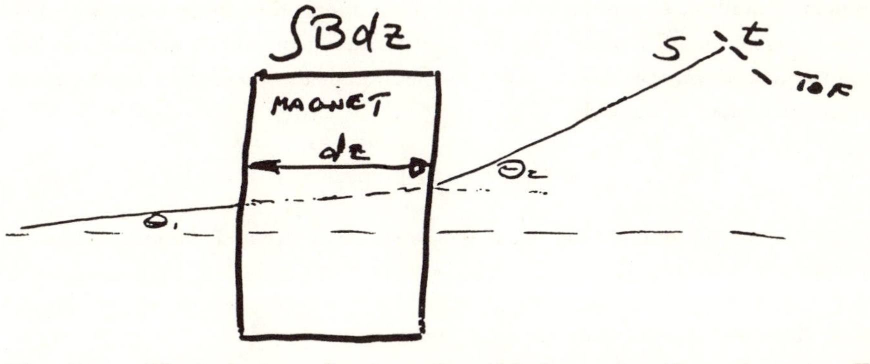

Fig. 15: The trajectory of a charged particle in a magnetic spectrometer. The magnetic field is normal to the page

leads to no change in the y component of momentum since B is in the y direction but leads to the coupled equations for P_x and P_z the x and z components of momentum:

$$\frac{dP_z}{dt} = -\frac{e}{c} v_x B_y \qquad \frac{dP_x}{dt} = \frac{e}{c} v_y B_z$$

Since B acts perpendicular to the momentum, the magnitide of the momentum is conserved, so it is sufficient to integrate the x–component equation:

$$\Delta P_x = -\frac{e}{c}\int_{\text{traj}} B_y v_z dt = -\frac{e}{c}\int B\, dz \qquad (6.1)$$

This leads to the usual description of a magnetic spectrometer as having a transverse momentum kick $= \int B dz$=300 MeV/c per Tesla meter, for unit charged particles. For highly charged particles, this quantity is simply multiplied by Z_p (i.e. $Z_p \times$ 300 MeV/c per Tm).

6.1 Momentum Resolution of a Magnetic Spectrometer

In the limit of small y–component of momentum, Eq. (6.1) reduces to

$$\Delta \sin\theta = \sin\theta_2 - \sin\theta_1 = \frac{\int B\, dz}{P} \qquad (6.2)$$

from which it should be apparent that $1/P$ is the quantity determined from a measurement of the difference in $\sin\theta$. Thus, the error in the quantity $1/P$ is related to the error in measurement of $\Delta \sin\theta$, which we may assume is normally distributed with standard deviation σ_{meas}:

$$\Delta(1/P) = \frac{\Delta P}{P^2} = \frac{\sigma_{\text{meas}}}{\int B\, dz} \qquad (6.3)$$

Measurement error usually dominates at high momentum, when the bending angle is small. At low momenta, multiple scattering usually dominates. For relativistic particles, the angular uncertainty due to multiple scattering (Eq. (5.4)) is inversely proportional to P, leading to the neat relation when plugged into Eq. (6.3):

$$\left.\frac{\Delta P}{P}\right|_{\text{ms}} = \frac{15\, MeV/c\, \sqrt{x/X_o}}{\int B\, dz} \qquad (6.4)$$

where the 21.2 MeV/c from Eq. (5.4) is divided by $\sqrt{2}$ to obtain the 15 MeV/c in Eq. (6.4), since only the component of multiple scattering in the bend plane affects the momentum resolution. Finally, since the uncertainties of multiple scattering and measurement error are uncorrelated so that they add in quadrature, the uncertainty in momentum measurement, the **momentum resolution**, is usually quoted in the form:

$$\frac{\Delta P}{P} = \sqrt{(a)^2 + (bP)^2} \qquad (6.5)$$

where the first term is due to multiple scattering in the spectrometer and the second term is from measurement error.

6.2 Resolution Smearing of a Momentum Spectrum

Suppose that x_o is a quantity to be measured, e.g. momentum, which is distributed with a steeply falling distribution, exponential for example:

$$dP(x_o) = f(x_o) \, dx_o = e^{-bx_o} dx_o$$

Further suppose that the true quantity x_o is measured with a Gaussian resolution function so that the result of the measurement is the quantity x, where

$$\mathcal{R}(x, x_o) \, dx = \text{Prob}(x)|_{x_o} \, dx = \frac{1}{\sqrt{2\pi\sigma^2}} \exp -\frac{(x-x_o)^2}{2\sigma^2} \, dx$$

The result for the measured spectrum is simply

$$f(x) \, dx = \int_{x_o=x-\infty}^{x_o=x+\infty} dx_o \, f(x_o) \, \text{Prob}(x)|_{x_o} \, dx$$

$$f(x) = \frac{1}{\sqrt{2\pi\sigma^2}} \int dx_o \, \exp -bx_o \, \exp -\frac{(x-x_o)^2}{2\sigma^2}$$

Complete the square:

$$f(x) = e^{-bx} e^{\frac{b^2\sigma^2}{2}} \times \frac{1}{\sqrt{2\pi\sigma^2}} \int_{x_o=x-\infty}^{x_o=x+\infty} dx_o \, \exp -\frac{(x - b\sigma^2 - x_o)^2}{2\sigma^2} \tag{6.6}$$

The result, since the Gaussian is normalized over ($-\infty, +\infty$), is simply

$$f(x) = e^{\frac{b^2\sigma^2}{2}} \times e^{-bx} = e^{-b(x - b\sigma^2/2)} \tag{6.7}$$

There are some important implications of this deceptively simple formula. It can be interpreted in two ways: the measured spectrum is shifted higher than the true spectrum by $\Delta x = b\sigma^2/2$; or equivlalently, the measured spectrum at a true quantity x_o is higher than the true spectrum by a factor $\exp(b^2\sigma^2/2)$. Also, the steeper is the spectrum (larger b), the larger is the effect of the resolution smearing. This is a consequence of the fact that as the spectrum becomes steeper, it is relatively less probable to get larger values of the quantity of interest from the distribution itself, compared to the fluctuations due to resolution.

As in most of these discussions, real life is considerably more complicated than the simple examples. Clearly, the momentum resolution is not Gaussian, with a constant σ for all momenta, as assumed above, but depends on the true momentum x_o as in Eq. (6.5). Thus, Eq. (6.6) doesn't integrate so neatly. In general a non-exponential tail is produced at higher (measured) momenta. Also, the momenta spectra of particles are not in general exponential. However, it remains true that the steeper is the spectrum, at any value of momentum, the larger is the effect of the momentum resolution.

6.3 Shift in a Spectrum due to the effects of Binning

In addition to the effect of momentum resolution in smearing a measured spectrum from the true distribution, unintended shifts may be introduced into measured (momentum) spectra by experimenters if the effects of binning are neglected. To clarify this point, again consider a steeply falling spectrum in a quantity x_o

$$dP(x_o) = f(x_o) \, dx_o.$$

One would like to measure the differential distribution $f(x_o)$. However the real world intrudes and what is actually measured is the integral of the distribution in small intervals (bins): e.g. for a bin of width Δ centered at the value x_o

$$F(x_o, \Delta) = \int_{x_o - \frac{\Delta}{2}}^{x_o + \frac{\Delta}{2}} f(x)\, dx$$

Typically, the differential distribution in a bin is estimated by dividing by the bin width:

$$\overline{f(x_o)} \equiv \frac{F(x_o, \Delta)}{\Delta}$$

Clearly, in the limit of zero bin width, $\Delta \to 0$, the measured distribution approaches the true distribution. For the case of wide bins, there is a reasonable procedure (not always followed) of plotting the value $\overline{f(x_o)}$ at a value of $\overline{x_o}$ shifted from the center of the bin, typically to a slightly lower value for a falling spectrum. **If this procedure is not followed, then the experimental spectrum will be higher than the true spectrum.** Especially, in the case of steeply falling spectra, not correcting for the effect of bin width can cause an apparently upward shift in the normalization.

To calculate the correct plotting point, we have to know the shape of the true spectrum, although an exponential is usually a good approximation in any individual bin. The plotting point $\overline{x_o}$ is chosen so that it lies on the true curve, or in mathematics:

$$\overline{f(x_o)} \equiv \frac{1}{\Delta} \int_{x_o - \frac{\Delta}{2}}^{x_o + \frac{\Delta}{2}} f(x)\, dx \equiv f(\overline{x_o})$$

For an exponential, $f(x) = A \exp(-bx)$, the solution is straightforward:

$$\overline{f(x_o)} = \frac{A f(x_o)}{b \Delta} \left[e^{\frac{b\Delta}{2}} - e^{\frac{-b\Delta}{2}} \right]$$

To get the correct solution, the exponentials must be expanded to second order in the small quantity:

$$e^z - e^{-z} = 2\left[z + z^3/3!\right] \simeq 2\, z\, e^{\frac{z^2}{3!}},$$

whereupon

$$\overline{f(x_o)} = A\, e^{-bx_o}\, e^{\frac{b^2 \Delta^2}{24}} = f(\overline{x_o})$$

and the place to plot the point is slightly shifted from the center of the bin

$$\overline{x_o} = x_o - \frac{(b\Delta)^2}{24}.$$

As a general rule, beware of any experimental measurement of a falling spectrum where the points are plotted at the centers of the bins!

7. Lorentz Transformations, Kinematics, Spectra of Decay Products

In addition to measuring the spectra of primary particles, experimentalists are forced quite often to deal with the spectra of unstable particles which can only be inferred from their decay products. Two examples are

$$\pi^0 \to \gamma + \gamma \qquad \Lambda^0 \to p + \pi^-$$

Lorentz transformations and Kinematics also play a real role in the daily life of an experimentalist, since we can not always position our equipment in the desirable rest frame.

7.1 Lorentz Transformations

Consider a particle of vector momentum $\mathbf{P}$, energy E, in a particular rest frame, and invariant mass m. The **four-vector** momentum p of this particle will be denoted:

$$p = (\mathbf{P}, iE)$$

in units where the speed of light c is taken as unity. The four-dot product of two four-momentum vectors p_1 and p_2 is denoted:

$$p_1 \cdot p_2 \equiv \mathbf{P_1} \cdot \mathbf{P_2} - E_1 E_2$$
$$= P_1 P_2 \cos\theta - E_1 E_2$$

where $\mathbf{P_1} \cdot \mathbf{P_2}$ is the dot product of the 3-momentum vectors, P_1 and P_2 are the moduli of the 3-vectors and θ is the angle between them. The squared modulus of a 4-vector is denoted

$$p^2 = p \cdot p = P^2 - E^2 = -m^2$$

and is invariant under a Lorentz Transformation.

We are often obliged to deal with particles in difference reference frames. Let a particle of invariant mass m have a four momentum

$$p^* = (\mathbf{P^*}, iE^*)$$

as measured in a coordinate system moving with a velocity β relative to the reference system in which the particle four momentum is

$$p = (\mathbf{P}, iE)$$

The **Lorentz Transformation** relates the momentum components in the reference frame to those measured in the moving coordinate system:

$$E = \gamma \left(E^* + \beta P_L^* \right)$$
$$P_L = \gamma \left(P_L^* + \beta E^* \right)$$
$$P_T = P_T^* \qquad (7.1)$$

where $P_L = P\cos\theta$, $P_T = P\sin\theta$, $P_L^* = P^*\cos\theta^*$, $P_T^* = P^*\sin\theta^*$ are the components of momentum, longitudinal and transverse to the direction of motion, in the respective frames; θ and θ^* are the angle of the particle relative to the direction of motion in the two frames, and

$$\gamma = \frac{1}{\sqrt{1-\beta^2}} \qquad (7.2)$$

Note that the transverse momentum, P_T, is conserved between the two frames, as is the **transverse mass**, $m_T = \sqrt{P_T^2 + m^2}$, while the energy and longitudinal momentum are **boosted**. The transformation of the angle relative to the direction of motion in the two frames is immediataly obtained from Eq. (7.1):

$$\tan\theta = \frac{P_T}{P_L} = \frac{\sin\theta^*}{\gamma(\cos\theta^* + \beta/\beta^*)} \tag{7.3}$$

where $\beta^* = P^*/E^*$ is the velocity of the particle in the moving frame.

The simplicity of the rapidity variable Eq.(2.1) becomes apparent when the Lorentz Transformation between the two frames is expressed in this variable. Let

$$y = \ln\left(\frac{E + P_L}{m_T}\right) = \frac{1}{2}\ln\left(\frac{E + P_L}{E - P_L}\right) \tag{7.4}$$

denote the rapidity of the particle in the reference frame, y^* be the rapidity of the particle measured in the moving frame and Y be the rapidity of the moving system:

$$Y = \frac{1}{2}\ln\left(\frac{1+\beta}{1-\beta}\right) \tag{7.5}$$

Then it follows from substituting Eq. (7.1) in Eq. (7.4) that

$$y = Y + y^* \tag{7.6}$$

The details are left as an exercise for the student.

7.2 Transformation to the rest system of a particle

The rest system of a particle is defined as the system in which the particle is at rest, i.e $P^* = 0$, $E^* = m$. It is then easy to see that the rest system of the particle moves with the particle velocity (as seen in the moving system):

$$E = \gamma m$$
$$P = \gamma\beta m$$

7.3 Two particle collisions—the lab and c.m. systems

The description of a two particle collision is a useful exercise in relativistic kinematics. In the laboratory system, an incident particle with momentum $\mathbf{P}_1$, energy E_1 and mass m_1 collides with a particle with mass m_2, at rest. The four vectors are:

$$p_1 = (\mathbf{P}_1, iE_1) \qquad p_2 = (0, im_2)$$

In the center of mass (c.m.) system the momenta of the particles are equal and opposite, and the four-vectors are:

$$p_1^* = (\mathbf{P}_1^*, iE_1^*) \qquad p_2^* = (-\mathbf{P}_1^*, iE_2^*)$$

The transformation between the lab and c.m. systems is given in terms of the four-vector total momentum of the system $p_1 + p_2$, which is conserved in a collision. The modulus of the total four-momentum is a Lorentz Invariant quantity, which is the same in all reference systems (before and after the collision because of four-momentum conservation):

$$-s \equiv (p_1 + p_2)^2 = -(E_1^* + E_2^*)^2 \tag{7.7}$$

It is clear that $\sqrt{s}$ is the total energy in the c.m. system, which is the same as the invariant mass of the c.m. system. In terms of the lab quantities:

$$s = m_1^2 + m_2^2 + 2E_1 m_2 \tag{7.8}$$

An important point to notice is that the Lorentz transformation of the sum of four-vectors is identical to the sum of the Lorentz transformations of the four-vectors. It is then immediately apparent, using the total four-momentum vector of the collision, that the c.m. rest frame has an invariant mass $\sqrt{s}$ and moves in the laboratory system (along the direction of $\mathbf{P_1}$) with a velocity corresponding to:

$$\gamma^{cm} = \frac{E_1 + m_2}{\sqrt{s}} \qquad \text{and} \qquad Y^{cm} = \cosh^{-1} \gamma^{cm}$$

Another useful quantity is Y^{beam}, the rapidity of the incident particle in the laboratory system

$$Y^{beam} = \cosh^{-1} \frac{E_1}{m_1}$$

Note that for equal mass projectile and target:

$$Y^{cm} = Y^{beam}/2$$

7.4 Two body decay of a heavy particle into light particles

This is a more advanced example of the transformation to a rest system, which has some interesting experimental applications. Consider a heavy particle which decays to two light particles. Examples would include:

$$J/\Psi \to e^+ e^- \qquad \pi^0 \to \gamma\gamma \qquad Z^0 \to \mu^+ \mu^-$$

To be specific, a particle of momentum $\mathbf{P}$, energy E, rest mass m, decays into two particles with $\mathbf{P_1}$, E_1, m_1 and $\mathbf{P_2}$, E_2, m_2. There are many questions we might wish to ask with this starting point, however we start out with a question near and dear to the heart of somebody with a photon detector. Suppose the heavy particle was travelling in a direction normal to the surface of a PbGl wall, heading toward the detector, when it decayed a distance L from the front surface. What is the distribution in the distance d between the two decay particles when they strike the front surface of the wall? (Would any theorist think up such a question?)

The two decay particles arrive at the wall at lateral distances d_1 and d_2 from where the point of impact of the parent would have been. Since it is a 2 body decay, all three particles lie in the same plane, so that the distance d between the 2 decay particles is simply

$$d = d_1 + d_2 = L(\tan\theta_1 + \tan\theta_2) \tag{7.9}$$

where θ_1, θ_2 are the decay angles of particles 1 and 2 in the lab system and L is the distance of the decay from the detector. In the rest system of the parent, the two particles have equal and opposite momenta, so that

$$x \equiv \cos\theta_1^* = -\cos\theta_2^*$$

and $\sin\theta_1^* = \sin\theta_2^* = \sqrt{1-x^2}$. This causes a few neat cancellations when Eq. (7.3) is substituted in Eq. (7.9) for both particles. In the case of equal masses for the two decay particles ($m_1 = m_2$), the exact result for d has the simple form:

$$\frac{d}{2L} = \beta\beta^* \frac{\sqrt{1-x^2}}{\gamma\left(\beta^2 - \beta^{*2} x^2\right)} \tag{7.10}$$

where β^* is the velocity of the decay particles in the rest system of the parent. The case $\beta^* = 1$ is particularly interesting since it involves massless particles in the final state, e.g. photons:

$$\frac{d}{2L} = \frac{\beta\sqrt{1-x^2}}{\gamma\left(\beta^2 - x^2\right)} \tag{7.11}$$

For $|x| \geq \beta$, $d \to \infty$ or is negative (impossible for an inherently positive quantity), which means that one of the photons goes backwards in the laboratory and therefore cannot hit the detector. This is an important reminder that a massless particle can not be turned around by a Lorentz Transformation.

We now consider the distribution in d for two cases: either the total energy of the parent ($E = E_1 + E_2$) is held constant; or the energy of one of the photons (E_1) is held constant. This will be a good example in the use of conditional probability. The energies of the two photons have the same constraint on $x = \cos\theta^*$ as used above, so that the ratio of the energies is easily computed, with the result:

$$r \equiv \frac{E_2}{E_1} = \frac{1-\beta x}{1+\beta x} \quad \text{and} \quad x = \frac{1}{\beta}\frac{1-r}{1+r} \tag{7.12}$$

For E fixed, we can ignore the case when one of the photons misses the detector, and the relativistic limit ($\beta \to 1$) of Eq. (7.11) may be taken simply:

$$\frac{d}{2L} = \frac{1}{\gamma\sqrt{1-x^2}} \tag{7.13}$$

For the case E_1 fixed, the variables to use are E_1 and $r = E_2/E_1$. In this case, care must be taken about the divergences, so that terms $\sim 1 - \beta^2$ can not be ignored. We solve Eq. (7.12) for $\sqrt{1-x^2}$ in terms of r, in the limit $\beta \to 1$:

$$\sqrt{1-x^2} = \frac{2\sqrt{r}}{1+r}$$

and substitute into Eq. (7.13), using the relation:

$$E = E_1 + E_2 = E_1(1+r) \qquad \gamma = \frac{E}{m} = \frac{E_1}{m}(1+r)$$

to obtain

$$\frac{d}{2L} = \frac{m}{2E_1\sqrt{r - \frac{m^2}{2E_1^2}}} \tag{7.14}$$

Note that the subtraction constant in the square root is due to a term $\sim 1 - \beta^2$, and corresponds to the value of $r_\infty = \frac{m^2}{2E_1^2}$, for which d diverges when E_1 is held fixed.

To complete the kinematics at fixed E_1, we note that the minimum value of r occurs when $x = 1$ and both decay photons are collinear with the parent: one going forward in the lab on the same trajectory as the parent and hitting the detector, while the other photon goes exactly backwards on the same trajectory, heading away from the detector. In this case, from Eq. (7.12),

$$r_{\min} = \frac{1-\beta}{1+\beta} = \frac{1-\beta^2}{(1+\beta)^2} \to \frac{1}{4\gamma^2} = \frac{m^2}{4E_1^2} \tag{7.15}$$

With the kinematics out of the way, we can now concentrate on finding the distribution in d.

The case E fixed is easy since it only depends on the angular distribution of decay particles in the parent rest frame, the distribution in $x = \cos\theta^*$. For fixed E there is a minimum separation of the two photons at the wall, which occurs for the symmetric decay, $x = 0$

$$\frac{d_{\min}}{L} = \frac{2m}{E} = \frac{2}{\gamma} \tag{7.16}$$

Use $d_{\min}$ in Eq. (7.13) to eliminate γ and obtain

$$\sqrt{1-x^2} = \frac{d_{\min}}{d}$$

or

$$x = \sqrt{1 - d_{\min}^2/d^2} \tag{7.17}$$

The integral probability that two photons land on the wall separated by a distance $s \leq d$ is just given by the integral probability that x lies between $\pm x(d)$, where $x(d)$ is given by Eq. (7.17). For the decay $\pi^0 \to \gamma + \gamma$, the decay is isotropic,

$$\frac{d\mathcal{P}}{dx} = \frac{1}{2}$$

so that

$$\mathcal{P}(s \leq d) = \mathcal{P}(-x(d) \leq x \leq x(d)) = x(d) = \sqrt{1 - d_{\min}^2/d^2} \tag{7.18}$$

For the case in which one of the decay photon energies, E_1, is held constant, the relationship between the spectra of the parent and the decay particles is required. This will be discussed in the following section, after which the distribution in d will be presented.

7.5 The spectra of decay particles—the parent-daughter factor

A parent particle of momentum P, energy E, rest mass m, decays in to two particles, as above. Furthermore, the differential probability to produce a parent particle with momentum P in range dP is given by the function:

$$d\mathcal{P}(P) = f(P)\,P\,dP$$

where for example we take $f(P)$ to be of the form

$$f(P) = A\,P^{-n}$$

so that

$$d\mathcal{P}(P) = A\,P\,P^{-n}\,dP \tag{7.19}$$

The conditional probability of finding a daughter with energy E_1, given a parent with momentum P, depends on the $x = \cos\theta^*$ distribution of the decay:

$$\left.\frac{\partial \mathcal{P}}{\partial E_1}\right|_P = \frac{\partial \mathcal{P}/\partial x}{\partial E_1/\partial x}$$

For equal mass decay particles,

$$E_1 = \frac{E}{2}(1 + \beta\beta^* x)$$

so that

$$\frac{\partial E_1}{\partial x} = \frac{E\beta\beta^*}{2} = \frac{P\beta^*}{2}$$

and

$$\frac{\partial P}{\partial E_1}\bigg|_P = \frac{2}{P\beta^*}\frac{\partial P}{\partial x}$$

For a uniform decay distribution

$$\frac{\partial P}{\partial x} = \frac{1}{2}$$

so that

$$\frac{\partial P}{\partial E_1}\bigg|_P = \frac{1}{P\beta^*}$$

The joint probability of finding a parent of momentum P and a daughter of energy E_1 is given by the rules of conditional probability:

$$\partial^2 P(P, E_1) = \partial P(P)\, \partial P(E_1)|_P = \frac{f(P)}{\beta^*} dP\, dE_1 \qquad (7.20)$$

The marginal probability distribution for E_1 is found by integrating over all values of P consistent with E_1. We evaluate the case for photons in the final state, $\beta^* = 1$, and only consider the relativistic limit, ignoring the difference between P and E. A more exact treatment has been given by Sternheimer[24].

$$\partial P(E_1) = \int_{E_{\min}=E_1(1+r_{\min})}^{\infty} AP^{-n} dP\, dE_1 = (1+r_{\min})^{-n+1} \frac{1}{n-1} A E_1 E_1^{-n} dE_1 \qquad (7.21)$$

When Eq. (7.21), the daughter spectrum, is compared to Eq. (7.19), the parent spectrum, we see that they are precisely the same form (for a power law) but the daughter spectrum is suppressed by a factor of $1/(n-1)$, for a spectrum falling with the $(n-1)$ power. This is called the **Parent-Daughter Suppression Factor**.

To continue the analysis, we now wish to find the spectrum of the second photon, E_2, for the energy of the other photon, E_1, held constant. The joint probability distribution of E_2 and E_1 is trivially related to the joint probability distribution of E and E_1, since

$$E = E_1 + E_2 \qquad \text{so that} \qquad \partial E_2|_{E_1} = \partial E|_{E_1}$$

The joint probability distribution comes directly from Eq. (7.20)

$$\partial^2 P(E_1, E_2) = f(E_1 + E_2)\, dE_1\, dE_2 \qquad (7.22)$$

We now use a famous rule of conditional probability to find the distribution of E_2 given E_1

$$P(A, B) = P(A) \times P(B)|_A$$

Thus the conditional probability for E_2, given E_1, is just the joint probability for both E_1 and E_2 divided by the marginal probability for E_1: or Eq. (7.22) divided by Eq. (7.21).

$$\partial P(E_2)|_{E_1} = \frac{\partial^2 P(E_1, E_2)}{\partial P(E_1)} = (1+r_{\min})^{n-1} \frac{n-1}{(1+r)^n} dr \qquad (7.23)$$

Note that this distribution **scales**—it is only a function of the ratio of the energies of the two photons. This equation can be integrated to find the probability of r in the range ($r_1 \leq r \leq r_2$):

$$P(r_1 \leq r \leq r_2) = \left(\frac{1+r_{\min}}{1+r_1}\right)^{n-1} - \left(\frac{1+r_{\min}}{1+r_2}\right)^{n-1} \qquad (7.24)$$

This probability is obviously correctly normalized over the range $(r_{\min} \leq r \leq \infty)$ where $r_{\min}(E_1)$ is given by Eq. (7.15).

7.6 Distribution of d for E_1 fixed

This problem can now be completed by rewriting Eq. (7.14) to relate the distance between the two photons to the ratio of their energies, with E_1 the energy of one of the photons being fixed:

$$r - \frac{m^2}{2E_1^2} = \frac{m^2/E_1^2}{d^2/L^2} \qquad (7.25)$$

The probability for d in any range is found by using Eq. (7.25) to find the corresponding values of r for the range and then using Eq. (7.24) to find the probability.

7.7 Why spend such effort on minute kinematic details?

These derivations and problems may seem a bit long-winded; but experimentalists spend most of their time dealing with such details rather than making great discoveries. Of course, if you do not spend enough effort coping with the details, you sometimes find a "discovery" which is nothing but a detail that you didn't work out correctly.

8. High Energy Physics in One Easy Lesson

One of the nice features of the search for the QGP is that it requires the integrated use of many disciplines in Physics: High Energy Particle Physics, Nuclear Physics, Relativistic Mechanics, Quantum Statistical Mechanics... An understanding of the properties of high energy $p-p$ and $p+A$ interactions is vital to the ability to distinguish the "ordinary physics" of relativistic nuclear interactions from the signatures of production of a new state of matter, the QGP. It is also possible that the "ordinary physics" may in itself be quite interesting.

8.1 Multiparticle Production in Nucleon-Nucleon $(p-p)$ Collisions

When high energy nucleons collide inelastically, the predominant mode of dissipating the energy is by multiple particle production. The produced particles are distributed relatively uniformly in rapidity, with limited transverse momemtum with respect to the collision axis, leading to a description of the process as "longitudinal phase space" (recall figure 4). The modern view of strong (or nuclear) interactions is that nucleons are composed of 3 valence quarks confined into a bound state by the strong force, Quantum Chromo Dynamics (QCD), which is mediated by the exchange of color-charged vector gluons. The self-interaction of the color-charged gluons, in sharp distinction to the behavior of the uncharged quanta of electrodynamics, is believed to provide the confinement property of QCD and the related property, known as asymptotic freedom, which is the reduction of the effective coupling constant at large momentum transfers. This leads to the perverse situation that the rare "hard scattering" events of inelastic hadron collisions, involving the scattering or production of constituents with large momentum transfer, can be treated very precisely in the framework of perturbative QCD; but that the vast majority of collisions are in the non-perturbative domain and are subject instead to a more qualitative description, based primarily on empirically observed regularities.

The general framework for the study of "soft" multiparticle physics was well in place by the early 1970's [26]. One of the important conceptual breakthroughs was the realization that the distribution of multiplicity for multiple particle production would not be Poisson unless the particles were emitted independently, without any correlation [27,28]. In that era single particle "inclusive" reactions[29] were extensively studied as was the distribution in the total multiplicity per collision, a multiparticle "exclusive" quantity[30,31].

8.2 Single Particle Inclusive Reactions

A single particle inclusive reaction involves the measurement of just one particle coming out of a collision, ignoring all others. The measurements are presented in terms of the (Lorentz) Invariant single particle inclusive differential cross section:

$$\frac{E d^3\sigma}{dp^3} = \frac{d^3\sigma}{P_T dP_T dy d\phi} = \frac{1}{2\pi} \mathbf{f}(P_T, y) \tag{8.1}$$

A uniform azimuthal distribution is usually assumed, so the integral over azimuth is simply 2π

$$\frac{d^2\sigma}{P_T dP_T dy} = \mathbf{f}(P_T, y) \tag{8.2}$$

The distributions in P_T are measured as a function of rapidity and integrated over P_T to find

$$\frac{d<n(y)>}{dy} = \frac{1}{\sigma_I} \int dP_T \, P_T \, d\phi \, E d^3\sigma/dp^3 = \frac{1}{\sigma_I} \int dP_T \, P_T \, \mathbf{f}(P_T, y) \tag{8.3}$$

This may be further integrated over all the kinematically possible rapidity to find the mean multiplicity per interaction, $<n>$:

$$<n> = \frac{1}{\sigma_I} \int dy \, dP_T \, P_T \, \mathbf{f}(P_T, y) \tag{8.4}$$

An important point to remember about inclusive single particle cross sections is illustrated in Eq. (8.4). Integrals of the single particle inclusive cross section are not equal to σ_I the interaction cross section, but rather equal to the mean multiplicity times the interaction cross section: $<n> \times \sigma_I$.

The inclusive charged particle P_T spectra, measured near the rapidity of the nucleon-nucleon c.m. system, y_{cm}^{NN}, (or 90° in the c.m. system in HEP jargon), is shown in figure 16. This figure[32] includes data over nearly the full available range in $\sqrt{s}$. The "high" P_T or "hard scattering" region, above 1 to 2 GeV/c, shows an enormous variation with c.m. energy, while the region below 1 GeV/c, the "soft" physics region which dominates the spectrum, remains essentially unchanged and is reasonably characterized over the full energy range by:

$$\mathbf{f}(P_T, y) = A(\sqrt{s}) \exp -6P_T .$$

This is the reason High Energy Physicists describe particle production as "longitudinal phase space". As the c.m. energy increased, the $<P_T>$ remains relatively constant, while the central plateau (figure 4) tends to expand to fill the available phase space in rapidity. Even at a fixed c.m. energy, the P_T and rapidity distributions are nearly independent. The $<P_T> = 2/6 = 0.333$ GeV/c at y_{cm}^{NN}, and decreases slightly, by less than 10% on the central plateau, as y moves away from y_{cm}^{NN}, then dropping to ~ 220 MeV/c near the projectile rapidity[33]. The mean transverse momentum as a function of rapidity is defined:

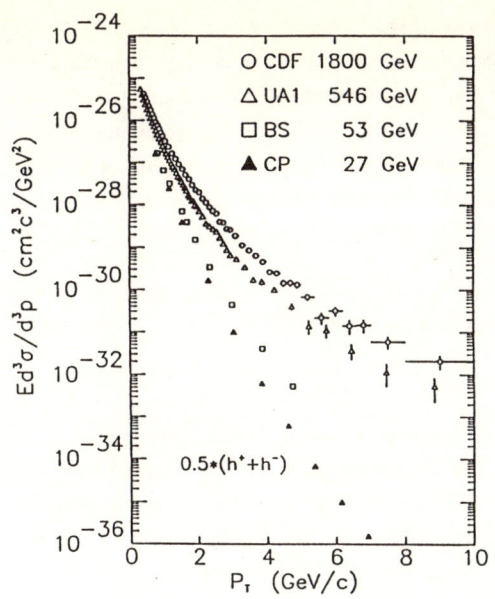

Fig. 16:
The transverse momentum dependence of invariant cross sections of charged-averaged hadrons $(h^+ + h^-)/2$, near y_{cm}^{NN}, in nucleon-(anti)nucleon collisions over the available range of c.m. energies. Typical data shown are from fixed target ($\sqrt{s} = 27$ GeV), ISR (53 GeV), CERN SPS collider (546 GeV), and FERMILAB collider (1800 GeV) [32].

$$<P_T>|_y = \frac{\int dP_T \, P_T \, P_T \, \mathbf{f}(P_T, y)}{\int dP_T \, P_T \, \mathbf{f}(P_T, y)} \quad (8.5)$$

The same equation applies to any rapidity region by integrating over rapidity first in both integrals. Note that in Lorentz invariant terms, this is the natural definition of $<P_T>$ so that the $|_y$ is not usually explicitly indicated.

8.3 A Warning—Beware the Seagull Effect

It has taken particle physicists many years to get comfortable with the the differential $dy = dP_L/E$ in the invariant cross section. (Relativistic Heavy Ion physicists are presently undergoing the same trauma.) Much effort was spent by particle physicists on detailed studies before it became clear that rapidity and P_T were the convenient independent variables for describing $p-p$ collisions. There is much work in the literature studying the dependence of the mean transverse momentum on P_L^* or Feynman x Eq(2.8). This creates an artificial variation of $<P_T>$ called the **Seagull Effect**[34], which in simple terms is a consequence of the fact that the transverse momentum cannot exceed the momentum.

The mean transverse momentum at fixed P_L^* is defined by using the **non-invariant** cross section as the probability distribution:

$$\begin{aligned}<P_T>|_{P_L^*} &= \frac{\int P_T \, P_T \, dP_T \, d^3\sigma/dP^3}{\int P_T \, dP_T \, d^3\sigma/dP^3} \\ &= \frac{\int P_T \, P_T \, dP_T \, \mathbf{f}(P_T, y)/E}{\int P_T \, dP_T \, \mathbf{f}(P_T, y)/E}\end{aligned} \quad (8.6)$$

This is equivalent to computing the $<P_T>$ of the distribution $\mathbf{f}/E$ where $\mathbf{f}$ is the invariant cross section. For large values of P_L^*, the energy E becomes equal to P_L^*, which for the purposes of Eq. (8.6) is a constant, so the true $<P_T>$ is obtained. For $x_F = 0$, the weighting distribution is

f/m_T which is obviously steeper in P_T than the unweighted f, resulting in a lower $<P_T>$. As x_F increases away from $x_F = 0$, the $<P_T>$ increases to its true value, producing a beautiful but irrelevant and confusing seagull drawing. (It may prove comforting to some to know that there was at least one thing on which Feynman's intuition was wrong!) This paragraph is placed here as a warning that the definition and plots of $<P_T>$ from the original particle physics work in the 1960's and 1970's may not mean the same thing as we understand them today.

8.4 Do particle physics spectra prefer P_T or m_T?

From the point of view of relativistic kinematics, m_T rather than P_T would seem to be the preferred variable, since under a Lorentz transformation Eq(7.1)

$$E^2 - P_L^2 = m_T^2 = P_T^2 + m^2 \tag{8.7}$$

is conserved. Although the P_T spectra of pions are generally (and conveniently) described as $\exp -6P_T$ (as above), in actual fact the data are better represented as a function of m_T. All spectra in P_T fall below an exponential for values of $P_T \sim m$, so that the spectra are better represented[35] by exponentials in m_T. (See Fig. 17). There is absolutely no change in the differential cross section for this change of variables, since $P_T \, dP_T = m_T \, dm_T$.

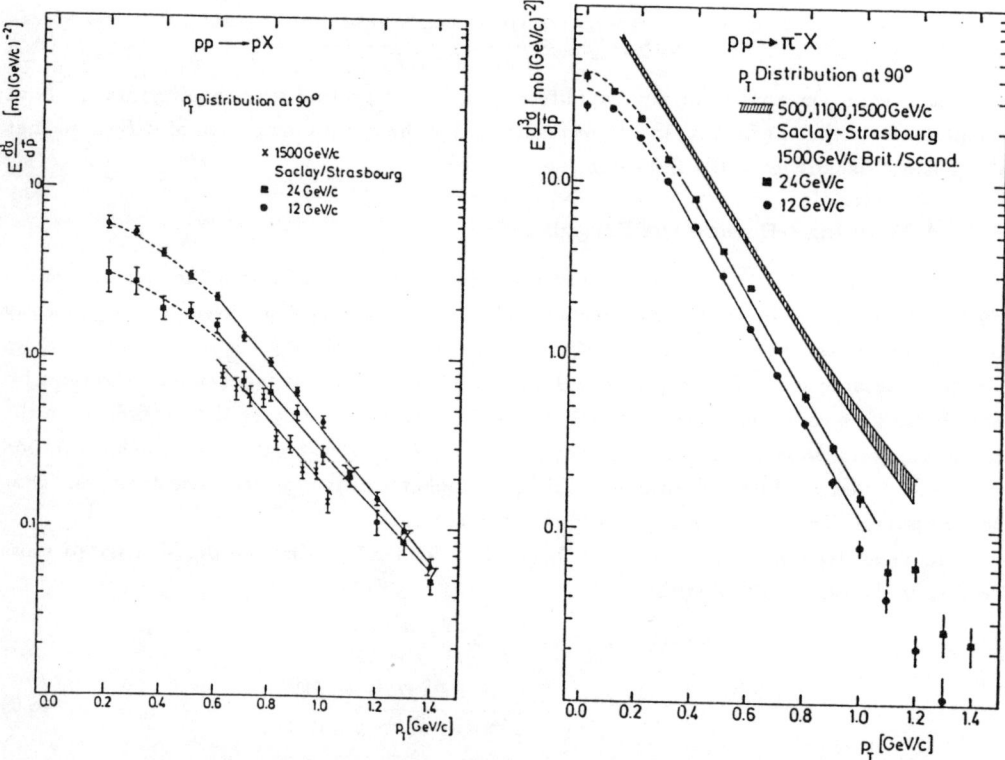

Fig. 17: Invariant cross sections at 90° in the c.m. system for proton and $\pi-$ production from $p-p$ collisions[35]. The solid lines are exponential fits in P_T, while the dashed lines are exponential in m_T.

8.5 $\sqrt{s}$ dependence of dn/dy

The multiplicity density in rapidity, dn/dy, (Eq. (8.3)) is one of the principal descriptive variables in both high energy and Relativistic Heavy Ion physics. The mean multiplicity in $p-p$ collisions increases logarithmaically with $\sqrt{s}$. In the early 1970's, it was thought[36] that this could be explained if dn/dy on the rapidity plateau reached a constant or limiting value, so that $<n>$, the integral of the distribution (Eq. (8.4)), would just increase as $\ln s$, the available rapidity range :

$$y^*(m)|_{\max} = \ln \frac{\sqrt{s}}{m} \qquad (8.8)$$

Although the ISR[36] produced the first real evidence for the rapidity plateau, another ISR[37] measurement was the first to show unambiguously that $dn/dy|_{y_{cm}^{NN}}$ was not a constant but rose steadily with increasing $\sqrt{s}$. The best present data[38] on the $\sqrt{s}$ dependence of the central density:

$$\rho(0) = \frac{1}{\sigma_I} \frac{d\sigma}{d\eta}\bigg|_{\eta^*=0} \qquad (8.9)$$

is shown in figure 18, where η is the pseudo-rapidity. When the lower energy data[35] are included, it becomes clear that the preferred fit over ther range $\sqrt{s} = 5 - 1000$ GeV is:

$$\rho(0) = 0.74 \left(\sqrt{s\,(GeV)}\right)^{0.210}$$

There is a clear, but very slow increase of $\rho(0)$ with $\sqrt{s}$. Note that the increase in central

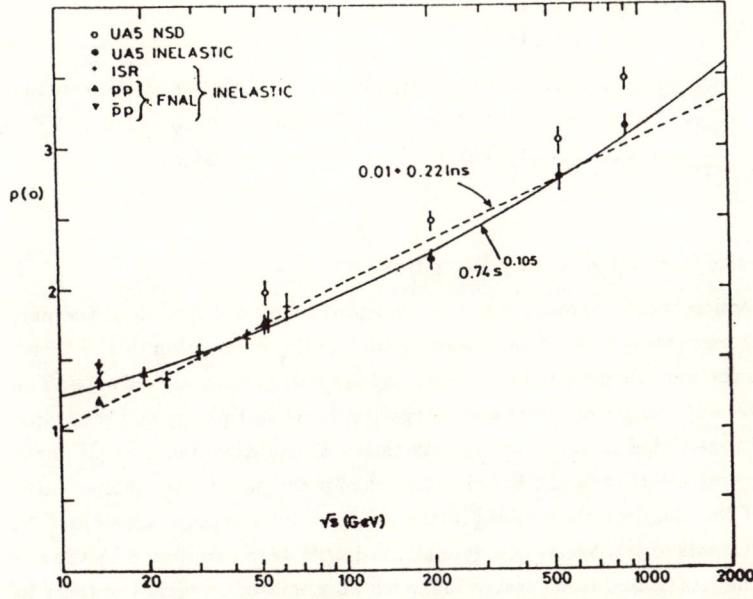

Fig. 18: Central density $\rho(0) = \frac{1}{\sigma_I}\frac{d\sigma}{d\eta}|_{\eta^*=0}$ plotted as a function of c.m. energy for data from Fermilab fixed target to Cern Collider energies. Fits to inelastic data using linear dependence in $\ln s$ or a power law dependence on s are shown[38].

multiplicity density in going from $\sqrt{s_{NN}} = 5.39$ GeV, which is the nucleon-nucleon c.m. energy for the BNL heavy ion program, to $\sqrt{s_{NN}} = 19.4$ GeV at CERN, is only a factor of 1.31.

8.6 Has the QGP already been found in $p - \bar{p}$ collisions ?

An experiment at the FERMILAB $p - \bar{p}$ collider, with the highest available $\sqrt{s} = 1.8$ TeV, has searched for production of the QGP by looking at the dependence on $<P_T>$ as a function of dn/dy on the central plateau[39]. The data show a striking increase of $<P_T>$ with dn/dy (see Fig. 19). This effect was also seen by UA1 at the CERN collider[40] and is expected in hydrodynamical modes[41], but is largely absent[42] at c.m. energies below 63 GeV. The key question from figure 19 is whether the rise in $<P_T>$ for $dN_c/d\eta \geq 25$ is evidence for the "Van Hove" signature of QGP production[11], or is merely a statistical fluctuation. This is left as an exercise for the student.

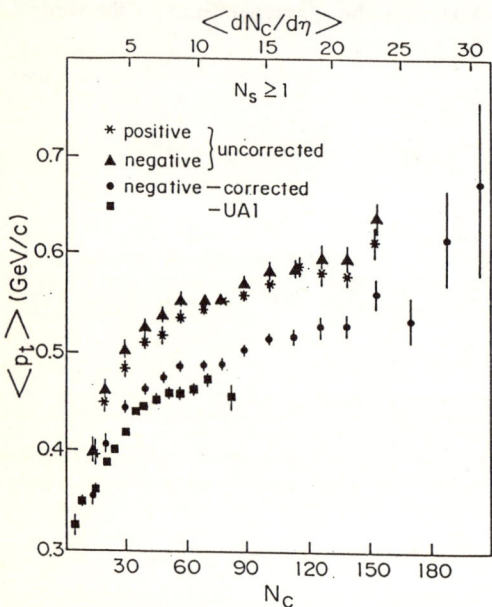

Fig. 19: $<P_T>$ vs. $dN_C/d\eta$ in the central region of $p - \bar{p}$ collisions at $\sqrt{s}$=1.8 TeV[39]. The UA1 data[40] are also shown.

8.7 Multiparticle inclusive reactions—E_T distributions

In the 1980's, multiparticle inclusive measurements, in which many but not all of the particles from an interaction are measured, became a leading tool in the description of the "soft" reactions which dominate the particle production process in high energy hadron collisions. The two principal multiparticle inclusive variables are the charged particle multiplicity and the transverse energy flow, taken in restricted intervals of rapidity rather than integrated over all phase space, These variables are very closely related, but it some time for this fact to be understood[43]. The multiplicity is one of the classical observables in the study of high energy collisions [44,45], while the original measurements of transverse energy distributions[46] were stimulated by the desire to detect and study the jets from "hard" scattering, with an unbiased trigger. Contrary to early expectations, the transverse energy is dominated by "soft collisions". The transverse energy is made up of a structureless cloud of low transverse momentum (~ 0.40 GeV/c) particles[47]. Jets are swamped[48].

The strong relationship of the multiplicity and transverse energy distributions is a consequence of the fact that the transverse momentum distribution for particle production is largely independent of the rapidity and multiplicity distributions [33]. Thus, an E_T measurement is simply an analog method of counting particles: each particle produced has roughly the same value of $E_i \sin \theta_i \simeq <p_T>$ [49].

The formal definition of the **transverse energy**, E_T, is

$$E_T = \sum_i E_i \sin \theta_i \quad \text{and} \quad dE_T(\eta)/d\eta = \sin\theta(\eta)\, dE(\eta)/d\eta \qquad (8.10)$$

where the sum is considered to be taken over all particles emitted on an event into a fixed but large solid angle. Following the original work of NA5 [46], the traditional solid angle was typically taken as the full azimuth, $\Delta\phi = 2\pi$, and c.m. pseudo-rapidity interval $\Delta\eta^* \simeq \pm 0.8$. It is important to note that E_T does not have a well defined property under Lorentz transformations. Relativistically preferable quantities have been discussed from time to time [50], but these are rarely used because the above definition of E_T is the most convenient for measurements in segmented calorimeters. In this case, the sum is over the energy E_i measured in the ith calorimeter cell, with average polar angle θ_i.

Three distinct varieties of "E_T" measurements have been reported in the literature. Each technique has its own systematic problems and biases, and the relationship of the quantity "E_T" measured in calorimeters to an idealized quantity (what a theorist would imagine) is not at all straightforward[51-52]. The first and "classical" method [46,48] uses a full azimuth hadron calorimeter, typically 5 to 8 hadron absorption lengths thick, which thus measures all hadrons regardless of whether they are charged or neutral. The second method uses a track chamber device to reconstruct the momenta of all charged particles and then to construct E_T^c, the transverse energy of charged particles, usually with the assumption that all the particles are pions [53,54]. The third method [55,56] uses an electromagnetic shower counter, typically 15 to 20 radiation lengths thick, to detect the energy of the photons from the decays of neutral mesons ($\pi^0 \to \gamma\gamma$ and $\eta^0 \to \gamma\gamma$) and has thus been called [55] "neutral transverse energy", E_T^0, or "electromagnetic" transverse energy [56], E_T^{em}. The shower counter can be a dedicated detector [55] or the electromagnetic section of a hadron calorimeter [56].

The now "classical" NA5 measurement[46] for 300 GeV $p-p$ collisions is shown in figure 20. This is the first measured E_T distribution in the present day usage of the terminology. The detector was essentially the same hadron calorimeter (ring calorimeter) as used in the NA35 experiment (figure 9), covering the full azimuth and c.m. polar angular interval $54° < \theta^* < 135°$ at 300 GeV incident energy.

9. Relativistic collisions involving nuclei

The study of relativistic collisions involving nuclei has a long tradition in high energy physics, dating from studies in the 1930's of cosmic ray interactions in photographic emulsions and in metal plates inside cloud chambers[44,45]. The subject revived in the early 1970's, when it was again realized that multiple particle production in nuclei is sensitive to the space-time structure of the fundamental nucleon-nucleon particle production process. Also, the concept of the nucleon as a composite system of quarks and gluons was being developed at this time, so there was a

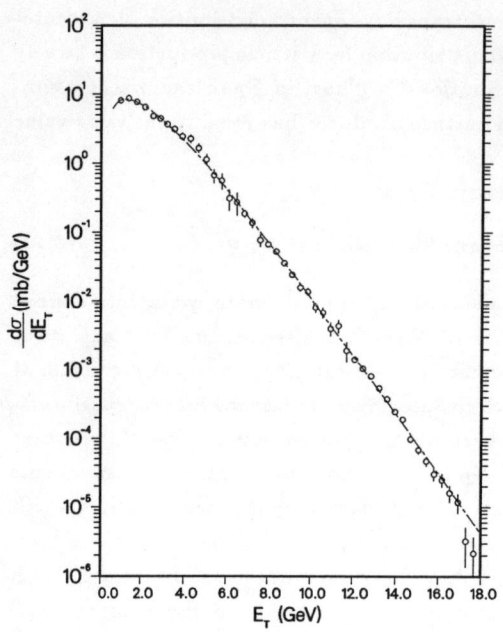

Fig. 20: The NA5 full azimuth E_T spectrum from $p-p$ collisions at 300 GeV incident energy [46] together with the fit to a gamma distribution with $p = 2.39 \pm 0.06$, $<E_T> = 2.2$ GeV, $\sigma_{pp}^{\text{det}} = 24 \pm 1$ mb, $\chi^2 = 62.4/43$ d.o.f.

desire to study the interactions of a real composite system (the nucleus) whose properties were well known [57,58]. Measurements of the interactions of ~ 100 GeV protons in nuclei were made at FERMILAB, and later at the CERN SPS, and produced results which proved to be much cleaner, and simpler to interpret, than expected.

9.1 Proton-Nucleus interactions at ~100 GeV

When a high energy proton passes through a nucleus, it can make several successive collisions. However, the charged particle multiplicity density, dn/dy, observed in proton+nucleus $(p+A)$ interactions is not simply proportional to the number of collisions, but increases much more slowly. The main features of $p+A$ interactions are strikingly illustrated in figure 21, which is the pseudorapidity distribution of relativistic charged particles ($v/c > 0.85$) from 200 GeV proton interactions in targets of various nuclear size [59]. The nuclear size is discussed in terms of $\bar{\nu}$, the average number of collisions, or more properly, absorption mean free paths encountered by an incident particle passing through a nucleus of atomic mass A:

$$\bar{\nu} = \frac{A\sigma_{hp}}{\sigma_{hA}}, \qquad (9.1)$$

where σ_{hp} and σ_{hA} are the absorption cross sections for the incident hadron on a nucleon and a nucleus, respectively [57,60]. The targets used covered the range from CH_2 to carbon to uranium, but unfortunately are not indicated on the figure. The most dramatic feature of figure 21 is that there is virtually no change in the forward fragmentation region ($\eta > 5.0$) with increasing A. By contrast, there is tremendous activity in the target region ($\eta \leq 0.5$). In the central region, $dn/d\eta$ increases with A and the peak of the distribution shifts backwards. The integral of the distribution (the average multiplicity) shows a linear increase with $\bar{\nu}$ independently of the identity of the incident hadron or the nuclear target (figure 22). All the data could be fit

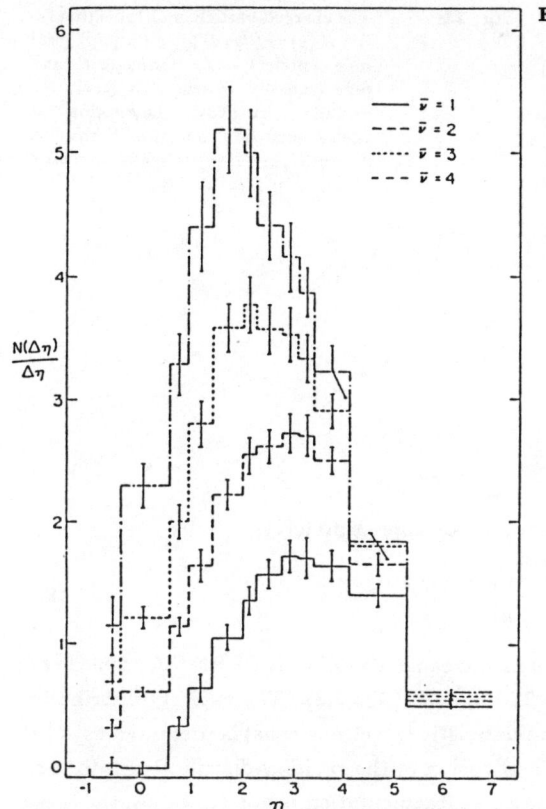

Fig. 21: Pseudorapidity distributions $dn/d\eta$ of relativistic charged particles for various values of $\bar{\nu}$ in 200 GeV/c proton-nucleus interactions [59].

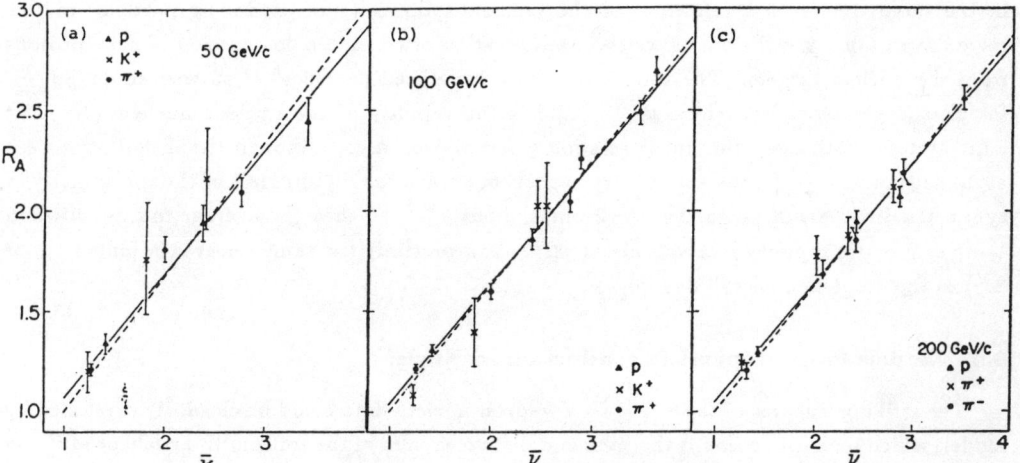

Fig. 22: Scaled average multiplicity R_A as a function of nuclear thickness. The data are for incident p, π^+, π^-, with momenta (a) 50 GeV/c, (b) 100 GeV/c, (c) 200 GeV/c. The solid lines are the result of the fit $R_A = a + b\bar{\nu}$, the dashed lines $R_A = 1 - b + b\bar{\nu}$, for each data set. All the data can be fit to a single linear relation, given in the text [59].

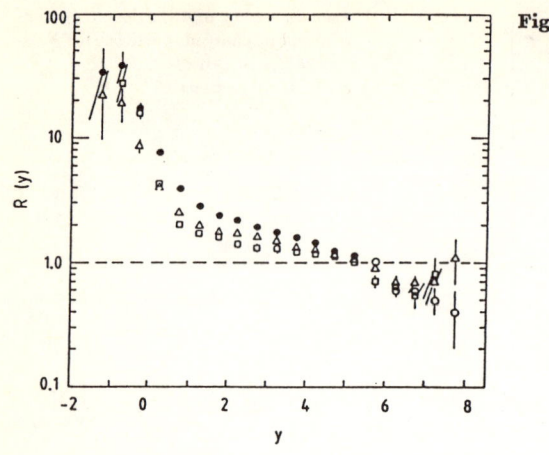

Fig. 23: The charged particle multiplication ratio, $R(y) = (d\sigma/dy)_{pA} / (d\sigma/dy)_{pp}$ for pXe (circles), pAr (triangles), and pNe (squares) interactions versus the rapidity y [61]. [Note: the caption has been copied from reference[61], however it seems that $R(y)$ is really the ratio of dn/dy and not $d\sigma/dy$.]

by the linear relation [60,59,57]
$<n>_{hA} / <n>_{hp} = (0.45 \pm 0.02) + (0.59 \pm 0.01)\bar{\nu}$, or approximately

$$R_A = \frac{<n>_{hA}}{<n>_{hp}} = \frac{1+\bar{\nu}}{2}. \qquad (9.2)$$

The same features are evident in data taken in a streamer chamber at CERN [61] for 200 GeV/c proton interactions on protons, Ne, Ar ($\bar{\nu} = 2.3$) and Xe ($\bar{\nu} = 3.3$). The rapidity distribution dn/dy for all charged particles (including non-relativistic target nucleons) is presented as $R(y)$, the ratio to $p-p$ interactions (figure 23). The division of the rapidity distribution into three distinct domains is clear from this plot: the projectile fragmentation region ($y > 5.0$), the target region ($y < 1.0$), and the central region ($1.0 < y < 5.0$). The dramatic change in slope of $R(y)$ in the target region is a reflection of the kinematic difficulty of producing particles in this region from a $p-p$ collision, as well as an indication of considerable emission of slow protons from the nuclear targets. The total number of "produced particles" [62] as well as $dn/dy|_{y_{cm}^{NN}}$ for all charged particles, where $y_{cm}^{NN} = 3.0$ is the rapidity of the nucleon-nucleon ($N-N$) c.m. system, both obey the linear relation given above. Also, although the distributions are asymmetric about y_{cm}^{NN}, the asymmetry cancels out when taking integrals of the multiplicity in symmetric intervals of up to $\Delta y = \pm 2$ units about y_{cm}^{NN}, so that the average multiplicities in these symmetric rapidity intervals about y_{cm}^{NN} show precisely the same linear relation(to 1% or 2%) as the "central density" $dn/dy|_{y_{cm}^{NN}}$.

9.2 Models for multiparticle production in Nuclei

The striking features of the ~ 100 GeV hadron-nucleus data could be elegantly explained by models which took into account the time and distance scales of the soft multiparticle production process [58]. A nucleus is considered to be rather transparent, so that a relativistic incident hadron can pass right through, and can make many successive collisions. Once a relativistic hadron interacts inside a nucleus, it becomes an excited hadron, and remains in that state inside the nucleus because the uncertainty principle and time dilation prevent it from fragmenting into particles until it is well outside the nucleus. This feature immediately eliminates the possibility of a cascade in the nucleus from the reinteraction of the secondary fragments. A further assumption is that the excited hadron interacts with the same cross section as an unexcited hadron and that the successive collisions do not affect the excited state or its eventual fragmentation products. This leads to the conclusion that the elementary process for particle production in nuclear collisions is the excited nucleon, and to the prediction that the multiplicity in nuclear interactions should be proportional to the total number of projectile and target participants, rather than to the total number of collisions.

The simplest expression of these ideas and assumptions is the Wounded Nucleon Model (WNM) [63], which describes a large body of data from high energy interactions with nuclei. In this model, the number of nucleons struck in a nuclear interaction is computed from the static (Glauber) nuclear geometry [64]; but a nucleon contributes only once to the production of particles no matter how many times (≥ 1) it is successively struck. The WNM was originally introduced to explain the behavior of the average multiplicity in nuclear interactions; and was later extended to explain multiplicity distributions [65] and E_T^c distributions [66] in limited intervals of rapidity. The behavior of the average multiplicity is neatly explained in this model: a nucleon-nucleon collision has two wounded nucleons, a nucleon-nucleus interaction has an average of $\bar{\nu}$ collisions, which corresponds to $(1 + \bar{\nu})$ wounded nucleons (the incident nucleon plus the $\bar{\nu}$ target nucleons), giving a ratio of the mean multiplicity in $p+A$ to $p-p$ interactions as $R_A = (1 + \bar{\nu})/2$. Recently, two of the leading models of soft collisions and fragmentation have been extended to nuclear collisions [67,68]; but the Wounded Nucleon Model, and its close relative, the Addidive Quark (really wounded projectile quark) Model (AQM) [69] retain their popularity because the geometry of the nuclear interaction can be separated from the dynamics of particle production, which can be taken directly from measured distributions [65,66,70,71].

9.3 "Hard" Collisions—Beware the "Cronin Effect"

Further constraints on the models of particle production in nuclei are provided by hard collisions such as lepton-pair production [72] (colloquially known as Drell-Yan [73]) and inclusive single particle production at large P_T [74]. In the constituent picture of hard scattering, the proton is already treated as a composite object, so the extension of this picture to the interactions of nuclei was thought to be straightforward. The transverse distance scale of the collision ($\simeq 1/P_T$) is so small that the constituents should act independently, thus giving a cross section proportional to the total number of constituents in the projectile and target, or proportional to $A_1 \times A_2$ for the interaction of two nuclei [75]. It was therefore somewhat surprising when measurements at FERMILAB in the period 1973-1977 indicated an "anomalous" nuclear enhancement (also known in HEP jargon as the "Cronin Effect"). The ratio of the inclusive cross sections at a given P_T in $p + A$ to $p - p$ interactions could be represented by a power law $A^{\alpha(P_T)}$. The exponent $\alpha(P_T)$ was greater than 1.0 at large P_T and varied with the $\sqrt{s}$, P_T, and the type of particle produced, leading to the description as "anomalous". After much work, over a period spanning 13 years [76], the deviation of the behavior from $A^{1.0}$ was reasonably well understood theoretically as being caused by smearing of the steeply falling hard-scattering P_T spectrum due to multiple scattering of the constituents in the nucleus [77,78].

Suffice it to say that there are no fully relativistic space-time models at this time that explain all facets of the data, and both the hard and soft collisions, although the subject is steadily advancing [79].

9.4 Reaction dynamics studies using E_T distributions.

The development of E_T distributions as a diagnostic tool to study the reaction dynamics of multiparticle production in relativistic nuclear collisions was started at CERN in 1980, when light ions were injected into the ISR. $\alpha-\alpha$ and $p-\alpha$ interactions were studied at nucleon-nucleon c.m. energies ($\sqrt{s_{NN}}$) of 31 GeV for $\alpha-\alpha$ and $\sqrt{s_{NN}} = 44$ GeV for $p-\alpha$. A subsequent run took place in 1983, with $\alpha-\alpha$, $p-\alpha$, $d-d$ and $p-p$ interactions all studied at $\sqrt{s_{NN}} = 31$ GeV. The AFS collaboration led the way in analyzing the charged multiplicity [65] and E_T^c distributions [54] in terms of the Wounded Nucleon Model (WNM) [66]. Excellent fits were obtained to the $\alpha-\alpha$ and $p-\alpha$ E_T^c distributions, in the symmetric rapidity range $|y| < 0.8$ in the nucleon-nucleon c.m. system, using $p-p$ data in the same interval to determine the spectrum of a "wounded nucleon", (see figure 24). The fit covers more than three orders of magnitude in cross section and extends to E_T^c values of 3 or 4 times the $<E_T^c>$.

The high luminosities of the second ISR light ion run allowed the BCMOR collaboration to extend measurements of E_T^0 distributions to over 10 orders of magnitude in cross section and to values of E_T^0 more than 10 times the average [70]. The spectrum of total neutral energy emitted in the central region was measured using an electromagnetic shower counter which detected, but did not separately resolve, the photons from the decays of neutral mesons ($\pi^0 \to \gamma\gamma$ and $\eta^0 \to \gamma\gamma$). A vertex with at least two charged tracks was reqired, and the spectra were corrected for the energy deposited by charged particles. The c.m. acceptance in which the neutral energy was detected covered 90% of 2π in azimuth with an average c.m. pseudorapidity acceptance inside this region of $|\eta| \leq 0.9$. Distributions in both the "transverse neutral energy" E_T^0 and the "total neutral energy" E_{TOT}^0 observed in the detector were obtained for $p-p$, $d-d$ and $\alpha-\alpha$ interactions all at the same $\sqrt{s_{NN}} = 31$ GeV (see figure 25). Note that the shapes of the spectra are essentially identical in both E_{TOT}^0 and E_T^0. This illustrates that, in a limited rapidity interval, the "energy flow" or the "transverse energy flow" are equivalently useful, since they are simply related by $<\sin\theta>$ in the interval ($\simeq 0.87$ for the data of figure 25), so long as the rapidity distribution of the transverse energy does not change over the spectrum.

One noteworthy feature of the spectra in figure 25 is that the $\alpha-\alpha$ data extend beyond the $p-p$ kinematic limit. This clearly indicates a multiple collision process. In the first published measurements of E_T distributions in nuclei, from $p+A$ interactions at FERMILAB [80], the A dependence of the spectra at a given E_T was parameterized as a power law, $A^{\alpha(E_T)}$, like the single particle inclusive spectra. It is evident that this parameterization makes no sense for the data in figure 25, since the ratio to $p-p$ collisions is undefined beyond the kinematic limit. The formalism of hard scattering is, once again, inappropriate for the soft multiparticle physics of E_T distributions; and analysis in terms of the WNM or AQM was thought to be more reasonable. The spectrum of a wounded nucleon (or quark) was determined from the measured $p-p$ distribution (figure 25), while the relative probability distribution of wounded nucleons or wounded projectile quarks was computed from the nuclear geometry of the α−particle [71]. Although the WNM successfully accounts for the increase of the observed integrated cross section and $<E_{TOT}^0>$ from the $p-p$ to the $\alpha-\alpha$ data, it leads to the wrong functional form for the high-energy tail of the $\alpha-\alpha$ distribution over a range of more than 6 orders of magnitude in cross section [70] (dot dash line). The Additive Quark Model [71] (solid line) seems to provide a much better description of the $\alpha-\alpha$ data, as does the DPM [67]; and it is tempting to speculate

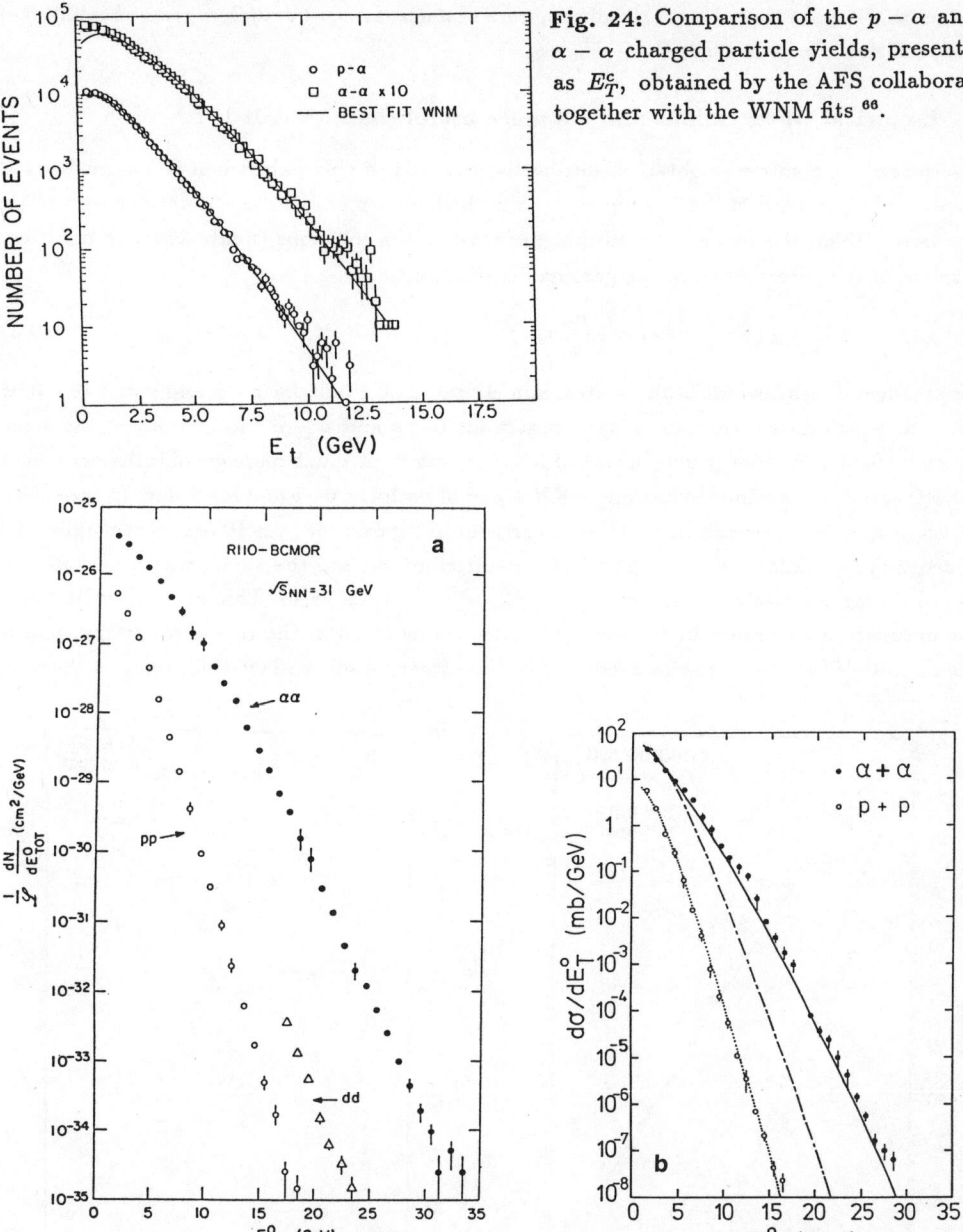

Fig. 24: Comparison of the $p-\alpha$ and $\alpha-\alpha$ charged particle yields, presented as E_T^c, obtained by the AFS collaboration, together with the WNM fits [66].

Fig. 25: The neutral energy spectra[70] for $p-p$, $d-d$, and $\alpha-\alpha$ interactions at $\sqrt{s_{NN}} = 31$ GeV. (a) Total neutral energy, E_{TOT}^0. (b) Transverse neutral energy, E_T^0 for $\alpha-\alpha$ and $p-p$ interactions together with the Wounded Nucleon Model fit (dot-dash line) and the Additive Quark Model fit (solid line). The dotted line is a fit of the $p-p$ data to a gamma distribution, with $p = 2.50$ and $<E_T^0> = 1.57$ GeV [71]. Only the statistical errors are shown on the figures. The systematic errors for the E_{TOT}^0 spectra are small. For technical reasons the systematic errors for the measurement of E_T^0 are larger [70].

whether this implies that the quark substructure of nucleons can be inferred from these studies of soft collision effects in nuclei[81].

9.5 Empirical Observations may be more useful than Models

A surprising result was obtained during the analysis of this data when it was realized [70] that a simple gamma distribution represented both the $p-p$ and the $\alpha-\alpha$ spectra as well as, if not better than, the models of multiple nucleon-nucleon collisions (figure 26a). In addition, the values of the parameter p of the gamma distribution:

$$<E> f(E) = \psi(z) = \frac{p}{\Gamma(p)}(pz)^{p-1} e^{-pz} \quad \text{where} \quad z = \frac{E}{<E>} \quad (9.3)$$

were practically identical for both spectra, namely $p = 2.50 \pm 0.06$ for $p-p$ and $p = 2.48 \pm 0.05$ for $\alpha-\alpha$, which meant that the E^0_{TOT} spectra for $p-p$ and $\alpha-\alpha$ interactions at the same nucleon-nucleon c.m. energy obey a sort of KNO scaling [82]. A much more graphic illustration of the KNO scaling is obtained by making a KNO-plot of both the $p-p$ and $\alpha-\alpha$ data (figure 26b). The E^0_{TOT} spectra plotted in this way are nearly indistinguishable over 10 orders of magnitude! In particular the tails of the $p-p$ and $\alpha-\alpha$ distributions are the same, when measured in units of the average value of the energy $<E^0_{TOT}>$, in each case. This appears to indicate some unexpected coherence in the multiple collisions involved in the $\alpha-\alpha$ scattering, and is quite opposite from the behavior predicted in the extreme-independent-collision hypothesis of

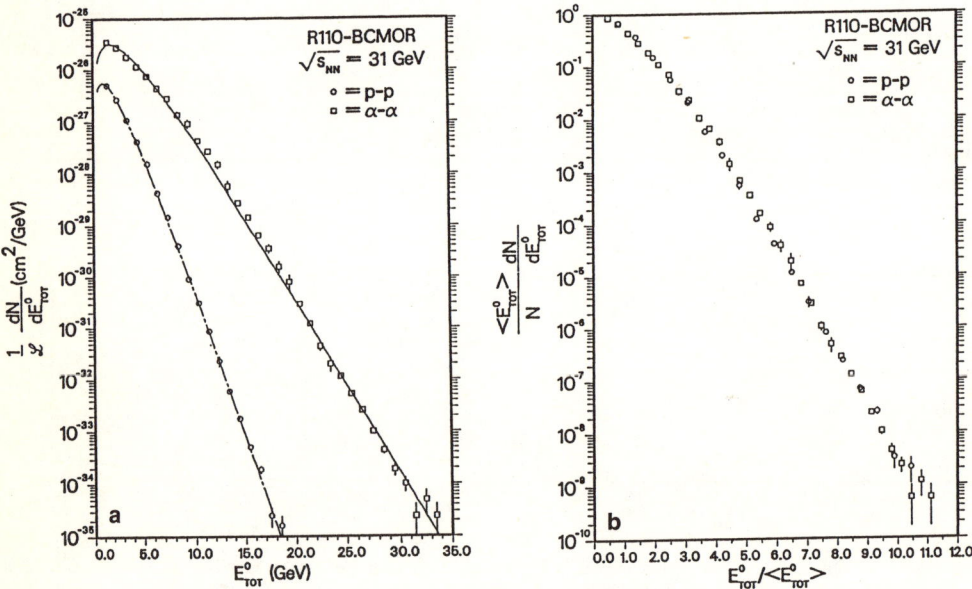

Fig. 26: The E^0_{TOT} spectrum of figure 25a for $\alpha-\alpha$ and $p-p$ interactions: (a) together with fits to a simple gamma distribution. The parameter $p = 2.50 \pm 0.06$ for the $p-p$ data (dot-dash line) and $p = 2.48 \pm 0.05$ for the $\alpha-\alpha$ data (solid line). (b) The same distributions replotted in the scaled energy variable, $z = E^0_{TOT} / <E^0_{TOT}>$, where $<E^0_{TOT}>$ is the average value of the energy for each distribution.

the WNM and the AQM, where the energy emitted on the independent collisions would be combined at random. Since the AQM also appears to fit the data, it is possible that the effect is simply an accident of the geometry of the multi-quark collisions. In order to address this question, 3 or 4 additional(!) orders of magnitude of cross section sensitivity would be required, so that the geometrical effect of the quarks would be exhausted and the dynamics would take over [83,84].

10. Details of the multiple-collision analysis.

The analyses of E_T distributions in light nuclei at the CERN ISR, by the AFS and BCMOR collaborations, proved to be complementary to the systematic studies of multiplicity in limited intervals of rapidity by the UA5 collaboration [85], and led to a whole new empirical formalism to for the description of soft multiparticle physics in proton-(anti)-proton collisions as well as in nuclei.

The multiparticle measures, E_T and multiplicity, are additive in the case of multiple collisions, either in the good sense, when they can be used to analyze the effects of multiple collisions in nuclei, or in the bad sense, when they are sensitive to the random pile-up of multiple events in time. In the extreme-independent-collision models of nuclear scattering, the effects of the nuclear geometry of the interaction can be calculated independently of the dynamics of particle production, which can be taken directly from experimental measurements. The calculation of the distribution in the number of basic elements of particle production — be they wounded nucleons, or wounded projectile and/or target nucleons or wounded projectile quarks, or number of $N-N$ collisions — is performed in the static (Glauber) approximation [64] by averaging over the impact parameter of the nucleus-nucleus scattering, usually by a simple Monte Carlo method [65,66,70,71]. Woods-Saxon densities [86]:

$$\rho(r) = \frac{1.0}{1.0 + e^{(r-c)/a_0}}, \qquad (10.1)$$

and in some cases the actually measured form-factors, are used for the spatial distributions of nucleons in the projectile and target nuclei. Sometimes, a hard core, or the minimum allowable distance between the centers of two nucleons in a nucleus, is also imposed. The only free parameter is these "geometrical" calculations is the effective $N-N$ inelastic cross section, σ_{NN} (or quark-nucleon cross section σ_{qN}), used to compute the absorption mean free path of a nucleon (or quark) in the nucleus. A collision is defined when the center of a nucleon (or quark) in the projectile intercepts the center of a nucleon in the target nucleus at a distance $r \leq \sqrt{\sigma/\pi}$.

The separation of the "geometry" from the dynamics allows experimental measurements to be used to derive the fundamental E_T spectrum of the elementary collision process, which is then used as the basis of the analysis of a nuclear scattering as the result of multiple indpendent collisions. Let $f_1(E)$ represent the differential probability for the emission of energy E in dE into the detector for the elementary collision process. Then $f_n(E_0) dE_0$, the probability of observing E_0 in dE_0 for n such collisions, independently overlapped, is given by the n-fold convolution of $f_1(E)$. This is easy to understand by writing $f_n(E_0)$ in the recursive form

$$f_n(E_0) = \int_0^{E_0} dE \; f_1(E) \, f_{n-1}(E_0 - E) \qquad (10.2)$$

where $0 \leq E_0 \leq n\sqrt{s_{NN}}$. Here, E_0 represents the energy emitted on n collisions. The first term inside the integral is the probability of emitting energy E on one collision, and the second term is the probability of emitting a total of $E_0 - E$ on $n - 1$ collisions.

It is important to realize that all the n-collision spectra are normalized to unity; only the shapes are determined. The spectra can be weighted according to the probability or cross section for n simultaneous collisions, as given by the "geometrical" calculation. Alternatively, the data can be used to fit for the n-collision cross sections σ_n, $n = 1, 2, \cdots, m$ as parameters:

$$\frac{d\sigma}{dE} = \sum_{n=1}^{m} \sigma_n f_n(E) = \sigma^{\text{det}} \sum_{n=1}^{m} r_n f_n(E) \qquad (10.3)$$

where r_n is the relative probability for n collisions and σ^{det} is the integral of the observed spectrum.

The analysis is greatly facilitated by using a gamma distribution to represent the energy spectrum of the elementary process:

$$f_1(E) = \frac{b}{\Gamma(p)} (bE)^{p-1} e^{-bE}. \qquad (10.4)$$

It should be noted that $p > 0$, $b > 0$, $f_1(E)$ is normalized to unity over the range $0 \leq E \leq \infty$ and that $\Gamma(p)$ is the gamma function of p, $= (p-1)!$ if p is an integer. The moments of the distribution are given in terms of the two parameters p and b:

$$\mu = p/b \quad \sigma = \sqrt{p}/b \quad \gamma_1 = 2/\sqrt{p} \quad \gamma_2 = 6/p \qquad (10.5)$$

where the moments are the mean (μ), the standard deviation about the mean (σ) and the first two cumulants (γ_1, γ_2). The gamma distribution is used because of its elegant convolution property. The n-fold convolution of the gamma distribution $f_1(E)$ is simply given by the function

$$f_n(E) = \frac{b}{\Gamma(np)} (bE)^{np-1} e^{-bE}, \qquad (10.6)$$

i.e. $p \to np$ and b remains unchanged. Furthermore, the convolution of a gamma distribution (parameters p, b) with another gamma distribution (parameters q, b) is simply a gamma distribution with parameters $p + q$, b. The convolution property of the gamma distribution allows the elementary spectrum to be "deconvoluted" from a measured spectrum, e.g. from $p - p$ collisions, and then re-convoluted to obtain the spectrum of the nuclear scattering.

This analysis is particularly straightforward in the case of E_T spectra of symmetric nuclei measured in a symmetric rapidity interval about y_{cm}^{NN}. The observed $p - p$ spectrum is treated as the probability function for the collision of two nucleons:

$$f_1(E) = \frac{1}{\sigma_{pp}^{\text{det}}} \frac{d\sigma_{pp}}{dE}, \qquad (10.7)$$

where σ_{pp}^{det} is the integral of the observed spectrum

$$\sigma_{pp}^{\text{det}} = \int_0^{\sqrt{s}} dE \; d\sigma_{pp}/dE. \qquad (10.8)$$

The spectrum is then fit to a gamma distribution for the parameters b and p. For instance, for the data of figure 25a, the fit to the $p-p$ E_{TOT}^0 spectrum is excellent (see figure 26a), with parameters $b = 1.41 \pm 0.01\,\text{GeV}^{-1}$,

$$p = 2.50 \pm 0.06, \qquad <E_{\text{TOT}}^0> = 1.77 \pm 0.3\,\text{GeV},$$
$$\sigma_{pp}^{\text{det}} = 13.1 \pm 0.3\,\text{mb}, \qquad \chi^2 = 24.6/15\,\text{d.o.f.}.. \tag{10.9}$$

Since the configuration is symmetric, the two wounded nucleons produce identical distributions in the detector, so that the spectrum of a wounded nucleon is derived from the proton-proton spectrum by a deconvolution into two equal spectra [87], $p \to \frac{1}{2}p$; and the WNM prediction for the $\alpha - \alpha$ data is obtained by summing over the convolutions of the wounded nucleon spectrum, weighted by the relative probabilities, r_n, from the "geometrical" calculation. The same principles apply to analyses done in terms of wounded projectile quarks, multiple $N - N$ collisions, etc.

Some technical points of the convolution analysis deserve to be emphasized. The asymptotic slope, b, is the same for all the convoluted spectra, while the mean value increases linearly with n, and the standard deviation about the mean increases as the $\sqrt{n}$. Once the energy is high enough so that the spectrum is dominated by the maximum number of collisions, the asymptotic slope becomes equal to the asymptotic slope of the elementary process. This is clearly illustrated by the slope of the WNM prediction in figure 25b, which is parallel to the $p - p$ spectrum. The slope of the $\alpha - \alpha$ data is quite different from the asymptotic $p - p$ slope and from the WNM prediction. This qualitative feature makes it possible to overcome the uncertainty in the geometrical calculation due to the value assumed for the only free parameter, σ_{NN}. The WNM fails qualitatively in figure 25b, independently of the details of the calculation [70]! The uncertainty of the calculation can become especially acute when, as in the case of figure 25, the observed cross-section in the detector for $p - p$ collisions is only 13 mb, or less than half of the actual cross section[43].

10.1 A new empirical description of soft multiparticle physics.

The use of gamma distributions to fit E_T spectra in $p - p$ collisions was driven by the convolution property for multiple collisions, and also by the fact that the gamma distributions fit the spectra rather well. For instance, the famous NA5 full azimuth E_T spectrum is nicely fit by a gamma distribution (recall figure 20). A similar regularity was discovered by the UA5 collaboration [85] in their study of multiplicity distributions in limited regions of rapidity, which are closely related to E_T distributions. A simple function, the negative binomial distribution (NBD), was found to provide an excellent representation of the multiplicity data. In a given pseudorapidity interval, with mean multiplicity $<n>$, the distribution of multiplicity, $P(n)$, can be fit by:

$$P(n) = \frac{-k!}{n!(-k-n)!}\left(\frac{-<n>}{k}\right)^n \left(1 + \frac{<n>}{k}\right)^{-(k+n)} \tag{10.10}$$

where k is a parameter. In the limit of large $<n>/k$, the NBD becomes a gamma distribution in the scaled variable $z = n/<n>$. Thus, a simple representation exists to describe "soft" multiparticle physics in terms of systematic variations of three parameters: the integrated cross section in a rapidity interval, the mean multiplicity (or $<E_T>$) in the interval, and the parameter k (or p) for the interval. At the present time, the underlying explanation for the success of these simple functions in describing the data remains a mystery [88].

11. A new era — relativistic heavy ion (RHI) interactions

A new era in the study of relativistic collisions involving nuclei opened up in the Fall of 1986 with the advent of beams of high energy heavy ions at CERN and at Brookhaven National Laboratory (BNL). At BNL, beams of ^{16}O and ^{28}Si have been provided at momenta of 14.6 GeV/c per nucleon; while at CERN, beams of ^{16}O and ^{32}S at 200 GeV per nucleon have been used. At both laboratories, some measurements were also made at lower incident momenta.

The main experimental challenge in the study of RHI interactions is to cope with the very high multiplicities produced (recall figure 3). In general, this is not a problem for the highly segmented calorimeters used in the study of E_T distributions, since these are analog devices. So long as the energy response of the detector is linear, and does not saturate, the problem of measuring E_T distributions in nucleus-nucleus interactions is no more difficult than in proton-proton collisions. For this reason, measurements of E_T distributions have been among the first results from this new field, and have played a leading role in the study of the reaction dynamics.

11.1 First results from CERN

The first result in this field was submitted for publication by the NA35 collaboration at CERN in November 1986, barely 2 months after the first 3 day test run [6]. The NA35 experiment consisted of a 2 by 1.2 by 0.72 m³ streamer chamber in a $1.5T$ superconducting magnet located upstream of a segmented calorimeter (recall figure 9). A contributing factor to the speedy success of this experiment was that all the NA35 calorimeters (except for the intermediate calorimeter covering the angular range $0.3° \leq \theta \leq 2.5°$, and used only as a passive absorber in this measurement) had been used in previous experiments (NA5 [46], NA24 [89]) so that their

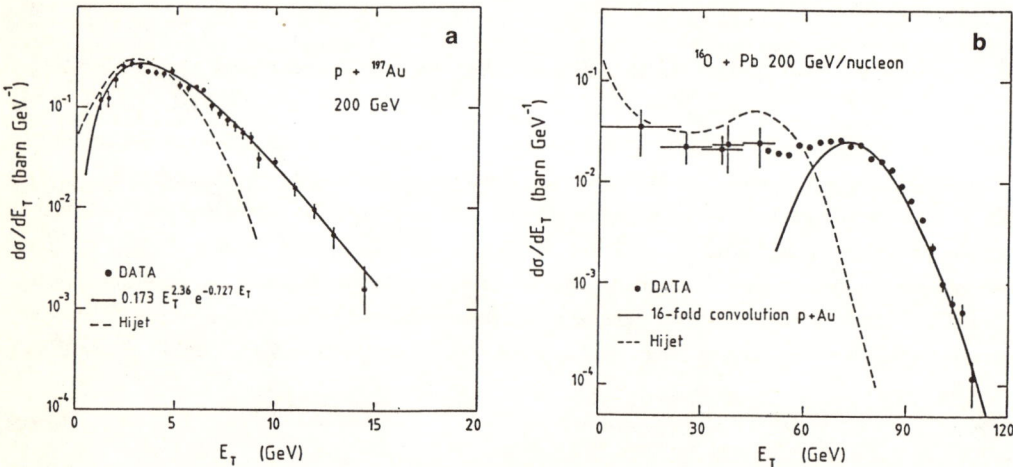

Fig. 27: The mid rapidity transverse energy distribution of $p + Au$ interactions (a) and $^{16}O + Pb$ interactions (b), all at 200 GeV/nucleon, measured in the same apparatus. The dashed curve gives the predictions of a Monte Carlo event generator Hijet [90] for both reactions. The solid curve on the $^{16}O + Pb$ data is the 16-fold convolution of the $p + Au$ fit. The analytical fit to the $p + Au$ data is given on the figure and is a gamma distribution with $p = 3.36$, $\sigma^{det} = 1.44$ barn, and $<E_T> = 4.6$ GeV.

behavior and calibration was well understood. The differential E_T distributions, as summed from the PPD and Ring Calorimeters, were obtained for the interactions of 200 GeV/nucleon ^{16}O in a Pb target and 200 GeV protons in an Au target, both measured in the same setup (see figure 27). The acceptance covered the lab-pseudorapidity interval $2.3 \leq \eta \leq 3.8$, which excluded the innermost ring of the Ring Calorimeter. At 200 GeV/nucleon incident energy, the nucleon-nucleon ($N-N$) c.m. energy is $\sqrt{s_{NN}} = 19.4$ GeV and the rapidity of the $N-N$ c.m. system is $y_{cm}^{NN} = 3.03$.

The results from this first ^{16}O test run were very striking. A most interesting and unexpected feature of this data was that the 16-fold convolution of the measured $p+Au$ spectrum beautifully reproduced the high energy edge of the $^{16}O + Pb$ spectrum (see figure 27).

11.2 Closely followed by BNL

These striking results were confirmed at BNL during the commisssioning period of the AGS-Tandem complex in November 1986. A small experiment was assembled using components of E802, one of the three major experiments at the facility[91], and measurements were made using a ^{16}O beam of momentum 14.6 GeV/c per nucleon [92]. The small experiment consisted of a full azimuth electromagnetic shower counter — an array of 96 blocks of lead-glass (PbGl) — placed 1 m downstream of the target. The array was 10 blocks wide by 10 high with a 2 by 2 block hole in the center for the beam to pass through. Each SF5 lead-glass block was 14.5 cm by 14.5 cm in area, and 17 radiation lengths thick (40 cm), viewed end on by a 5 inch photomultiplier. These blocks had been used previously in the CCR and successor experiments at the CERN ISR [55,70].

The PbGl array measured the electromagnetic energy emitted in a laboratory polar angular interval $10° \leq \theta \leq 32°$, corresponding to the laboratory pseudorapidity interval $1.25 < \eta < 2.44$ or roughly -0.5 to +0.7 in the $N-N$ c.m. system. At 14.6 GeV/c per nucleon incident momentum, the $N-N$ c.m. energy is $\sqrt{s_{NN}} = 5.4$ GeV and the rapidity of the $N-N$ c.m. system is $y_{cm}^{NN} = 1.72$. The observed energy was denoted E_{TOT}^0, keeping with the ISR[55,70] notation, since the PbGl responds primarily to neutral mesons, π^0 and η^0, which are detected via their two-photon decay. Relativistic charged hadrons ($v/c > 0.8$) also emit Cerenkov light in the PbGl, equivalent to approximately 500 MeV per particle [94], and contribute, in this particular configuration, on the average, about 50% of the observed energy [92,93]. The response of the PbGl to Cerenkov light is linear to $< 1\%$, whatever the source, so that no attempt was made to correct E_{TOT}^0 for the Cerenkov light from charged hadrons. The relative scale variation of E_{TOT}^0 is better than $\pm 1\%$ for the ^{16}O data and $\pm 5\%$ for the proton data. The absolute scale of E_{TOT}^0 was set by calibration in an electron beam.

In figure 28, the spectrum of energy observed in the PbGl is shown for ^{16}O interactions on Au, Cu, and mylar ($C_5H_4O_2$) targets and for positive hadron (80% proton) [92] interactions on an Au target, all measured at 14.6 GeV/c per nucleon in the same experimental arrangement. The PbGl detector covers a restricted pseudorapidity interval near y_{cm}^{NN}. Because of the restricted acceptance, the transverse and total energy recorded in the detector are highly correlated. If the transverse energy density in pseudorapidity is constant over the detector, then the transverse energy and total energy observed in the detector are simply related by a contant factor of $< \sin \theta > = 0.29$, which seems to be the case[81]. Because the segmentation of the PbGl was

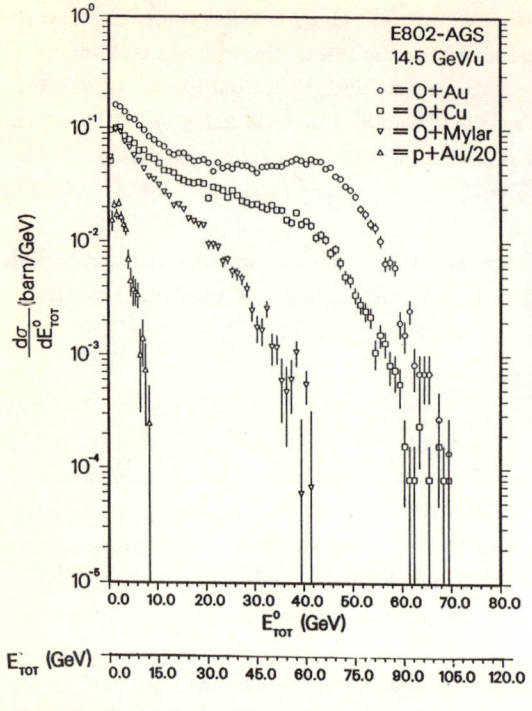

Fig. 28: Differential cross sections in E^0_{TOT}, the energy recorded in the PbGl, for ^{16}O interactions on Au, Cu and Mylar ($C_5H_4O_2$) targets and for proton interactions on an Au target. The $p+Au$ cross section has been divided by a factor of 20 for clarity of presentation. The two lower scales on the figure correspond approximately to the total energy and transverse energy falling on the detector and have been estimated with an average correction $E^0_T \simeq 0.29 \, E^0_{TOT}$, $E_T \simeq 3/2 \, E^0_T$, as discussed in the text [92,93].

relatively crude in this small experiment, the transverse resolution is difficult to unfold, so the E^0_{TOT} spectrum is used for further discussion.

The E^0_{TOT} spectrum for $^{16}O+Au$ shows an initial fall-off and then a broad plateau, a peak centered at 40 GeV and then a sharp drop-off until the yield runs out at ~ 70 GeV (see figure 28). The $^{16}O+Cu$ data also show evidence of considerable energy emission, even though the maximum thickness of a Cu nucleus is only $\sim 2/3$ that of Au. It is of particular interest that the edges of the $^{16}O+Cu$ and $^{16}O+Au$ spectra become virtually identical above 50 GeV if the Cu cross section is multiplied by a factor of ~ 6.

These features can be described by a simple, geometrical model, motivated by the original NA35 observation that the high energy edge of the $^{16}O+Pb$ spectrum is just the 16-fold convolution of the $p+Au$ spectrum. This idea was extended to provide a description of the entire $^{16}O+Cu$ and $^{16}O+Au$ spectra in E802 [92,93,95]. The observed $p+Au$ spectrum was fit to a gamma distribution (figure 29) and then convoluted from 1 to 16 times, with weights for the n-fold convolutions obtained from a geometrical calculation which averaged over the impact parameter of the nucleus-nucleus scattering to obtain the distribution in the number of projectile nucleons which interact at least once in the target. An $N-N$ inelastic cross section of 30 mb was used in the calculation, corresponding to an absorption mean free path of ~ 2.2 fm. Excellent representations of both the $^{16}O+Au$ and $^{16}O+Cu$ spectra were obtained from this "wounded-projectile-nucleon" (WPN) model (see figure 30). It then appears that the peak in the $^{16}O+Au$ spectrum and the identical shape of the high-energy edges of the $^{16}O+Au$ and $^{16}O+Cu$ spectra arise from events in which all 16 projectile nucleons interact.

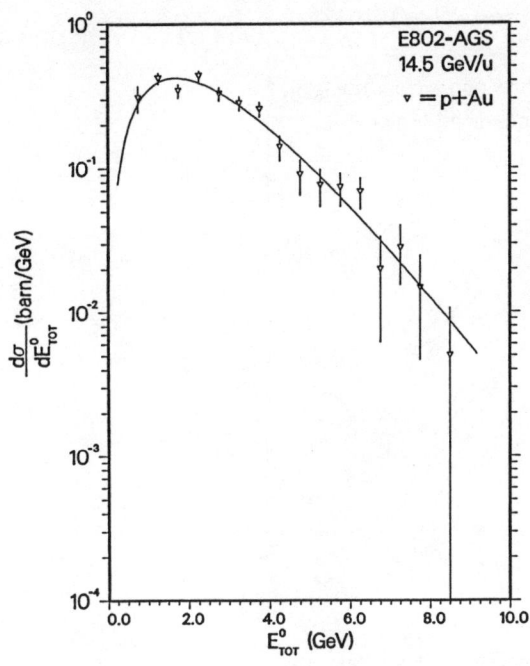

Fig. 29: Fit [92,93] of the $p + Au$ spectrum to a single gamma distribution (solid curve) with $\sigma^{det} = 1.44 \pm 0.77$ barn, $b = 0.95 \pm 0.09 GeV^{-1}$, $p = 2.60 \pm 0.24$, with average energy $<E^0_{TOT}> = p/b = 2.73 GeV$.

Although the WPN model seems quite reasonable and fits the E^0_{TOT} data rather well, the analysis contains some potential inconsistencies. The observed $p + Au$ spectrum, which is necessarily averaged over all impact parameters, was used for the 16-fold convolution, whereas, it would seem that a "centrally averaged" $p + Au$, or $p + Cu$ spectrum would be more appropriate. Also, the WPN model is in serious contradiction with the understanding of relativistic nuclear interactions gleaned from proton-nucleus collisions at higher energies, where nuclei appear to be relatively transparent.

A simple resolution of these inconsistencies is to presume that an incoming 14.6 GeV/u nucleon loses most of its energy in the first few collisions. The subsequent lower energy collisions are unlikely to produce energy observable in the PbGl (neutral-mesons). In this way the energy emitted in a $p + A$ interaction would not depend much on impact parameter or nuclear size, and it is appropriate to use the experimental $p + Au$ spectrum to analyze the $^{16}O + Cu$ and $^{16}O + Au$ data. The maximum energy observed in these reactions occurs when all sixteen projectile nucleons interact: i.e., for events in which the impact parameter is sufficiently small to allow complete overlap of the $^{16}O + Au$ or $^{16}O + Cu$ nuclei. The cross sections for maximum energy deposit in Au and Cu are then in the ratio of these impact parameters squared, $[R(Au) - R(^{16}O)]^2 / [R(Cu) - R(^{16}O)]^2$, a factor of ~ 5, which is close to the measured factor of ~ 6.

The observation that the maximum energy emitted in $^{16}O + Cu$ interactions is essentially the same as in $^{16}O + Au$ interactions, together with the success in reproducing the observed ^{16}O spectra from the $p + Au$ spectrum by a simple geometrical model, was cited as compelling evidence by the E802 collaboration [92] that ^{16}O projectiles at 14.6 GeV/c per nucleon can be sufficiently stopped in Cu so that pion emission effectively ceases.

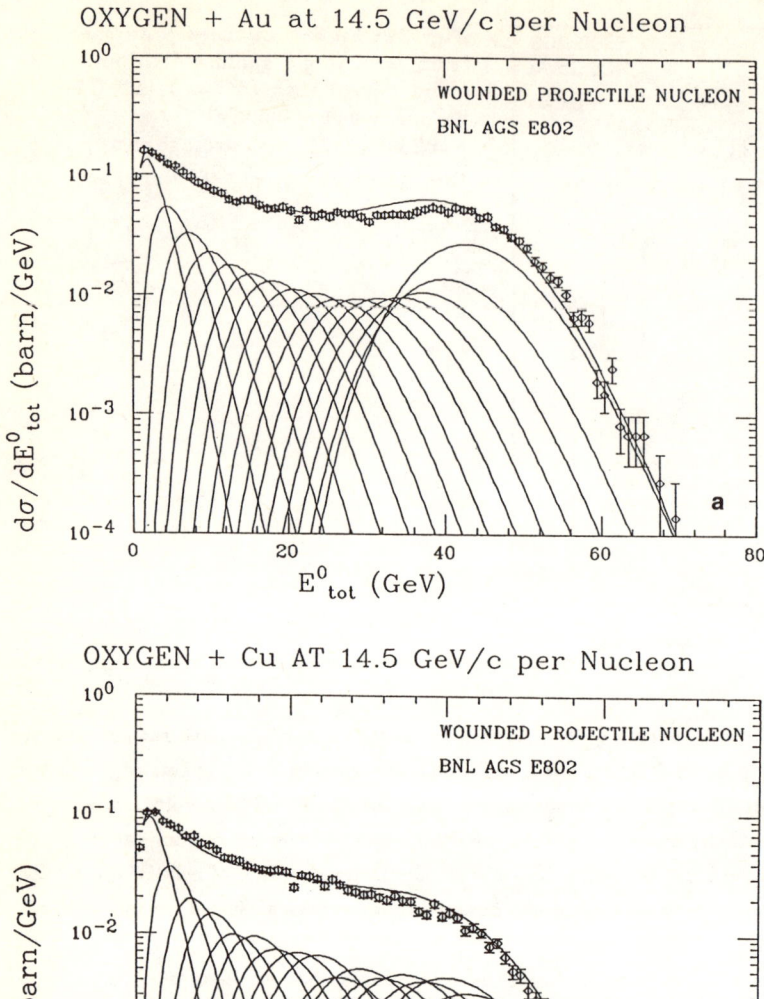

Fig. 30: The E^0_{TOT} spectra of figure 28 for ^{16}O interactions on Au (a) and Cu (b) together with the sum of 1 to 16-fold convolutions of the measured $p + Au$ spectrum weighted according to the probability for 1,2,...16 of the projectile nucleons to interact in the target. The individual components of the sum are also shown. The curves have been normalized to the observed cross-section for each target, $\sigma^{det} = 3.27$ barn for $^{16}O + Au$ and $\sigma^{det} = 1.66$ barn for Cu [92,93].

11.3 Further E_T results at BNL and CERN

Additional results in support of "energy stopping" at BNL energies are provided by measurements from the full setup[91] of E802, for ^{28}Si interactions in a variety of targets. For this series of measurements, the PbGl was composed of 245 six inch square SF5 blocks, located 3 m downstream of the target, covering a laboratory polar angular interval of 8.0° to 32.0°, or $1.3 < \eta_{lab} < 2.5$ in pseudorapidity. This is roughly the same η coverage as the small experiment, with 3 times finer segmentation. However, only 1/2 of the azimuth was covered; the other half of the azimuth being occupied by a magnetic spectrometer with full particle identification[96].

Because of the excellent segmentation in the full setup, measurements of the energy flow in the PbGl can be presented interchangeably in terms of either the total energy, E_{TOT}^{PbG}, or the transverse energy E_T^{PbG} observed in the detector. (Previously, these quantities had been denoted E_{TOT}^0 and E_T^0). These variables are highly correlated: the distribution in $E_T^{PbG}/E_{TOT}^{PbG} \equiv <\sin\theta>_{eff}$ is Gaussian, with an rms width of $\sim 10\%$ of the mean value of 0.285 ± 0.003. The systematic variation of $< E_T^{PbG}/E_{TOT}^{PbG} >$ as a function of E_{TOT}^{PbG} is less than 5 %

Although the data analysis is not yet complete, conclusions on "energy stopping" can be drawn already from the "trigger" distributions, figure 31, for ^{28}Si interactions on ^{27}Al, ^{64}Cu, ^{108}Ag and ^{197}Au targets. These distributions have not been target out subtracted, and may differ in absolute scale by 5 % compared to previous data[93]. However, the relative scale is better than 1 % for all targets; and the target empty correction is negligible for $E_T^{PbG} > 2$ GeV, and is the same for all the targets, which are 3 % of an interaction length for this figure. The upper edges of the $Si + Cu$ and $Si + Ag$ spectra differ from the $Si + Au$ spectrum by roughly 15 %, and 5 % in E_T^{PbG}, whereas the target nuclear thicknesses differ by 40 % and 20 % respectively.

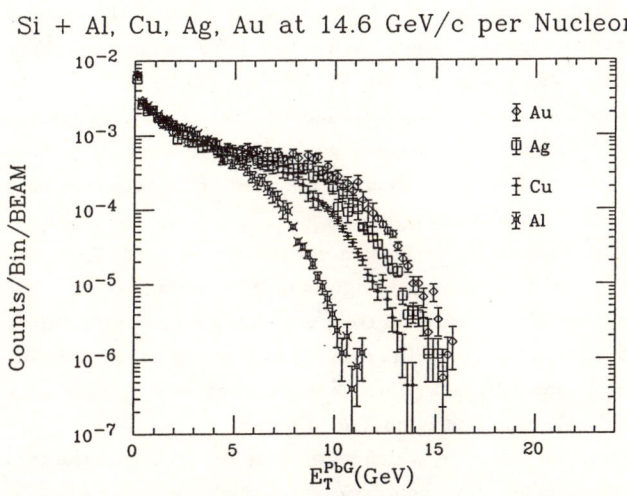

Fig. 31: Preliminary E802 E_T^{PbG} spectra in the 1/2 azimuth PbGl for interactions of 14.6 GeV/c per nucleon ^{28}Si projectiles in targets of Au, Ag, Cu and Al. Note that the target-out background has not been subtracted in this figure; and that the data are not in cross section units, but are in units more natural for a "trigger" counts per bin per incident beam particle[105].

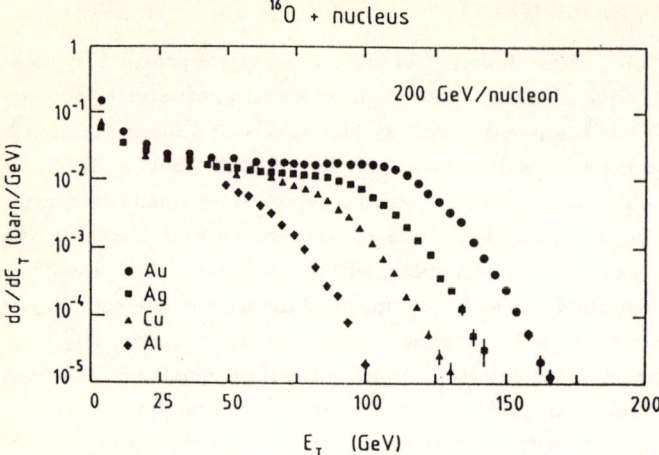

Fig. 32: Transverse energy spectra for ^{16}O on Au (circles), Ag (squares), Cu (triangles) and Al (diamonds) at 200 GeV/nucleon as seen in the NA35 acceptance [98].

It should be noted that E_T spectra are very sensitive to the region of pseudorapidity in which they are measured[97,91]. However, E_T spectra measured in the central rapidity region at AGS energies appear to differ qualitatively from comparable spectra measured at CERN.

The NA35 E_T spectra from the full ^{16}O run in Fall 1986 [98] are shown in figure 32 for four targets, Au, Ag, Cu and Al. The acceptance for these data covers the central pseudorapidity range $2.28 < \eta < 3.94$, where the $dE_T/d\eta$ distribution is expected to be relatively constant. All the spectra are characterized by a "rapidly varying cross section for peripheral collisions, followed by a long plateau which ends in a steep fall-off" [98]. The upper edges of the spectra show a substantial increase in E_T with increasing target nuclear size. Particularly striking in comparison to the E802 data (figures 28, 31) is that the NA35 data (figure 32) indicate $\sim 35\%$ more energy emission for $^{16}O + Au$ interactions compared to $^{16}O + Cu$. The spectra in electromagnetic transverse energy, E_T^{em} were also obtained by NA35 from the PPD (recall Fig. 9). The NA35 distributions in (total) transverse energy, E_T, and the electromagnetic component only, E_T^{em}, show essentially the same characteristics.

Some observations on the systematic errors of E_T measurements can be made from the NA35 data. A comparison of the $^{16}O + Au$ E_T spectrum in figure 32 [98] with the $^{16}O + Pb$ E_T spectrum in figure 27 [6] shows $\sim 40\%$ more E_T in figure 32, compared to figure 27, with only a 10% increase in acceptance. These data were taken by the same group, in the same detector, at the same incident momentum, only 2 months apart. This discrepancy exceeds considerably the systematic error on the overall E_T scale, which is estimated to be 15% .

In spite of systematic differences between data sets, which happens to be obvious in the case of NA35, many useful conclusions can be drawn from such data by making comparisons within sets of measurements which all share the same overall systematic effects. Two comparisons emphasized by the NA35 collaboration are the ratio of E_T emission at 60 GeV/nucleon incident energy to that at 200 GeV/nucleon, and the ratio of E_T^{em} emission in ^{32}S and ^{16}O interactions at 200 GeV/nucleon in an Au target.

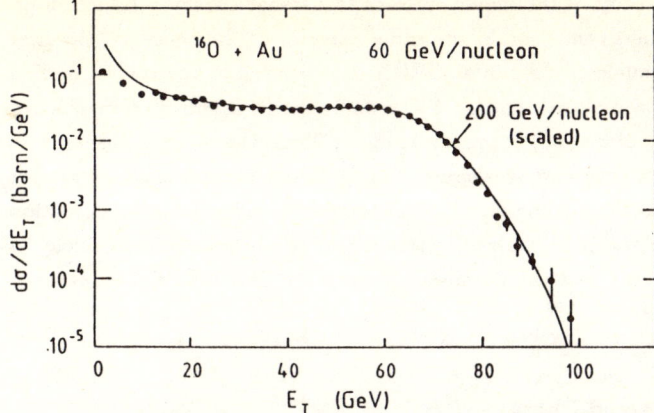

Fig. 33: NA35 transverse energy spectrum for $^{16}O + Au$ for 60 GeV/nucleon (circles) together with a fit (solid line) to the corresponding spectrum at 200 GeV/nucleon scaled down in E_T by a factor of 0.58 [98].

The E_T and E_T^{em} spectra were obtained by NA35 at 60 GeV/u as well as at 200 GeV/u. It should be noted that the NA35 calorimeters were moved for the 60 GeV/u run, so that the pseudorapidity acceptance covered $2.08 < \eta < 3.72$, in an attempt to remain relatively centered about $y_{cm}^{NN} = 2.43$ at $\sqrt{s_{NN}} = 10.7$ GeV. A spectacular illustration of the simple scaling of the NA35 E_T spectra for the two different c.m. energies is given in figure 33, where the $^{16}O + Au$ E_T spectrum at 60 GeV/nucleon is compared to the corresponding 200 GeV/nucleon spectrum scaled down in E_T by a constant factor of 0.58. The excellent agreement in spectral shapes in figure 33 tends to imply that only the $< E_T >$ changes with $\sqrt{s_{NN}}$ over this range, with the gamma distribution shape parameter p remaining constant (recall figure 27), i.e. another example of KNO scaling[30].

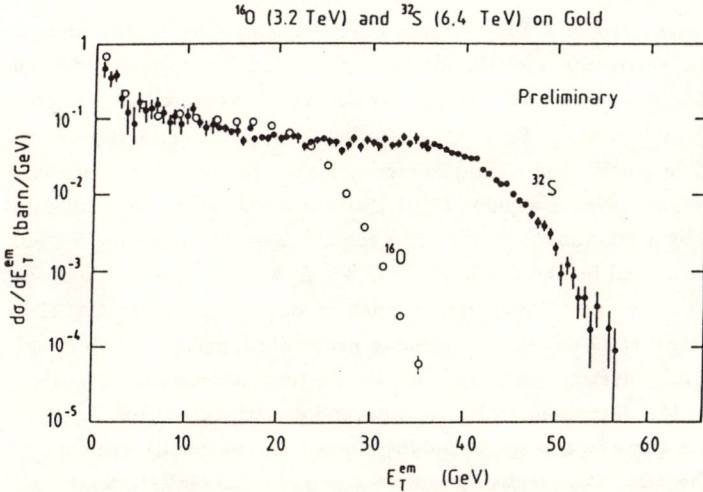

Fig. 34: Differential cross section $d\sigma/dE_T^{em}$, seen in the NA35 PPD for $^{32}S + Au$ (solid circles) with the corresponding spectrum for $^{16}O + Au$ (open circles) for comparison at 200 GeV/nucleon. The ^{32}S spectrum is preliminary [98].

Another first by NA35 was a study of the dependence of transverse energy production on projectile mass at fixed incident energy/nucleon in the same detector [98]. In figure 34, the first data from the Fall 1987 200 GeV/nucleon ^{32}S run at CERN is compared to the corresponding $^{16}O+Au$ data. The transverse electromagnetic energy E_T^{em} seen by the PPD shows an increase of $\sim 77\%$ for central collisions, for a doubling of the projectile mass. Thus, the original observation that the upper edge of the $^{16}O + Pb$ spectrum was simply the 16-fold convolution of the $p+Au$ spectrum [6] (recall figure 27) breaks down for the ^{32}S projectile [98]. This latter observation (figure 34) is more in accord with the previous understanding of the interactions of nuclei in this energy range, where transparency was the dominant feature [58]. The lack of transparency at lower energies was also not unexpected [99].

12. Measurements of the Forward Energy—E_F

The correlation of forward and transverse energy was exploited, primarily by the WA80 Collaboration at CERN[101], to analyze the E_T distributions in $A+A$ interactions without recourse to $p-p$ or $p+A$ data. The usefulness of forward energy (E_F) measurements rests on the assumption that the forward calorimeter detects only projectile spectators, thus making the forward energy an extremely sensitive probe of geometry of the $A+A$ collision. Thus, the solid angle of the forward calorimeter is set to a very small forward cone around the beam direction $\eta \gtrsim Y^{beam}$. The ideal aperture would allow the zero-degree-calorimeter (ZDC) to measure the full kinetic energy of the projectile, in the case of no interaction, and to measure zero energy for any nucleon in the projectile that suffered an inelastic collision. Thus the energy recorded in the ZDC should be proportional to the number of non-interacting nucleons ("spectators") in the projectile. In the real world the situation is somewhat more complicated, since some of the interacting nucleons may suffer very little transverse deflection, and remain within the ZDC accceptance; or some of the spectator nucleons may have unexpectedly large transverse deflection and fall outside of the ZDC.

The energy distributions measured in a ZDC offer a different sensitivity to the nuclear collision geometry than E_T measurements; and the interpretation of E_F spectra would seem more straightforward in the ideal case, since they should only depend on whether the projectile nucleons interact and not how or how many times they interact. The ZDC spectra from the WA80 Collaboration [101] for ^{16}O interactions at 60 and 200 GeV/nucleon are shown in figure 35a. The full beam kinetic energy should appear at 960 (3200) GeV for the 60 (200) GeV/nucleon data but has been suppressed by a "minimum-bias" trigger requirement that less than 88% of the full projectile energy be measured by the ZDC. The $^{16}O + ^{12}C$ data at 200 GeV/nucleon show essentially no cross section for events depositing a small amount of energy in the ZDC because, even in the most central collision, there are many projectile "spectators" since the target nucleus is too small to fully overlap the projectile. By contrast, a pronounced peak is seen at small ZDC energies in the data from $^{16}O + Au$ interactions, where full overlap does occur on central collisions, and there is a large probability for all the projectile nucleons to interact. The acceptance of the ZDC was effectively more restrictive at 60 GeV/nucleon, and closer to the ideal, because the WA80 calorimeters remained fixed for the two incident energies so that the ZDC acceptance ($\eta > 6.0$) effectively moved considerably forward of $Y^{beam} = 4.85$

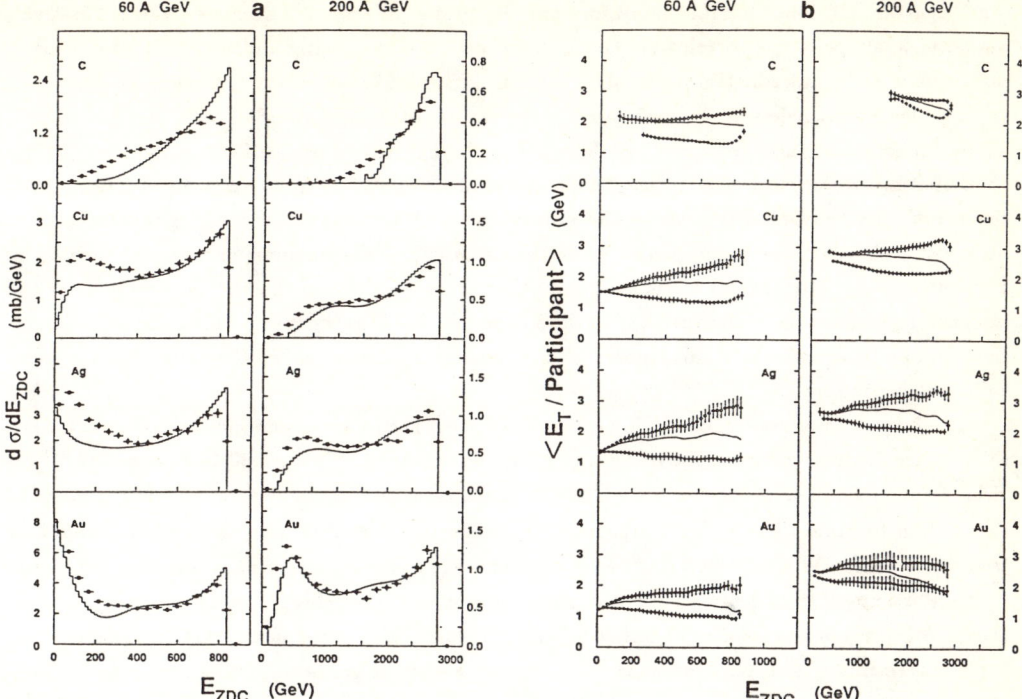

Fig. 35: (a) Energy spectra measured in the WA80 zero-degree-calorimeter (filled circles) in ^{16}O induced reactions. The histograms give the results of the Lund Model (FRITIOF) [68,101]. (b) Determination, by WA80, of the average values of E_T/participant as a function of the energy measured in the zero-degree-calorimeter. The pseudorapidity range used in the E_T determination is $1.6 < \eta < 5.5$. The solid line and the open circles represent different nuclear geometry calculations (see [101] for details).

at 60 GeV/nucleon. The peak in the $^{16}O + Au$ data becomes more pronounced and moves to zero E_{ZDC}, at 60 GeV/nucleon, indicating the absence of projectile spectators for a large fraction of the interactions. An increase of events with low values of E_{ZDC} is also observed for the Ag and Cu targets. The authors of WA80 compare their data to the Lund-FRITIOF Monte Carlo calculation [68] (solid line on figure 35a), and it is amusing to note that the agreement is not particularly good for the lighter targets, where "there is a clear tendency for FRITIOF to underestimate the cross section for small values of E_{ZDC}" [101]. This tendency is more apparent at 60 GeV/nucleon.

12.1 Correlations between transverse energy E_T and forward energy E_F.

A new diagnostic tool in relativistic heavy ion interactions has been provided by the study of the correlation between forward and transverse energy emission. At first glance, the correlation appears to be a trivial consequence of energy conservation: the energy not observed in the forward direction will be observed as "transverse energy" [102]. Early measurements[98,101] clearly demonstrated the anti-correlation between E_T and E_{ZDC}; and confirmed the intuitive notion that "central collisions" exhibit small forward energies and large transverse energies. However, the detailed shape of the correlation curve, as a function of projectile and target atomic masses, can provide stringent constraints on the reaction dynamics of collisions in nuclei.

The energy recorded in an ideal zero-degree-calorimeter corresponds to the number of projectile "spectators", and thus directly determines the number of projectile "participants". The participating projectile nucleons may make more than one $N - N$ collision in the target nucleus, with the number of successive collisions being proportional to the nuclear thickness. The relationship between wounded projectile nucleons and $N - N$ collisions, or wounded projectile nucleons and wounded target nucleons, is a characteristic, simply, of the nuclear geometry of the interaction, and is easily computed. The data then may be directly compared to the geometrical calculations to see which, if any, model of the fundamental element of E_T emission in nuclear collisions is preferred.

The forward-transverse correlation measurements at the incident energies available at CERN seem to track very well with the nuclear participant geometry. The WA80 Collaboration [101] divides their $< E_T >$–E_{ZDC} correlation data by the corresponding nuclear participant geometry calculation to obtain the $< E_T/\text{Participant} >$ integrated over their pseudorapidity acceptance (see figure 35b). While the 200 (60) GeV/nucleon data appear to be consistent with a constant $< E_T/\text{Participant} >$ of 2 (1.3) GeV, the conclusion is not compelling, due to the large systematic uncertainty of the "nuclear geometry" calculations. However, a new analysis of these data, including results of 200 GeV/nucleon ^{32}S interactions in the same targets, appears to be compatible with a constant value of $< E_T/\text{Participant} > \simeq 2$ GeV, in the pseudorapidity interval $2.4 < \eta < 5.5$, for all targets and both projectiles at 200 GeV/nucleon [103]. This preliminary result would be striking corroboration of the WNM for $< E_T >$ at CERN energies, if it holds up; and an interesting extension would be to see whether the detailed shapes of the E_T spectra also support a WNM explanation.

Unlike the results at CERN energies, the forward-transverse energy correlation measurements of the E802 collaboration [105,104], for 14.6 GeV/c per nucleon ^{28}Si interactions do not simply follow a nuclear geometry calculation, either wounded projectile nucleons as for E^0_{TOT} (recall figure 30) or the WNM. No doubt, the forward-transverse energy correlation provides enhanced sensitivity to the different models of E_T production in relativistic nuclear collisions; and will continue to be exploited. However, this is a new technique, unique to the field of Relativistic Heavy Ion interactions; and the analysis and interpretation of the data (in the real world) is more complicated than in the idealized case, and has taken longer than anticipated to complete.

13. Pseudorapdity distributions, $dE_T/d\eta$ and the issue of "nuclear stopping"

The issue of "nuclear stopping" continually surfaces in discussions of the data from relativistic heavy ion interactions. The degree of "nuclear stopping" is an important element in estimates of the temperature, energy density and baryon density of the large volume of hot, compressed, nuclear matter remaining after a "central" nuclear collision. The formation of a new state of matter, the Quark Gluon Plasma (QGP) is expected to take place when the temperature exceeds a critical value $T_c \gtrsim 250$ MeV, or the baryon or energy density exceeds the value 1.0 to 3.0 GeV/fm^3. [2,3]

Pseudorapidity distributions of transverse energy emission play an important role in the discussions of "nuclear stopping" because one of the simplest concepts of stopping is that the projectile and target "participants" merge into a hadronic "fireball", which then decays isotropically [6,51,98,101,97]. This would be quite distinct from the normal "longitudinal phase space" behavior of the "soft" multiparticle production processes that predominate in high energy collisions, in which produced particles have limted transverse momentum ($<p_T> \simeq \frac{1}{3}$ GeV/c) and very broad and flat rapidity distributions.

13.1 The Fireball Model—I

The "fireball model" contains three distinct assumptions.

- The number of participants is calculated assuming a head-on, collision in which the projectile interacts with all nucleons in a "cylinder cut through the center of the target nucleus" [97]. For spherical nuclei, with B nucleons in the projectile and A in the target ($B \leq A$), the number of projectile participants is $n^P_{\text{part}} = B$ and the number of target participants is given by

$$n^T_{\text{part}} = A\left(1 - \left(1 - (B/A)^{\frac{2}{3}}\right)^{\frac{3}{2}}\right). \tag{13.1}$$

- The center-of-mass system and the available c.m. kinetic energy T^{fb}_{cm} are calculated as if a solid projectile, of mass B nucleons, containing the full beam energy, interacted with a solid target of mass n^T_{part} nucleons.

- The total available c.m. energy T^{fb}_{cm} is dissipated by isotropic emission in the fireball rest frame.

Thus in the fireball model, the relevant symmetry point is not the rapidity of the $N-N$ c.m. system, y^{NN}_{cm}, but the rapidity of the $B + n^T_{\text{part}}$ fireball, y^{fb}_{cm}. Also, the "isotropic" assumption has a very neat and striking consequence: the pseudorapidity distribution of transverse energy emission, in the fireball model, is a universal curve about the symmetry point, $\eta = y^{fb}_{cm}$, for all projectiles, targets, and incident energies (see figure 36).

FIREBALL MODEL PSEUDORAPIDITY DISTRIBUTION

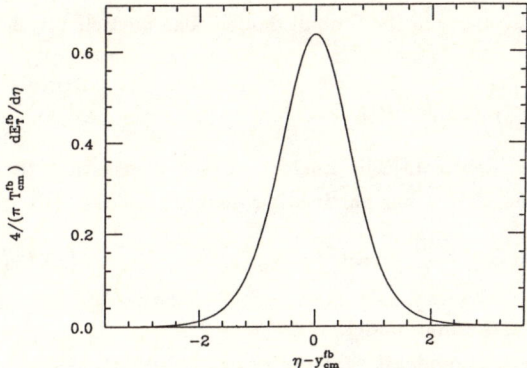

Fig. 36: Universal pseudorapidity distribution of $dE_T/d\eta$, in the fireball model, relative to the rapidity of the fireball center of mass.

It is instructive to work out this distribution from the isotropic probability for energy emission in the fireball c.m. system:

$$\frac{d\mathcal{P}(\theta^*, \phi^*)}{d\Omega^*} = \frac{1}{4\pi} = \frac{d\mathcal{P}(\theta^*, \phi^*)}{\sin\theta^* \, d\theta^* \, d\phi^*} \tag{13.2}$$

Thus,

$$\frac{d\mathcal{P}}{\sin\theta^* \, d\theta^*} = \frac{1}{2} \tag{13.3}$$

The probability is evidently normalized. The available energy distribution is:

$$\frac{dT}{d\theta^*} = T_{cm}^{fb} \frac{d\mathcal{P}}{d\theta^*} = \frac{T_{cm}^{fb}}{2} \sin\theta^* \tag{13.4}$$

Now, use the definition of E_T :

$$\frac{dE_T}{d\theta^*} \equiv \sin\theta^* \frac{dT}{d\theta^*} = \frac{T_{cm}^{fb}}{2} \sin^2\theta^* \tag{13.5}$$

Integrate over θ^* to find the total $E_T^{\text{integrated}}$

$$E_T^{\text{integrated}} = \int_0^\pi \frac{dE_T}{d\theta^*} d\theta^* = \int_0^\pi d\theta^* \frac{T_{cm}^{fb}}{2} \left(\frac{1 - \cos 2\theta^*}{2}\right) = \frac{\pi}{4} T_{cm}^{fb} \tag{13.6}$$

The pseudorapidity distribution of E_T is determined from Eq. (13.5)

$$\frac{dE_T}{d\eta} = \frac{dE_T/d\theta^*}{d\eta/d\theta^*} \tag{13.7}$$

where

$$\eta = -\ln\tan\theta^*/2 \qquad \frac{d\eta}{d\theta^*} = -\frac{1}{\sin\theta^*} \tag{13.8}$$

so that

$$\frac{dE_T}{d\eta} = -\frac{T_{cm}^{fb}}{2} \sin^3\theta^* \tag{13.9}$$

whereupon, from Eq (2.5)

$$dE_T^{fb}/d\eta = \frac{T_{cm}^{fb}}{2} \operatorname{sech}^3\left(\eta - y_{cm}^{fb}\right) \tag{13.10}$$

A cultural division between nuclear and particle physicists arises when it comes to discussing the pseudorapidity distribution of charged multiplicity in the fireball model. The particle physicist says that $< P_T >$ is constant, so that

$$dn^{fb}/d\eta = \frac{1}{< P_T >} dE_T^{fb}/d\eta$$

i.e. the $dn/d\eta$ and $dE_T/d\eta$ distributions are identical. The nuclear physicist says that the temperature is constant, which implies that the $< E >$ per particle is constant, so that

$$dn^{fb}/d\eta = \frac{1}{< E >} dE^{fb}/d\eta = \frac{T_{cm}^{fb}}{2 < E >} \operatorname{sech}^2\left(\eta - y_{cm}^{fb}\right) \tag{13.11}$$

This latter point of view is correct in the strict thermal model.

Another, less unique, conseqence of the fireball model is that the symmetry point, y_{cm}^{fb}, the position of the maximum $dE_T/d\eta$, should move systematically backwards in a $B + A$ interaction, as the target atomic mass A increases. The prediction of the shift in the position of $dE_T/d\eta|_{\text{max}}$

is quantitatively precise in the fireball model; but is qualitatively predicted in all multiple collision models, since the rapidity of an excited nucleon will decrease, as its effective mass or transverse momentum increases, even without any loss of energy [106].

Although the fireball model will turn out to have little applicability in the regime of relativistic heavy ion collisions presently under study, its predictions serve as a guide on how to characterize and evaluate measured pseudorapidity distributions. It is convenient to characterize the pseudorapidity distributions by $\eta|_{max}$ and $dE_T/d\eta|_{max}$, the position and value of the maximum of the distribution, and by $\Delta\eta_{FWHM}$, the full width of the distribution at half maximum. For most pseudorapidity distributions, the integral of $dE_T/d\eta$ over all η is well approximated by

$$\int_{all\ \eta} d\eta\ dE_T/d\eta \equiv E_T^{integrated} \simeq dE_T/d\eta|_{max} \times \Delta\eta_{FWHM}. \qquad (13.12)$$

In the fireball model, $dE_T/d\eta|_{max} = T_{cm}^{fb}/2$, $E_T^{integrated} = \frac{\pi}{4}T_{cm}^{fb}$, and the FWHM of the distribution is universal: $\Delta\eta_{FWHM}=1.41$. For the "particle physicist" the multiplicity distribution has the same FWHM as the E_T distribution; for the "nuclear" physicist, $dn/d\eta$ has $\Delta\eta_{FWHM}=1.75$.

13.2 The Fireball model fails—$dn/d\eta$ isn't universal

A direct comparison of pseudorapidity distributions for ^{16}O interactions at 14.5 GeV/c per nucleon, 60 GeV/nucleon and 200 GeV/nucleon is available, for charged particle multiplicity, from E808 [109], an emulsion experiment at the AGS and CERN. The distributions of shower particles ($v/c > 0.7$) measured in central collisions of ^{16}O with $AgBr$ are shown in figure 37. There is approximate scaling in the target fragmentation region and the distributions show a longitudinal expansion to larger η as the projectile energy increases. These distributions appear to be very symmetric about y_{cm}^{NN}, for all 3 incident energies, and do not indicate any backward shift of $\eta|_{max}$, as seen in comparable results from $p+A$ interactions [59] (recall figure 21), although perhaps the $\bar{\nu} = 2.4$ for the emulsion [107] is not sufficiently large for the effect to be visible. The principal disagreement with the fireball model comes from the systematic increase of $\Delta\eta_{FWHM}$ from 2.4 to 3.0 to 3.6 with increasing projectile energy, and the fact that all of these values are considerably wider than the universal fireball model prediction of 1.41, or 1.75.

A measurement of the pseudorapidity distribution was obtained by E802 from the small early setup with a 14.6 GeV/c per nucleon ^{16}O beam[91] (recall figure 28). Although the segmentation of

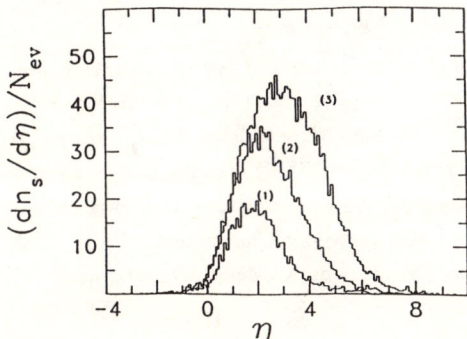

Fig. 37: Normalized pseudorapidity distributions of shower particles ($v/c > 0.7$) from central $^{16}O + Ag/Br$ interactions in an emulsion at three energies (1) 14.6, (2) 60, and (3) 200 GeV/nucleon. The central selection corresponds to $\sim 30\%$ of the Ag/Br interaction cross section[109].

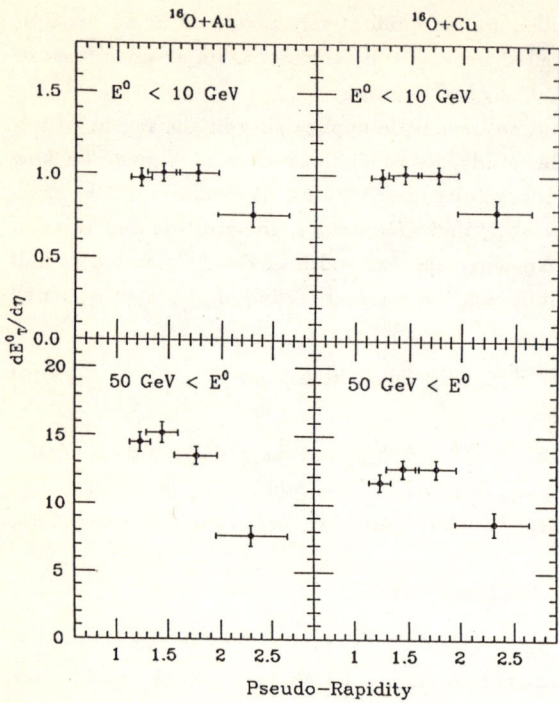

Fig. 38: The pseudorapidity dependence of the transverse energy recorded in the E802 full azimuth PbGl detector for 14.5 GeV/c per nucleon ^{16}O interactions in Au, and Cu targets [91].

the detector in η was rather crude, the transverse energy flow, $dE_T^0/d\eta$ could be obtained in four broad η intervals, for windows of total energy, E_{TOT}^0, observed in the PbGl detector (see figure 38). The PbGl detector responds only to relativistic particles ($v/c > 0.8$), so these distributions should be directly comparable to the emulsion data in figure 37. The PbGl distributions are roughly gaussian with $\Delta\eta_{FWHM} \simeq 2.0$ for the $^{16}O + Au$ and $^{16}O + Cu$ distributions, in agreement with the emulsion results. For peripheral collisions, $E_{TOT}^0 < 10 GeV$, the distributions are symmetric about $y_{cm}^{NN}=1.7$. For the Cu data, there is the hint of a shift backward of the peak for "central collisions", $E_{TOT}^0 > 50 GeV$, to 1.6. For the $^{16}O + Au$ data, the peak shifts steadily backwards with increasing centrality (E_{TOT}^0), from ~ 1.6 at low values to ~ 1.4 for the most "central collisions". These data show that the flow in energy gets more transverse with increasing target atomic mass and with increasing observed total energy E_{TOT}^0, particularly for the Au target. This is indicative that there is some, albeit small, effect of more than one collision of the projectile in the target nucleus, even at AGS energies.

At CERN, NA34 presented[51,100,108] a relatively comprehensive set of $dE_T/d\eta$ measurements, for fixed values of E_T in the range $-0.1 < \eta < 2.9$, for $p + Pb$, $^{16}O + W$ and for $^{32}S + W$ (see figure 39). These data are reasonably consistent with the characteristic shape from $p - p$ minimum bias collisions and with the emulsion data (recall figure 37). It is difficult to discern a difference in these distributions as a function of centrality (E_T) or projectile. The data at 200 GeV/nucleon appear to show $\Delta\eta_{FWHM} \simeq 3.0$, and and appear to peak backwards of y_{cm}^{NN}; but it is difficult to make definitive conclusions about $\eta|_{max}$ with this data. Measurements, with full pseudorapidity coverage, are expected in the near future.

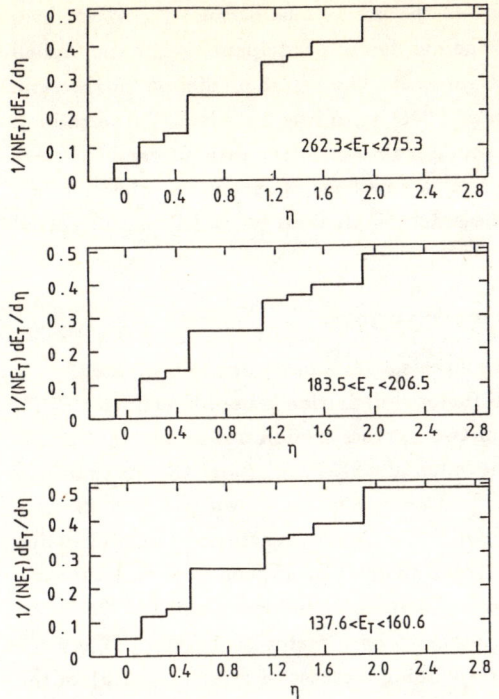

Fig. 39: NA34-Helios $dE_T/d\eta$ normalized by the integrated E_T in the acceptance $-0.1 < \eta < 2.9$ for $^{32}S + W$ at incident energy of 200 GeV/nucleon [100].

13.3 A success—the relative endpoints of the E_T distributions

Although, the fireball model utterly fails to represent the shape of any of the pseudo-rapdity distributions shown above, the WA80 Collaboration presents their measurements of $dE_T/d\eta|_{max}$ for central collisions of ^{16}O projectiles at 60 and 200 GeV/nucleon as a ratio to the fireball model prediction [101]. The "mid-rapidity energy stopping" ratio, S_{mid}, was defined as the ratio of $dE_T^{cent}/d\eta|_{max}$, the measured value of the maximum in the pseudorapidity distribution for central collisions, to $T_{cm}^{fb}/2$ is the fireball model prediction. Amazingly, the fireball model prediction scaled out the target dependence very well, at both incident energies, and also scaled out the projectile dependence at 200 GeV/nucleon. The upper edges of the spectra for ^{16}O interactions in ^{27}Al, ^{64}Cu, ^{108}Ag, and ^{197}Au, could be simply characterized as $S_{mid} \simeq 35\%$ (26%) at 60 (200) GeV/nucleon for all the targets. The fireball model prediction for the ratio of $dE_T^{cent}/d\eta|_{max}$ in $^{32}S + Au$ to $^{16}O + Au$ interactions is 1.77, which is close to the value observed by NA35[98], and not far from the values $\simeq$1.65 observed by NA34[100] and $\simeq$ 1.60 by WA80[101].

13.4 But it's probably just the Wounded Nucleon Model

The success of the fireball model in representing the target (and projectile) dependences of the upper edges of the E_T spectra is striking in comparison to its utter failure in representing the shape of the pseudorapidity distributions. This is probably because the target scaling and the projectile scaling in the fireball model correspond very closely to the the predictions of the Wounded Nucleon Model (WNM), in which E_T scales simply as the total number of participating nucleons. There are slight differences in the predictions because the WNM uses an $N-N$ absorption mean free path to calculate the number of participants, while the fireball model simply assumes that all the central nucleons interact. The WNM prediction for the ratio of the upper edges of the E_T spectra in $^{32}S+Au$ and $^{16}O+Au$ reactions is 1.77, the ratio of the number participants for central collisions. In the fireball model, the ratio of participants is 1.65, while the ratio of $T_{cm}^{fb}/2$ is 1.77. It is clear that a new level of systematic precision will have to be achieved by the experiments, before these fine distinctions in models can be sorted out.

14. Systematic Comparison of all experiments–a surprise

The discussion of pseudorapidity distributions suggests a systematic way to compare the E_T distributions from all the experiments. The NA34–Helios [51] definition is used to estimate E_T^{cent}, the E_T observed in "central collisions" (a factor of two in cross section down from the "knee" of the distribution), for all experiments. Then, the value of $dE_T^{cent}/d\eta|_{max}$ at the maximum in the pseudorapidity distribution can be estimated[43]. The results are shown in figure 40, as a function of the nucleon-nucleon c.m.energy $\sqrt{s_{NN}}$ for $^{16}O+Au\,(W)$ reactions. The broken line on the figure is $dn/d\eta|_{y_{cm}^{NN}}$ for $p-p$ collisions [38] (recall figure 18). If geometry were the only nuclear effect, then the $\sqrt{s_{NN}}$ dependences would be identical in the $p-p$ and the $^{16}O+Au$ reactions. However, $dE_T^{cent}/d\eta|_{max}$ in $^{16}O+Au$ increases by a factor of 3 from $\sqrt{s_{NN}}=5.4$ to 19.4 GeV, which is much greater than the corresponding increase of a factor of 1.31 of the multiplicity density in $p-p$ collisions. The relative increase of $3/1.3=2.3$ is roughly equal to $(1+\bar{\nu})/2\simeq 2.4$ for Au, which is consistent with the observation of "energy stopping" at the AGS, in contrast to the full "transparency", at $\sqrt{s_{NN}}=19.4$ GeV, where all the successive collisions in an Au nucleus are likely to be equally effective in producing transverse energy.

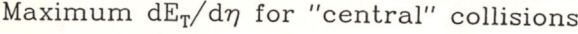
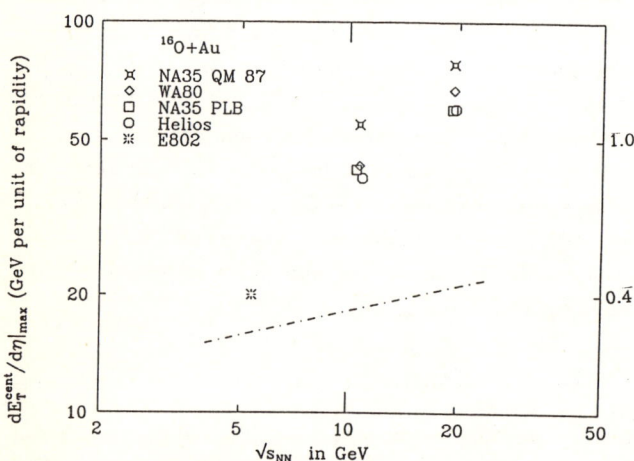

Fig. 40: Comparison of estimated [43] values at the maximum in the pseudorapidity distribution, $dE_T/d\eta|_{max}$, for central collisions, in all relativistic heavy ion experiments, plotted against nucleon-nucleon c.m. energy $\sqrt{s_{NN}}$. The broken line is $dn/d\eta|_{y_{cm}^{NN}}$ for $p-p$ collisions[38]

Although, the estimates of E_T^{cent} are clearly dominated by systematic effects, the most notable being between the two NA35 measurements[6,98], a similar conclusion on the relative $\sqrt{s_{NN}}$ dependence of nucleus-nucleus and $p-p$ collisions can be drawn from the emulsion data[109] in figure 37. $dn/d\eta|_{max}$ increases by a factor of 2.4 from $\sqrt{s_{NN}}$ =5.4 to 19.4 GeV, an increase of a factor of 1.8 relative to $p-p$, which is again close to the value of $(1+\bar{\nu})/2 \simeq 1.7$ for $AgBr$.

It is evident from these considerations, that systematic comparision of the data from nucleus-nucleus, $p+A$, and $p-p$ interactions at AGS and CERN energies can lead to quantitative conclusions, without detailed recourse to particular models.

15. Measurements of inclusive spectra of identified charged particles

The "energy stopping" observed at the AGS, may raise hopes that a baryon rich system has been created with a density approaching 3-5 times the normal nuclear density. The particle composition, particularly production of strange particles, has been proposed as a signature for the QGP[16,17]. Thus, a measurement of the particle composition and momentum spectra could provide essential input for understanding the reaction dynamics of nucleus-nucleus collisions in this stopping regime. The first results for momentum distributions of identified charged particles from such collisions in this energy regime have recently been obtained in the E802 magnetic spectrometer[96,110] (recall figure 12).

Semi-inclusive transverse momentum distributions of $\pi^\pm$, $K^\pm$, and p have been measured as a function of rapidity for central collisions (hence the **semi-inclusive** nomenclature) of 14.6 GeV/c per nucleon ^{28}Si with an ^{197}Au target of $\approx 3\%$ interaction length. The characterization of central collision events is obtained by requiring a high charged-particle multiplicity (upper 7% of the distribution) in the target-multiplicity array (TMA) $(-1.3 \leq \eta \leq 3.0)$.[91,96]

The experimental apparatus consists of a rotatable magnetic spectrometer with 25 msr solid angle, covering about 0.6 units of rapidity for pions at mid-rapidity at a given angle setting. Particle tracking is provided by 40 planes of projective drift chambers, divided evenly on each side of a bending magnet with a maximum integrated field of 1.5 Tm. The momentum resolution, which is limited by multiple scattering, is $\Delta P/P \leq 1.4\%$. Particle identification is provided by measuring particle flight time between a thin plastic scintillator beam counter (timing resolution σ=40 ps) and a 160 slat plastic scintillator array located behind the tracking system at 6.5 m from the target.[111] The quality of the particle identification is shown in the scatter plot (Fig. 41) of inverse momentum versus time-of-flight for charged particles at the pion mid-rapidity setting of the spectrometer. The combined timing resolution of 75 ps (1σ) provides adequate K-π separation up to 2.2 GeV/c. In this momentum range the mean separation between particle types is at least 5 times the σ for a given group. The only contamination of particle types occurs for the pions, for which there is a small contribution due to electrons. The electrons are completely separable up to $P = 0.7$ GeV/c; based on extrapolating the electron/pion ratio, the electron contamination at larger P is no greater than 5%.

15.1 Relativity really works!

In figure 41, the measured quantity $1/P$ is plotted against the measured time of flight (recall figure 15). The characteristic shape of the curves for the different particles is a pretty illustration of relativistic kinematics. Recall that the momentum of a particle of mass m is

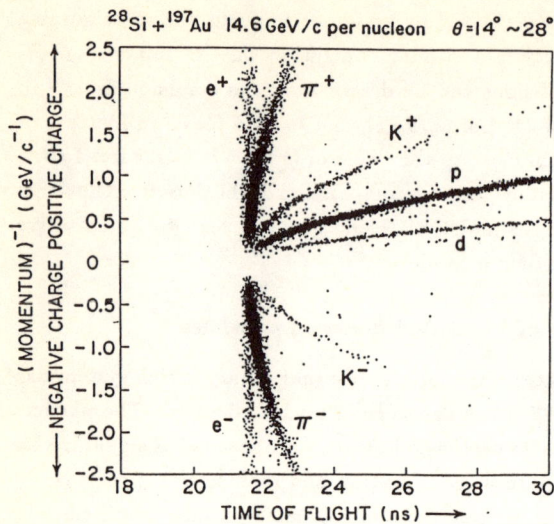

Fig. 41: Scatter plot of inverse momentum versus time of flight, for positively and negatively charged particles emitted from Si+Au interactions.

$$P = \gamma\beta m \qquad \gamma = \frac{1}{\sqrt{1-\beta^2}}$$

The velocity is determined from the measured time of flight t over the known distance s

$$\beta = \frac{s}{ct}$$
$$t = \frac{s}{c}\left(\frac{1}{\beta}\right) \qquad (15.1)$$

so that

$$\frac{1}{P} = \frac{\sqrt{\left(\frac{1}{\beta}\right)^2 - 1}}{m} = \frac{\sqrt{\left(\frac{c}{s}t\right)^2 - 1}}{m} \qquad (15.2)$$

For low values of β (long times), the non-relativistic limit is obtained, and the relationship between momentum and velocity is linear: $1/P$ is linearly proportional to $(1/\beta)$, or to the time t. For large values of $\beta \to 1$, the γ dependence takes over, and the linear dependence of $1/P$ on t curves over and becomes a square root dependence, for low values of t, as indicated in Eq. (15.2) and nicely shown in figure 41.

15.2 Results for the P_T spectra and $dn(y)/dy$

The compactness of the spectrometer allows a measurement of the kaon momentum spectra down to 0.5 GeV/c; below that, the fraction that survives decay is too small ($\sim$20%). Note that the fraction of kaons surviving after a proper time t_P (in the kaon rest frame), corresponding to $t = \gamma t_P$ in the lab system is

$$\mathcal{P}_{\text{surv}} = \exp -t_P/\tau = \exp -\frac{s}{c\tau}\left(\frac{1}{\gamma\beta}\right) = \exp -\left(\frac{s}{c}\frac{m}{P}\Big/\tau\right) \qquad (15.3)$$

where τ is the lifetime of the particle (as measured in its rest frame).

The track-reconstruction efficiency was determined from a manual reconstruction of 200 events and from Monte-Carlo simulations. The single particle position resolution of 150 μm (1σ)

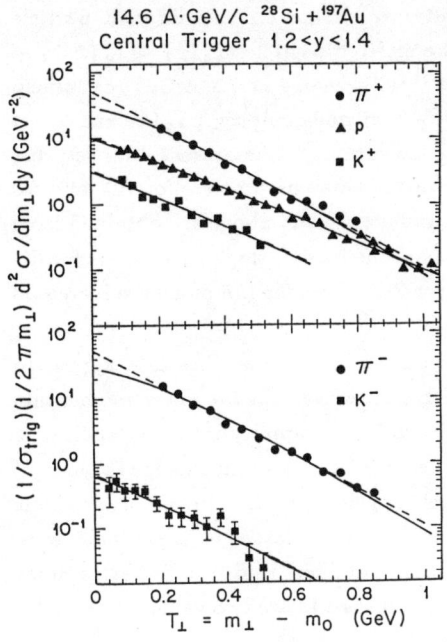

Fig. 42: Invariant cross section per central trigger as a function of $T_\perp$ for $\pi^\pm$, $K^\pm$, and protons for the rapidity interval $1.2 < y < 1.4$. The dashed lines show the exponential fits to the data; the solid curves show fits to Boltzmann distributions. The uncertainties in the data are either shown or are of the size of the drawn points.

and pulse pair resolution of 2 mm, together with the large redundancy in the tracking system, allows a track reconstruction efficiency of 85±5%. This efficiency is only weakly dependent on multiplicity, namely a loss of ≤2.5% for each additional track for the observed range of multiplicities. At mid-rapidity, the average multiplicity in the spectrometer for events where there is at least one track is 1.5, with 99% of events having multiplicity ≤ 6. Decay and acceptance corrections have been performed on a track-by-track basis, assuming that a particle will be lost if it decays. Monte-Carlo calculations, however, indicate that some muons from pion decay will be mis-identified as pions, and thus be accepted as valid tracks. The inclusion of these particles in the pion spectra causes a negligible change in the slope of the pion momentum distribution and increases the pion integrated yield by no more than 10% from its correct value.

The measured momentum distributions for pions, kaons, and protons in the rapidity interval $1.2 < y < 1.4$ are shown in Fig. 42. The error estimates are statistical uncertainties only; there is no additional uncertainty in the relative normalization, but the overall normalization is uncertain to ±10%. The distributions are shown in terms of the invariant cross section, per trigger, as a function of the transverse kinetic energy $T_T = m_T - m_o$, where $m_T = \sqrt{P_T^2 + m_o^2}$ is the transverse mass, P_T is the transverse momentum, and m_o is the rest mass. For each rapidity interval, fits were performed using three possible types of distributions: $A_o exp(-m_T/T_o)$, the Boltzmann form $A_B m_T exp(-m_T/T_B)$, and $A_p exp(-P_T/T_p)$, where the parameters A and T are determined for each interval. As in $p-p$ collisions, the momentum distributions are found to be better described by exponentials in m_T than in p_T. The quality of the fits to the present data for all particle species is good for both the exponential and Boltzmann dependence on m_T. The best fit values of $T_o (MeV)$ for the "mid-rapidity" interval shown in figure 42 are: π^+ 162 ± 10, π^- 161 ± 10, K^+ 203 ± 15, K^- 175 ± 25, p 215 ± 5.

The measured values of T_o (or T_B) at mid-rapidity are similar for the different particle species, but progressively increase with the mass of the emitted particle: $T_o(\pi) \leq T_o(K^+) \leq T_o(p)$. This order is different from the Bevalac results[112] with beams at 2.1 A·GeV/c, namely $T_o(\pi) \leq T_o(p) \leq T_o(K^+)$. For $p-p$ collisions at 12 GeV/c at mid-rapidity[35], $T_o(\pi)$ and $T_o(p)$ are both equal to 150 MeV to within the 2% statistical uncertainty. The present data set, due to limited statistics, allows only a poor determination of the slope parameters for K^- and for only a limited range in y for K^+. The inverse slope parameters given above are obtained from fits to data in the same T_T range. If one wishes to compare the inverse slope parameters for fits to data in the same m_T range, then the extracted T_B (or T_o) values for the pions are increased by 10-20 MeV.

The rapidity dependences of T_o and T_B for π^- are shown in Fig. 43. The data appear to indicate a variation in inverse slope of about 40 MeV with y. However, the fits to the momentum distributions are made in different ranges of m_T for the different rapidity intervals, the range being shown in the upper panel of Fig. 43. It is estimated that a systematic change in slope of 10-20 MeV may occur over the rapidity range as a result of this restriction in m_T range. This restriction has the effect of decreasing the T_B (or T_o) in Fig. 43 at larger y. If the data were fit with the same m_T range for each rapidity interval, it would likely lead to a flatter rapidity distribution of the inverse slopes. For reference, indicated on Fig. 43 are two values of rapidity:

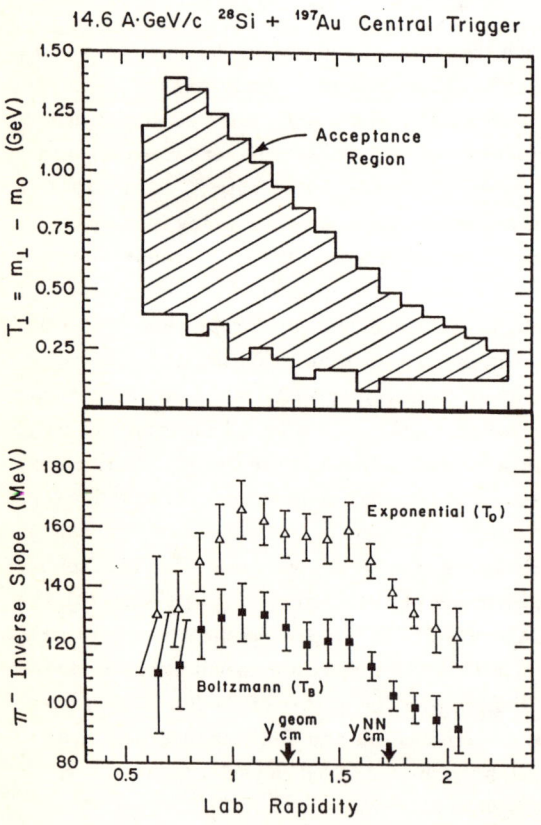

Fig. 43: The upper panel shows the range in $T_\perp$ for which the pion data are fitted to obtain inverse slope parameters in each rapidity interval. The lower panel shows the inverse slope parameters for the π^- distributions, for both exponential and Boltzmann distribution fits. The uncertainties include an estimate of the systematic errors. The participant c.m. rapidity $y_{cm}^{geom} = 1.25$ and nucleon-nucleon c.m. rapidity $y_{cm}^{NN} = 1.72$ are indicated.

y_{cm}^{NN}, the rapidity of the nucleon-nucleon c.m. system; and y_{cm}^{geom} i.e. $y_{cm}^{fb} = 1.25$, the rapidity of a system of ≈ 103 nucleon participants, composed of the incident Si and a core of 75 Au nucleons swept out by the Si nucleus in a head–on collision. **The Fireball Model strikes again!**

The rapidity distribution, dN/dy is obtained by integrating the transverse momentum distributions over m_T, using all 3 forms of the distribution to bound the systematic errors in the extrapolation. The values depend on the chosen method of extrapolation, but agree within $\pm 13\%$. The dN/dy distributions, obtained by integrating the Boltzmann form, are shown in Fig. 44 for protons, kaons and pions. The distributions for π^+ agree with those for π^- within the statistical uncertainty of 4% and are broadly peaked. The shape of the $\pi^\pm$ distribution also agrees very well with the $dE_T^0/d\eta$ distribution in the PbGl for central collisions in Au (recall figure 38). The distribution for protons rises dramatically near the target rapidity, consistent with the expected contribution of target spectator protons. By comparing the integrated spectra one also obtains the particle yield ratios. For the mid-rapidity interval $1.2 < y < 1.4$, the integrated ratios are 19.2±3% for K^+/π^+ and 3.6±0.8% for K^-/π^-, including the systematic error in the extrapolation to $T_T = 0$. These values are somewhat smaller than those given in a preliminary report[96]; the difference arises solely from the extrapolation to $T_T = 0$ used here, instead of the low p cutoff of 0.5 GeV/c used in the preliminary version[96].

The K^-/π^- ratio for $p-p$ collisions at the same rapidity is $\approx 2.4 \pm 2.0\%$, the large uncertainty arising from the low yield of K^-. Because of this large uncertainty one cannot determine whether this ratio is different in heavy-ion collisions. The π^+/π^- ratio is 1.00 ± 0.04, whereas it is ≈ 1.6 for $p-p$ reactions[35]. This result presumably reflects the contribution of $n-n$ and $n-p$ interactions as well as $p-p$ interactions in $Si+Au$ interactions. The value for K^+/π^+ is much larger than the corresponding ratio of 4-8% measured in $p-p$ collisions[113]. Analysis of data for $p+A$ reactions indicates a K^+/π^+ ratio that is intermediate between the $p-p$ and $Si+Au$ results, albeit for a somewhat different kinematic range[113]. An enhancement of K^+ production, called the K^+ distillation effect[114], has been predicted to occur if either very high baryon density matter or a quark-gluon plasma is formed[16,17]. However, more mundane mechanisms are also possible. For example, the rescattering of the reaction products $\pi^+ n \to K^+ \Lambda$ could provide a mechanism for the K^+ enhancement[96,115].

15.3 The Fireball Model II— for single particle inclusive spectra

A simple prediction for the rapidity dependence of dn/dy is given by the single source Boltzmann distribution[116], a more elegant version of the fireball model. The solid curve on Fig. 44 is a fit to the pion distributions, with parameters $y_o = 1.46$ and $T = 125 MeV$, where y_o is the rapidity of the thermal source with temperature T. For a thermal distribution one expects a common temperature T for all species of emitted particles, provided they freeze out at the same time. No such dependence is seen in the above data. A variety of possible mechanisms have, however, been proposed to postulate different effective temperatures for different species....

It is once again a useful exercise, to work out the predictions of the single source Boltzmann model for inclusive single particle distributions. Assume that particle emission is isotropic with a thermal distribution:

$$d^3n/dp^3 = \frac{C}{e^{(E-\mu)/T} \pm 1}$$

where the + sign is for particles which obey Fermi-Dirac statistics and the - sign for Bose-

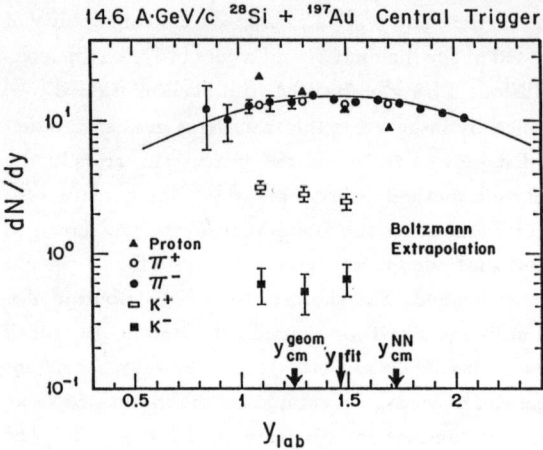

Fig. 44: Rapidity distributions dN/dy per trigger for $\pi^\pm$, $K^\pm$, and protons, assuming a Boltzmann form to extrapolate to $T_\perp = 0$. The errors shown are relative errors only; there is an overall uncertainty of $\pm 10\%$ due to the absolute normalization. The values for π^+ and π^- are the same within the measured uncertainties for all rapidity intervals. The solid curve shows the fit to Eq. (15.8), for a source located at $y_{fit} = 1.46$ and $T = 125$ MeV.

Einstein statistics. C is a constant and μ is called the chemical potential[116]. Usually, the limit of a Maxwell-Boltzmann distribution is taken:

$$d^3n/dp^3 = A\, e^{-E/T} \qquad (15.4)$$

where $A = C e^{\mu/T}$ is taken as a constant for each particle species [117]. The total number of particles (of a given species), N, is obtained by integrating d^3n/dp^3 over all space and over all particle energies above the minimum energy for a given mass particle, $E = m$.

$$N = A \times 4\pi m^2 T\, K_2(m/T) \qquad (15.5)$$

where K_2 is a Macdonald or modified Hankel function.

For single particle inclusive distributions, the fireball model gives distinctive predictions for the number of particles per unit rapidity and the "effective Temperature" as a function of rapidity. Starting with the non-invariant probability:

$$d^3n/dp^3 = A\, e^{-E/T}$$

the invariant probability is obtained by multiplying both sides of the equation by E:

$$E\, d^3n/dp^3 = A\, E\, e^{-E/T} \qquad (15.6)$$

Expressed in terms of rapidity and m_T,

$$E = m_T \cosh y \qquad \text{and} \qquad d^3p = m_T\, dm_T\, dy\, d\phi$$

and the differential distribution is obtained:

$$\frac{E\, d^3n}{dp^3} = \frac{1}{2\pi} \frac{d^2n}{m_T dm_T dy} = A\, m_T\, \cosh(y - y_o)\, \exp\left(\frac{-m_T \cosh(y - y_o)}{T}\right) \qquad (15.7)$$

where y_o is the rapidity of the fireball rest system. In simple terms, Eq. (15.7) says that the transverse momentum distribution at fixed rapidity has the form $E\,d^3n/dp^3 \simeq m_T\,e^{-m_T/T_B}$ where $T_B = T/\cosh(y - y_o)$. Thus, the fireball model predicts the form of the P_T distribution and the variation of the effective temperature $T_B(y)$ with rapidity. Furthermore, the equation may be integrated over m_T to obtain dn/dy:

$$\frac{dn}{dy} = 4\pi\,A\,T\,e^{-m_o/T_B(y)}\left[T_B^2(y) + m_o\,T_B(y) + m_o^2/2\right] \tag{15.8}$$

where again,

$$T_B(y) = T/\cosh(y - y_o) \tag{15.9}$$

y_o is the rapidity of the thermal source of temperature T and m_o is the rest mass of the particle produced. The predicted variation of dn/dy with rapidity (Eq. (15.8)) is slightly more complicated than that of the effective temperature $T_B(y)$. However, both should exhibit a maximum at the rapidity of the source y_o. Note also, that in the limit $m_0 \to 0$, $dn/dy \to dn/d\eta$ in agreement with the previous treatment Eq (13.11).

15.4 Further results on K^+/π^+

Preliminary results have recently been obtained in the E802 spectrometer from a run, in November 1988, with 14.6 GeV/c protons on a variety of nuclear targets. The ratio of the inclusive cross sections (no central trigger), $d\sigma/dy$, as a function of rapidity, has been measured for $p + ^{197}Au$ and $p + ^9Be$ reactions (see figure 45). Note that the inclusive cross section is plotted: thus, if dn/dy stayed the same for both reactions, the cross sections would be in the ratio $A^{2/3}$, as shown on the figure. The data for $\pi^\pm$ and K^- look rather similar to the ~ 100 GeV data[61] (recall figure 23). The ratio for protons increases dramatically toward the target, faster even than the ratio of the total number of target nucleons available. What is particularly surprising about the data is that the K^+ ratio exhibits nearly precisely the same dramatic

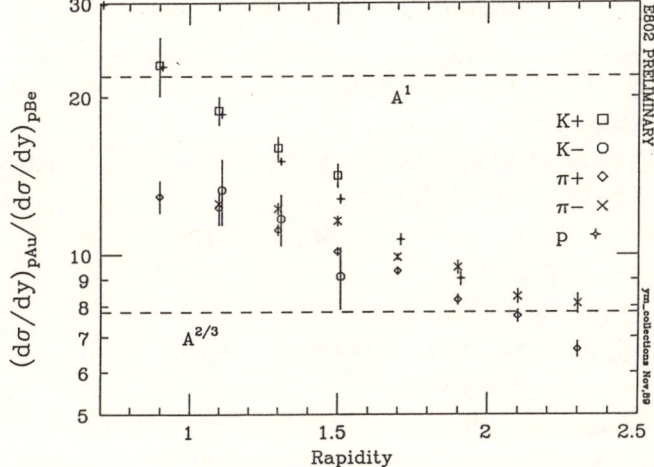

Fig. 45: Preliminary E802 results for the ratio of the inclusive cross sections (no central trigger) $d\sigma/dy$ as a function of rapidity for $p + ^{197}Au$ and $p + ^9Be$ reactions at 14.6 GeV/c.

increase toward the target as the protons, indicating a common origin for both K^+ and proton production. Could this, at last, be an indication of the importance of rescattering in the AGS energy range[96,115,118].

16. Conclusion

I have tried to leave the impression that this is a new and still dynamic field, with much ongoing activity and many unexplored facets. Regrettably, I have had to totally ignore many aspects of this broad field, one case in point being the effects of the large Z of relativistic heavy ions, the Coulomb reactions (the domain of E814[97] at the AGS), and another being identical particle Interferometry, already observed at CERN, and just now being seen in the large aperture E802 spectrometer.

The present round of experiments has been largely descriptive and exploratory, establishing the framework for understanding the interactions of Relativistic Heavy Ions. With the Au and Pb beams expected in 1992 at BNL and CERN, and with RHIC expected to be approved for construction in 1991, the future looks bright.

I find it particularly intriguing (and challenging) that **virtually all of the signatures of the QGP discussed above, have already been observed** in the first round of exploratory experiments. Naturally, I do not believe that the QGP has yet been discovered. In the case of the real QGP, all the signatures would have to turn on and turn off in a predictable, reproducible controllable and unified way. There is still much work to be done!

References

1. W.J. Willis in Proceedings of the XVI International Conference on High Energy Physics, Chicago-Batavia, Ill., USA, 1972, eds. J.D. Jackson and A. Roberts (NAL, Batavia, Ill. 1973), Vol. 4, pp 321-331. Note particularly the suggestion attributed to G. Cocconi on page 323 that "new states of matter might be produced when
hadrons could be concentrated at high densities for a sufficiently long time".

2. Proceedings of the Sixth International Conference on Ultra-Relativistic Nucleus-Nucleus Collisions–Quark Matter 1987, eds. H. Satz et al., Z. Phys. C- Particles and Fields 38 (1988) 1; and previous proceedings in this series;
see also J.C. Collins and M.J. Perry, Phys. Rev. Lett. 34 (1975) 1353;
N. Cabibbo and G. Parisi, Phys. Lett. 59B (1975) 67.

3. R. Anishetty, P. Koehler and L. McLerran, Phys. Rev. D22 (1980) 2793.

4. U. Heinz et al., Phys. Rev. Lett. 58 (1987) 2292.

5. G. Baym and S.A. Chin, Phys. Lett. 62B (1976) 241;
G. Chapline and M. Nauenberg, Phys. Rev. D16 (1977) 450;
H.Satz, Ann. Rev. Nucl. Part. Sci. 35 (1985) 245.

6. NA 35 Collaboration, A. Bamberger et al., Phys. Lett. B184 (1987) 271.

7. W. Thomé et al., Nucl. Phys. B129 (1977) 365.

8. J. D. Bjorken, Phys. Rev. D27 (1983) 140.

9. P. Hut, Nucl. Phys. A418 (1984) 301c.
10. L. Van Hove, Nucl. Phys. A461 (1987) 3c.
11. L. Van Hove, Phys. Lett. 118B (1982) 138.
12. L. Van Hove, Z. Phys. C21 (1983) 93.
13. R. Hwa and K. Kajantie, Phys. Rev. D32 (1985) 1109;
    K. Kajantie, J. Kapusta, L. McLerran and A. Mekjian Phys. Rev. D34 (1986) 2746.
14. G. Goldhaber, S. Goldhaber, W. Lee and A. Pais, Phys. Rev. 120 (1960) 300.
15. M. Gyulassy, Phys. Rev. Lett. 48 (1982) 454.
16. J. Rafelski and B. Muller, Phys. Rev. Lett. 48 (1982) 1066.
17. T. Matsui, B. Svetitsky and L. McLerran, Phys. Rev. D34 (1986) 2047.
18. T. Matsui and H. Satz, Phys. Lett. B178 (1986) 416.
19. W. M. Geist, Phys. Lett. B211 (1988) 233.
20. John. F. Donoghue and K. S. Sateesh, Phys. Rev. D38 (1988) 360.
21. A more complete treatment of this subject is found in the following references:;
    Review of Particle Properties, Phys. Lett. B204 (1988);
    Richard Fernow, *Introduction to experimental particle physics*, (Cambridge, 1989);
    R. M. Sternheimer,*Methods of experimental physics*, ed. L. C. L. Yuan and C. S. Wu, (Academic Press, New York, 1963), vol. 5A, p 1.
22. Y. S. Tsai, Rev. Mod. Phys. 46 (1974) 815.
23. C. W. Fabjan and T. Ludlam, Ann. Rev. Nucl. Part. Sci. 32 (1982) 335.
24. R. M. Sternheimer, Phys. Rev. 99 (1955) 277.
25. L. Van Hove, preprint CERN-TH.4353/85, to be published in the Proceedings of the Neils Bohr Centenary Symposium, Copenhagen, 4-7 October 1985.
26. For example see A.H. Mueller in Proceedings of the XVI Int. Conf. on High Energy Physics, Chicago-Batavia, Ill., USA, 1972, eds. J.D. Jackson and A. Roberts (NAL, Batavia, IL 1973), Vol. 1, pp 347-388;
    M. Jacob, ibid., Vol. 3, pp 373-458;
    L. Van Hove, Phys. Reports 1C (1971) 347.
27. A.H. Mueller, Phys. Rev. D4 (1971) 150;
    see also L. Caneschi, Nucl. Phys. B35 (1971) 406.
28. L. Foa, Phys. Reports 22 (1975) 1;
    see also R. Panvini in Proceedings of the XVI Int. Conf. on High Energy Physics, Chicago-Batavia, Ill., USA, 1972, eds. J.D. Jackson and A. Roberts (NAL, Batavia, IL, 1973), Vol. 1, pp 330-346.
29. H. Bøggild and T. Ferbel Ann. Rev. Nucl. Part. Sci. 24 (1974) 451.
30. Z. Koba, H.B. Nielsen and P. Olesen, Nucl. Phys. B40 (1972) 317.
31. P. Slattery, Phys. Rev. D7 (1973) 2073.
32. F. Abe et al., Phys. Rev. Lett. 61 (1988) 1819.

33. e.g. see G. Giacomelli and M. Jacob, Phys. Repts. 55 (1979) 1;
    P. Capiluppi et al., Nucl Phys B79 (1974) 189, ibid. B70 (1974) 1.
34. H. Bøggild, et al., Nucl Phys B57 (1973) 77.
35. V. Blobel, et al., Nucl Phys B69 (1974) 454.
36. For the flavor of the period, see Maurice Jacob's Eulogy for the ISR in CERN report 84-13, (CERN, SIS, Geneva 1984).
37. K. Guettler, et al., Nucl Phys B116 (1976) 77.
38. UA5 Collaboration, G.J. Alner et al., Z. Phys. C 33 (1986) 1.
39. T. Alexopoulos, et al., Phys. Rev. Lett. 60 (1988) 1622.
40. UA1 Collaboration, G. Arnison et al., Phys. Lett. 118B (1982) 167.
41. G.N. Fowler et al., Phys. Lett. 145B (1984) 407.
42. ABCDHW Collaboration, A. Breakstone et al., Phys. Lett. 132B (1983) 463.
43. M. J. Tannenbaum, Int. J. Mod. Phys. A. 4 (1989) 3377.
44. W.B. Fowler et al., Phys. Rev. 95 (1954) 1026. In the very early days it was not at all clear that a nucleon-nucleon collision could lead to the production of more than one meson;
    See R.E. Marshak, *Meson Physics* (McGraw-Hill, New York,1952), chapter 8. Also see reference [45].
45. U. Camerini, W.O. Lock and D.H. Perkins in *Progress in Cosmic Ray Physics*, Volume I, ed. J.G. Wilson, (North-Holland, Amsterdam,1952), pp 1-34.
46. C. DeMarzo et al., Phys. Lett. 112B (1982) 173.
47. C. DeMarzo et al., Nucl. Phys. B211 (1983) 375.
48. UA1 Collaboration, G. Arnison et al., CERN-EP-82/122, presented to the XXI International Conference on High Energy Physics, Paris, 1982 (unpublished);
    See also UA1 Collaboration, G. Arnison et al., Phys. Lett. 107B (1981) 320.
49. In detail the situation is somewhat more complicated. The many subtleties and details of "soft" physics [33] have been glossed over in this discussion but they do not change the essential relationship between E_T and multiplicity.
50. W. Ochs and L. Stodolsky, Phys. Lett. 69B (1977) 225;
    H. Fesefeldt, W. Ochs and L. Stodolsky, Phys. Lett. 74B (1978) 389;
    W. Ochs, Physica Scripta 19 (1979) 127.
51. HELIOS Collaboration, T. Åkesson et al., Z. Phys. C38 (1988) 383.
52. R. Wigmans, Nucl. Instrum. Meth. A259 (1987) 389.
53. AFS collaboration, H. Bøggild et al., CERN-EP-82/104, submitted to the XXI International Conference on High Energy Physics, Paris, 1982 (unpublished).
54. AFS Collaboration, H. Gordon et al., Phys. Rev. D28 (1983) 2736.
55. COR Collaboration, A.L.S. Angelis et al., Phys. Lett. 126B (1983) 132;
    see also J. de Physique 43 (1982) C3-134, figure 8.
56. B.C. Brown et al., Phys. Rev. D29 (1984) 1895.

57. P.M. Fishbane and J.S. Trefil, Phys. Rev. D3 (1971) 238;
    see also A. Dar and J. Vary, Phys. Rev. D6 (1972) 2412.

58. P.M. Fishbane and J.S. Trefil, Phys. Rev. D9 (1974) 168, Phys. Lett. 51B (1974) 139;
    K. Gottfried, Phys. Rev. Lett 32 (1974) 957;
    A.S. Goldhaber, Phys. Rev. D7 (1973) 765;
    see also A. Białas and W. Czyż, Phys. Lett. 51B (1974) 179;
    B. Andersson and I. Otterlund, Nucl. Phys. B88 (1975) 349.

59. J.E. Elias et al., Phys. Rev. D22 (1980) 13;
    C. Halliwell et al., Phys. Rev. Lett. 24 (1977) 1499.

60. W. Busza et al., Phys. Rev. Lett. 34 (1975) 836.

61. C. De Marzo et al., Phys. Rev. D29 (1984) 2476; ibid. D26 (1982) 1019. Note that the streamer chamber and many of the collaborators are from the NA5 experiment, reference [46].

62. "Produced particles" are defined in reference [61] as the particles below 600 MeV/c which are not identified as protons, and all particles with momenta larger than 600 MeV/c. The produced particles are treated as pions.

63. A. Białas, A. Błeszynski and W. Czyż, Nucl. Phys. B111 (1976) 461.

64. R. Glauber in *Lectures in Theoretical Physics*, Volume I (1958), eds. W.E. Brittin and L.G. Dunham, (Interscience, New York, 1959), pp 315-414; R.J. Glauber in *High-Energy Physics and Nuclear Structure*, ed. S. Devons, (Plenum, New York, 1970) pp 207-264. See also R. Glauber and G. Matthiae, Nucl. Phys B21 (1970)123, and references therein.

65. AFS Collaboration, T. Åkesson et al., Phys. Lett. 119B (1982) 464.

66. H. Brody, S. Frankel, W. Frati and I. Otterlund, Phys. Rev. D28 (1983) 2334;
    see also AFS collaboration, reference [54]; and B. Callen et al., Proc. Quark Matter 1984, ed. K. Kajantie (Springer, Berlin 1985), pp 133-142.

67. Dual Parton Model: see A. Capella and J. Tran Thanh Van, Z. Phys. C38 (1988) 177;
    Also see: K. Werner, ibid. C38 (1988) 193;
    J.P. Pansart, Nucl. Phys. A461 (1987) 521c;
    J. Ranft, Phys. Lett. B188 (1987) 379.

68. Lund Model: see B. Andersson, G. Gustafson, B. Nilsson-Almqvist, Nucl. Phys. B281 (1979) 289;
    Also see: B. Andersson, G. Gustafson, Z. Phys. C3 (1980) 223;
    B. Andersson, G. Gustafson, G. Ingelman, T. Sjöstrand, Phys. Rep. 97 (1983)31.

69. A. Białas and E. Białas, Phys. Rev. D20 (1979) 2854;
    A. Białas, W. Czyż and L. Lesniak, Phys. Rev. D25 (1982) 2328.

70. BCMOR Collaboration, A.L.S. Angelis et al., Phys. Lett. 168B (1986) 158; 141B (1984) 140;
    see also, M.J. Tannenbaum et al., Proc. Quark Matter 1984, ed. K. Kajantie (Springer, Berlin 1985), pp 174-186.

71. T. Ochiai, Z. Phys. C35 (1987)209; Phys. Lett. B206 (1988) 535, and references therein.

72. J.H. Christenson, G.S. Hicks, L.M. Lederman, P.J. Limon, B.G. Pope and E. Zavattini, Phys. Rev. Lett. 25 (1970) 1523; Bull. Am. Phys. 15 (1970)579.
73. S.D. Drell and T-M. Yan, Phys. Rev. Lett. 25 (1970) 316.
74. J.W. Cronin et al., Phys. Rev. D11 (1975) 3105;
    D. Antreasyan et al., Phys. Rev. D19 (1979) 764, and references therein.
75. J. Pumplin and E. Yen, Phys. Rev. D11 (1975) 1812;
    P.M. Fishbane and J.S. Trefil, Phys. Rev. D12 (1975) 2113.
76. e.g. see BCMOR Collaboration, A.L.S. Angelis et al., Phys. Lett. B185 (1987) 213, and references therein.
77. M. Lev and B. Petersson, Z. Phys. C21 (1983) 155.
78. U.P. Sukhatme and G. Wilk, Phys. Rev. D25 (1980) 1978;
    see also J.H. Kuhn, Phys. Rev. D13 (1976) 2948.
79. e.g. see S. Frankel and W. Frati, Nucl. Phys. B308 (1988) 699;
    S.J. Brodsky and A.H. Mueller, Phys. Lett. B206 (1988) 685;
    K. Werner, BNL-41500, to be published in Phys. Rev. D;
    Y. Iga, R. Hamatsu, S. Yamazaki, H. Sumiyoshi, Z. Phys. C38 (1988) 557;
    G. Gatoff, A.K. Kerman, and D. Vautherin, Phys. Rev. D38 (1988) 96;
    A. Białas and M. Gyulassy, Nucl. Phys. B291 (1987) 793.
80. B. Brown et al., Phys. Rev. Lett. 50 (1983) 11;
    AFLMPRW Collaboration, H.E. Miettinen et al., Nucl. Phys. A418 (1984) 315c;
    see also C. Bromberg et al., Phys. Rev. Lett. 42 (1979) 1202.
81. M. J. Tannenbaum, in *Proceedings of Hadronic Matter in Collision*, Tucson, AZ, USA, 6-12 October 1988, ed. P. Carruthers, J. Rafelski (World Scientific, Singapore).
82. Recall that KNO scaling originally was defined as "scaling in the mean" of the charged multiplicity distribution as a function of c.m. energy $\sqrt{s}$. See reference [30].
83. Note that the AFS collaboration, using a full hadron calorimeter, did not initially confirm the fit to a simple gamma distribution or the KNO scaling, but preferred the WNM fit to their data (see B. Callen et al. [66]). After extensive analysis of the energy response of the calorimeter, the corrected E_T spectra in $|\eta| \leq 0.7$ now fit the simple gamma distribution, and are relatively close to KNO scaling, with $p = 2.78 \pm 0.01$ for the $p-p$ data and $p = 2.48 \pm 0.08$ for the $\alpha - \alpha$ data [84].
84. S. Frankel, in *Proceedings of Hadronic Matter in Collision*, Tucson, AZ, USA, 6-12 October 1988, ed. P. Carruthers, J. Rafelski (World Scientific, Singapore);
    B.W. Callen, thesis, (Univ. of Pennsylvania, Philadelphia, PA, 1988).
85. UA5 Collaboration, G.J. Alner et al., Phys. Rep. 154 (1987) 247.
86. e.g. see M.A. Preston and R.K. Bhaduri, *Structure of the Nucleus* (Addison Wesley, Reading, MA, 1975).
87. Note that in the case of an asymmetric rapidity interval, the projectile and target can be assigned different weights, so that the deconvolution is ϵp and $(1 - \epsilon)p$ for the projectile and target wounded nucleons respectively, where ϵ and $(1-\epsilon)$ are the fractions of the total

energy contributed by the projectile and the target wounded nucleons in a given rapidity interval. Alternatively, more complicated algorithms can be used; e.g. see M. Kutschera, J. Hüfner and K. Werner, Phys. Lett. B192 (1987) 283. See also S. Frankel and W. Frati, reference [79].

88. e.g. see A. Giovannini and L. Van Hove, Z. Phys. C30 (1986) 391;
    P. Carruthers and C.C. Shih, Phys. Lett.165B (1985) 209, and references therein;
    K. Fiałkowski, Phys. Lett. 169B (1986) 436;
    see also P.A. Carruthers, M.Plümer, S. Raha and R.M. Weiner, Phys. Lett. B212 (1988) 369;
    E.R. Nakamura and K. Kudo, ibid. 381.

89. C. De Marzo et al., Phys. Rev. D36 (1987) 8,16.

90. T.W. Ludlam, BNL Report 51921 (1985) 373.

91. eg. see M.J. Tannenbaum, Nucl. Phys. A488 (1988) 555c.

92. E802 Collaboration, T. Abbott et al., Phys. Lett. B197 (1987) 285.

93. E802 Collaboration, L.P. Remsberg, M.J. Tannenbaum et al., Z. Phys. C38 (1988) 35.

94. D.E. Alburger et al., Nucl. Instrum. Methods A254 (1987) 88.

95. A similar analysis for the NA35 data [6] was performed by J. Ftacnik et al., Phys. Lett. B188 (1987) 279.

96. E802 Collaboration, Y. Miake, G.S.F. Stephans, et al., Z. Phys. C38 (1988) 135;
    See also E802 Collaboration, T. Abbott, et al., BNL-43459, Submitted to Nucl. Instr. and Meth. A (1989).

97. E814 Collaboration, P. Braun-Munzinger et al., Z. Phys C38 (1988) 45.

98. NA35 Collaboration, W. Heck et al., Z. Phys. C38 (1988) 19.

99. R. Anishetty, P. Koehler and L. McLerran, Phys. Rev. D22 (1980) 2793;
    S. Date, M. Gyulassy and H. Sumiyoshi, Phys. Rev. D32 (1985) 619.

100. Helios Collaboration, T.Åkesson et al., CERN-EP/88-121, Phys. Lett. B214 (1988) 295.

101. WA80 Collaboration, S.P. Sorensen et al., Z. Phys. C38 (1988) 3, 51;
     WA80 Collaboration, R. Albrecht et al., Phys. Lett. B199 (1987) 297.

102. P.V. Landshoff and J.C. Polkinghorne, Phys. Rev. D18 (1978) 3344.

103. WA80 Collaboration, G.R. Young et al., presented at the 7th International Conference on Ultra-Relativistic Nucleus-Nucleus Collisions (Quark Matter '88), September 1988, Lenox, Mass. To appear in the proceedings.

104. E802 Collaboration, E. Duek et al., "Projectile Energy Degradation at 14.5 GeV/u", submitted to 3rd International Conference on Nucleus-Nucleus Collisions, St. Malo, France, June 1988, (BNL-41447). See also reference [91].

105. E802 Collaboration, M.J. Tannenbaum et al., Proceedings of the 23rd Rencontre de Moriond, *Current Issues in Hadron Physics*, ed. J. Tran Thanh Van, (Editions Frontieres, Gif-sur-Yvette, France, 1988).

106. Recall that $\cosh y = E/m_T$; and see also Frankel and Frati, reference [79].

107. e.g. see W. Busza and R. Ledoux, Ann. Rev. Nucl. Part. Sci. 38 (1988) 119.
108. T. Åkesson et al., Z. Phys. C38 (1988) 397; see also Nucl. Phys. A447 (1985) 475c.
109. L.M. Barbier et al., Phys. Rev. Lett. 60 (1988) 405.
110. E802 Collaboration, T. Abbott, et al., submitted to Phys Rev Letters, (1989).
111. T. Sugitate et al. , Nucl. Inst. and Meth. A249 (1986) 354.
112. S. Nagamiya and M. Gyulassy, *Advances in Nuclear Physics*, (Plenum, New York, 1984), vol. 13, p. 201..
113. J. V. Allaby et al., CERN Preprint 70-12 (1970);
     D. Dekkers et al., Phys. Rev. 137 (1965) B962;
     U. Becker et al. , Phys. Rev. Lett. 37 (1976) 1731.
114. C. Greiner, D -H. Rischke, H. Stöcker, and P. Koch, Phys. Rev. D38 (1988) 2797.
115. R. Mattiello, H. Sorge, H. Stöcker and W. Greiner, Phys. Rev. Lett. 63 (1989) 1459.
116. S. Das Gupta and A. Z. Mekjian, Physics Reports 72 (1981) 131.
117. In principle, in a strict thermal model, C should be taken as a constant and the A parameters for the different particle species should vary according to their chemical potentials $A_i = C\, e^{\mu_i/T}$.
118. G. Baym, P. Braun-Munzinger and V. Ruuskanen, Phys. Lett. B190 (1987) 29;
     A.D. Jackson and H. Bøggild, Nucl. Phys. A470 (1987) 669.

Physics of the Quark–Gluon Plasma

J.C. Parikh

Physical Research Laboratory, Navrangpura, Ahmedabad 380009, India

1. Programme of the Lectures

The purpose of these lectures is to introduce, some of the important physical concepts and theoretical approaches, to study matter in the quark gluon phase. The subject has grown immensely in the past ten years and hence it would not be possible for me to discuss all aspects of it in these lectures.

I propose to begin by a brief review of QCD and its implications at high temperature and density. Basically, it will motivate the study of quark gluon plasma. Next, I will discuss the thermodynamic and collective properties of the plasma making use of finite temperature perturbative QCD. Some essential aspects of the method would be described without formal proofs. The difficulties encountered with such an approach would be highlighted and the usefulness of non-perturbative but classical studies would be emphasized. The study of collective properties of the plasma, using classical colour hydrodynamics would be discussed with a focus on non-abelian non-perturbative effects. Following this, the lectures will contain an elementary account of space-time evolution of the plasma in ultrarelativistic heavy ion collisions. Finally, I will also review some of the signatures, that have been proposed, to detect the existence of matter in this state in the laboratory. It should be pointed out that I will not discuss the Monte-Carlo simulation of QCD on the lattice. All the same, I will quote the results of such simulations without providing any details of the calculations.

2. QCD and QCD at Finite Temperature

2.1. A Brief Review of QCD

Quantum Chromodynamics (QCD) is currently the fundamental theory for understanding strong interaction physics. The strongly interacting particles which are observed in the laboratory are the baryons and the mesons and they are known to have more elementary constituents called the quarks. The quarks have spin 1/2, baryon number 1/3, fractional electric charge (2/3e, -1/3e) and come in different flavours (up, down, strange, charm, bottom and top(?)). They also carry a "colour" charge (an internal dynamical variable) such that each quark (antiquark) can be in one of three colour charge states labelled r, b, g, ($\bar{r}, \bar{b}, \bar{g}$). QCD Lagrangian is derived on the assumption of invariance of the system to local unitary transformations in the 3-dimensional internal colour space (local SU(3) symmetry) of the quarks. In such a (non-abelian gauge) theory quarks interact by the exchange of massless spin 1 particles called gluons which themselves carry the colour quantum number (there are 8 colour charge states for the gluon) and interact with each other. The QCD Lagrangian as determined by the symmetry principle of SU(3) local gauge invariance contains one unknown dimensionless parameter, the coupling constant g. The Lagrangian is

$$L_{QCD} = \bar{q}(i\gamma^\mu \partial_\mu - m)q - g(\bar{q}\gamma^\mu T_a q) A_\mu^a - \frac{1}{4}F_{\mu\nu}^a F_a^{\mu\nu} \qquad (2.1)$$

where the spinor quark field q(x) is a column vector with 3 components in colour space, m is the quark mass, A_μ^a ($\mu = 1, \ldots 4$; $a=1, \ldots 8$) are the gluon (gauge) fields with 4 Lorentz components and 8 colour components. T_a ($a=1,\ldots 8$) are the generators of the SU(3) group having the commutation relation

$$[T_a, T_b] = if_{abc}T_c \quad (a,b,c=1,2,\ldots\ldots 8) \qquad (2.2)$$

where f_{abc} are the completely antisymmetric structure constants.

$$F_{\mu\nu}^a = \partial_\mu A_\nu^a - A_\mu^a + gf_{abc}A_\mu^b A_\nu^c \qquad (2.3)$$

is the gluon field tensor. In Eq. (2.1) the first term represents free quarks, the second term is the interaction of quarks with gluons and the last one containing only gluons has the kinetic energy and the cubic and the quartic self interaction terms of the gluons. It is worth emphasizing that the self interaction of coloured gluons is responsible for the difference between Quantum Electrodynamics (QED) and QCD. The two most significant features of this theory are (i) asymptotic freedom, and (ii) confinement.

Asymptotic freedom implies that with increasing momentum transfers or with decreasing distances the interaction between the quarks becomes weaker -- i.e. $g^2(Q^2) \to 0$ as $Q^2 \to \infty$ or $g^2(r) \to 0$ $r \to 0$. The expression for the "running" coupling constant $g^2(Q^2)$ (order 1-loop) in QCD is

$$g^2(Q^2) = \frac{g^2(\mu^2)}{1 + \frac{g^2(\mu^2)}{12\pi}(33-2N_f)\ln(\frac{Q^2}{\mu^2})} \qquad (2.4)$$

Note that for $Q^2 > \mu^2$ and the number of quark flavours $N_f < 16$, $g^2(Q^2) \to 0$ as $Q^2 \to \infty$. It is instructive to contrast this behaviour with the one in QED where the "running" coupling constant is

$$g^2_{QED}(Q^2) = \frac{g^2_{QED}(\mu^2)}{1 - \frac{g^2_{QED}(\mu^2)}{3\pi}\ln(\frac{Q^2}{\mu^2})} \qquad (2.5)$$

Note that $g^2_{QED}(Q^2) \to \infty$ as $Q^2 \to \infty$. This result of QCD (Eq. (2.4)) can explain from a fundamental point of view (instead of the parton model) the deep inelastic scattering of leptons from nucleons where it is observed that the three quarks within a nucleon behave essentially as free particles. Due to the smallness of the ("running") coupling constant at small distances, perturbative calculations provide corrections to the lowest order parton model results which can be verified experimentally. This is a major achievement of the theory. Perturbative QCD has also been applied to the reaction processes $e^+ + e^- \to$ hadrons. More generally, one concludes that short distance phenomena in strong interactions can be studied using perturbative methods. The scale ($\mu = \Lambda_{QCD}$ in Eq. (2.4)) which defines short distance can be determined only by experiments and it is estimated that $\Lambda_{QCD} \sim$ 100-200 MeV. Thus, for $Q^2 >> \Lambda^2_{QCD}$ we have weak coupling. The corresponding length scale is > 1 fm.

While this is very encouraging, it is simultaneously necessary to show theoretically that, a quark (or the colour it carries) remains confined within the hadron, since no free quark has been observed in experiments. The confinement property thus implies that the interaction between the quarks, due to the exchange of gluons, ought to become increasingly attractive with increasing distance. Further, if colour is confined within a hadron then the hadron cannot have any net colour charge -- i.e. it is a colour neutral object (SU(3) singlet). This means that the colour of the quarks (antiquarks and gluons) must combine to give a net colourless hadron. Clearly, in order to theoretically deduce from QCD, confinement (long distance-infrared) effects, one must use non-perturbative methods. It has been demonstrated in lattice QCD calculations (which are non-perturbative) that there is a linearly rising (and hence confining) potential between a (heavy) static quark and an antiquark. At a more phenomenological level a sum of Coulomb and linear potential has been used to successfully describe the charmonium and bottomium spectra. These model calculations and their fit to the data validate, in some respects, the lattice results. Finally, the observation of jets in $e^+ + e^- \to$ hadrons reactions as well as in p+p collisions, and their analysis in terms of flux tubes and Field Feynman (1978) model, provides further support for the idea of confinement. These theoretical and experimental results make one believe in asymptotic freedom and confinement properties of QCD. While they are satisfactory in many respects, one would like to have a much greater physical understanding of the non-abelian gauge theory, including its vacuum, for deeper insights into these features.

2.2. QCD at High Density and Temperature

An important consequence of QCD under extreme conditions of density and temperature is the "existence" of quark-gluon plasma. QCD predicts that nuclear matter at densities a few hundred times the nuclear density (~ 0.15 GeV/fm^3) or at very high temperatures (kT $\sim$ 200 MeV) will not continue to be in the hadronic form, but would become a "soup" of quarks and gluons. The quarks and the gluons are now not confined within the hadronic volume but are free to move in a much larger (macroscopic) volume. Actually, a phase transition from the (confined) hadronic matter to a (deconfined) quark matter (quark-gluon plasma, QCD plasma) is theoretically expected. This is schematically shown in Fig. 2.1.

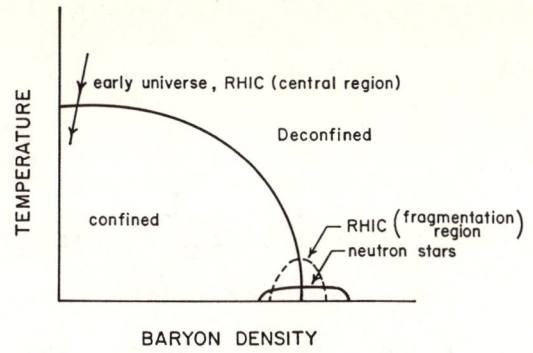
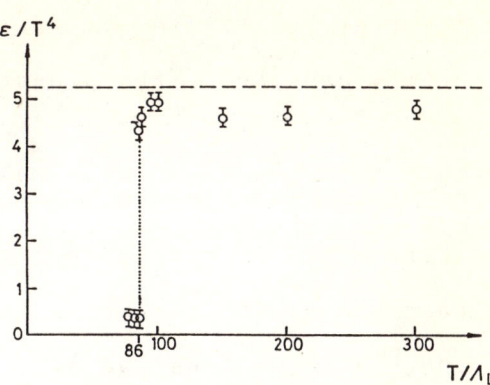

Fig. 2.1. A schematic diagram of the theoretically predicted phase structure of strongly interacting matter

Fig. 2.2. The energy density of SU(3) Yang-Mills matter as a function of temperature. The dashed line shows the corresponding ideal gas value (from ref.4).

Lattice simulations of QCD at finite temperature indicate that there is a first order phase transition in the pure SU(3) gauge thoery (no quarks in the system). The result of one such calculation is shown in Fig. 2.2 (Celik et al 1983). Observe the rapid rise in the internal energy density near $T/\Lambda_L \sim 86$ indicating a first order phase transition. The critical temperature is predicted to be $T_c \sim 200$ MeV. Lattice calculations including the quarks are as yet not completely clear regarding the question of phase transition. It appears that if the quark mass is infinite there is a first order phase transition in 3 flavour QCD. For other values of quark masses the first order transition seems to get weaker.

There are simple thermodynamic and phenomenological arguments which provide some understanding of the phase change. Consider an ideal gas of relativistic pions at temperature $T (T \gg m_\pi)$. If we neglect the mass then the energy density of the pions is:

$$\varepsilon_\pi = g_\pi \times (\frac{\pi^2}{30}) T^4 = 3 \times (\frac{\pi^2}{30}) T^4 \qquad (2.6)$$

where the factor $g_\pi = 3$ represents the 3 charge states of the pions. If there is a deconfinement transition then instead of pions we will have an ideal gas of quarks (q), antiquarks ($\bar{q}$) and gluons (g). The energy density (ε_q) is given by

$$\begin{aligned}\varepsilon_q &= g_q \times (\frac{\pi^2}{30}) T^4 + B \\ &= (2 \times 8 + 7/8 \times 2 \times 2 \times 2 \times 3)(\frac{\pi^2}{30}) T^4 + B \\ &= 37 (\frac{\pi^2}{30}) T^4 + B\end{aligned} \qquad (2.7)$$

On the r.h.s. in Eq. (2.7) g_q is the total number of degrees of freedom for the ($q\bar{q}g$)-system and B is the bag constant which phenomenologically reflects the effects due to the vacuum. Its value is taken as $B^{1/4} \approx 190$ MeV. In g_q the first term is the contribution from gluons - 2 for the spin polarization and 8 for the colour. The second term is the contribution due to (massless) q and $\bar{q}$, where the 3 factors of 2 represent particle-antiparticle, spin and flavour degrees of freedom and the factor 3 is for the colour.

The pressure is related to the internal energy density (when $\mu = 0$) by the equation

$$P = Ts - \varepsilon \qquad (2.8)$$

where s is the entropy density which is given by

$$s = g_\alpha \left(\frac{4\pi^2}{90}\right) T^3 \qquad (2.9)$$

where g_α is either g_π or g_q. Then,

$$P_\pi = g_\pi \left(\frac{\pi^2}{90}\right) T^4 = \frac{1}{3} \epsilon_\pi \qquad (2.10)$$

and

$$P_q = g_q \left(\frac{\pi^2}{90}\right) T^4 - B$$
$$= \frac{1}{3}(\epsilon_q - 4B) \qquad (2.11)$$

Due to the bag pressure term B, the pion gas is the favoured state ($P_\pi > P_q$) at low temperatures, whereas at high temperatures, the deconfined state would be the preferred one ($P_q > P_\pi$). The critical temperature T_c in the model is obtained by equating the pressure in the two phases (Eqs. (2.10) and (2.11) at $T = T_c$. We therefore get

$$B = (g_q - g_\pi)\left(\frac{\pi^2}{90}\right) T_c^4 \qquad (2.12)$$

With the above value of the bag constant B the critical temperature is $T_c \sim 136$ MeV. Also at $T = T_c$ we have

$$\frac{\epsilon_q(T_c)}{\epsilon_\pi(T_c)} = \left(4\frac{g_q}{g_\pi} - 1\right)/3 \qquad (2.13)$$

Equation (2.13) clearly shows that the sudden jump observed at $T = T_c$ in the lattice calculations is a consequence of the increase in the number of degrees of freedom (or entropy) at $T = T_c$.

In view of these considerations, it appears that, in nature, this state of matter may have been the dominant component in early universe, when the temprature $T > T_c \approx$ 200 MeV. It is also likely that, for heavy neutron stars, the density in central core may be high enough so that, matter is again in (cold) QCD plasma phase. In terms of laboratory experiments, there is the tantalizing prospect of producing and detecting this form of matter in ultra relativistic nucleus nucleus collisions. Conservative estimates suggest that, at energies $E/A > 100$ GeV in heavy ion collisions the plasma may be formed. It is very important to carry out these experiments not only for their astrophysical implications but also more fundamentally as a crucial test of QCD.

3. Perturbative QCD at Finite Temperature

3.1. Perturbation Theory at Finite Temperature - General Discussion

The "running" coupling constant in QCD is small for large momentum transfers Q^2 (see Eq. (2.4)) and hence perturbation theory is useful for studying short distance phenomena in QCD. For the high temperature and high density QCD plasma, one again believes that the coupling constant will be small, supporting the application of perturbative methods to study the plasma properties. It ought to be appreciated that, without asymptotic freedom, it would have been very difficult to imagine working out the properties of high density and high T nuclear matter, which is interacting strongly. Note that, near $T = T_c \sim 200$ MeV, the validity of perturbative approach is doubtful, as it is close to the QCD scale parameter Λ_{QCD} = 100-200 MeV.

The basic quantities that one evaluates in perturbative quantum field theory are the Green's functions. The 2-point (Green) function for gauge field operators is defined by the vacuum expectation value

$$iG^{ab}_{\mu\nu}(x,x') = \delta^{ab}<0|T(A^a_\mu(x)A^b_\nu(x'))|0> \qquad (3.1)$$

where the operators $A^a_\mu(x)$ in Heisenberg representation are related to ones in Schrödinger representation by the relation

$$A^a_\mu(x) = e^{iHt} A^a_\mu(\vec{x})e^{-iHt} \qquad (3.2)$$

H is the total Hamiltonian of the system. In Eq. (3.1) T is the time ordering operator so that field operators at earlier times are to the right of those at later times. $G_{\mu\nu}$ is essentially a correlation function between two field operators at different spacetime locations. More generally, one defines an n-point function involving the time-ordered product of n field operators. These are usually evaluated in momentum space to a given order in coupling constant by using Feynman graphs and rules. All physical observables can then be deduced from the Green's function.

At finite temperature, one makes use of the analogy between imaginary time and inverse temperature β, to introduce temperature Green's function

$$G^{ab}_{\mu\nu}(x,\tau,\vec{x}'\tau') = -\delta^{ab}<T_\tau(A^a_\mu(\vec{x}\tau)A^b_\nu(\vec{x}'\tau'))> \qquad (3.3)$$

where

$$A_\mu^a(\vec{x}\tau) = e^{H\tau} A_\mu^a(\vec{x}) e^{-H\tau} \tag{3.4}$$

with τ real and T_τ the τ ordering operator. If $\tau =$ it is pure imaginary, Eq. (3.4) reduces to Eq. (3.2). The bracket $\langle \ \rangle$ on r.h.s. of Eq. (3.3) denotes, thermodynamic expectation value -- i.e. it is a trace over all states in the Hilbert space weighted by the thermodynamic density $\rho_\beta = e^{\beta(\Omega - H + \mu N)}$ where $\beta = 1/T$. In a more explicit way Eq. (3.3) may be written as

$$G_{\mu\nu}^{ab}(\vec{x}\tau,\vec{x}'\tau') = -\delta^{ab} \text{Tr}[\rho_\beta(T_\tau(A_\mu^a(\vec{x}\tau)A_\nu^b(\vec{x}'\tau')))] \tag{3.5}$$

With these definitions, the machinery of perturbation theory developed for T=0 can be applied to systematically evaluate Green's functions at finite T. The crucial difference is that in the T≠0 case, integration over all internal "time" variables τ_i (in Feynman diagram) is restricted to the interval $0 < \tau_i < \beta$, unlike the T=0 case where the range is $-\infty < t_i < \infty$. Furthermore, the range of the variables τ and τ' can also be limited to the interval $0 < \tau, \tau' < \beta$. It is then easy to show that the Green's functions have a periodicity with period β for each "time" variable

$$G_{\mu\nu}^{ab}(\vec{x}\,0, \vec{x}'\tau') = \pm G_{\mu\nu}^{ab}(\vec{x}\beta,\vec{x}'\tau')$$

$$G_{\mu\nu}^{ab}(\vec{x}\tau, \vec{x}'\,0) = \pm G_{\mu\nu}^{ab}(\vec{x}\tau, \vec{x}'\beta) \tag{3.6}$$

where the + and − signs on the r.h.s. are respectively for

bosons and fermions. In momentum space, these properties of the τ variable imply a discrete sum over frequencies so that,

$$\int d\tau \int d^3x \rightarrow \sum_{\omega_n} \int d^3k \qquad (3.7)$$

where

$$\omega_n = \begin{cases} \pm \dfrac{2n\pi}{\beta} & \text{(bosons)} \\ \pm \dfrac{(2n+1)\pi}{\beta} & \text{(fermions)} \end{cases} \qquad n=0,1,2,\ldots \qquad (3.8)$$

These constraints on ω_n follow from Eq. (3.6).

Applying these considerations to QCD plasma, one can evaluate the thermodynamic potential Ω, by calculating appropriate Feynman diagrams and obtain the equilibrium properties of the plasma. In order to determine the collective time-dependent properties, one has to use the linear response theory and relate it to the thermal Green's functions. We shall first discuss the calculation of thermal equilibrium properties and a little later the time dependent ones.

3.2. Thermodynamic Properties of QGP

We have already seen that (Sec.2.2) using the Bag model and an ideal relativistic gas of $q,\bar{q}$ and g one can obtain the internal energy and the pressure for the QGP and the pionic gas. It also allows one to deduce the transition temperature T_c. The lowest order contribution to Ω is the ideal gas value

$$\Omega_{(o)} = -\frac{\pi^2 T^4}{45} [8 + 7/4 \times 3 \times 2] \tag{3.9}$$

for a plasma with 2 flavours for the quarks.

One can include the effect of interactions by evaluating Feynman diagrams order by order. The lowest order correction is of $O(g^2)$ and arises from the two loop graphs. The gluonic terms alone are shown in Fig. 3.1. Combining all the contributions of $O(g^2)$ one obtains (Kapusta 1979)

$$\Omega_{(2)} = \Omega_{(o)} + \frac{g^2 T^4}{144} (8)(3 + 5/4 \times 2) \tag{3.10}$$

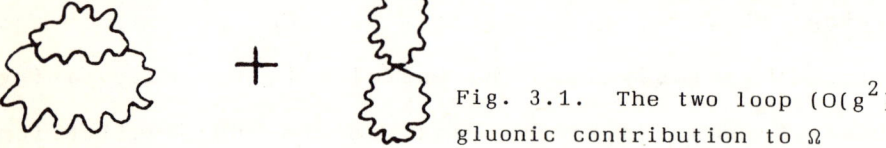

Fig. 3.1. The two loop ($O(g^2)$) gluonic contribution to Ω

It may therefore seem that, if necessary, one can evaluate higher order terms and hence generate a perturbative expansion for Ω in powers of g^2. This naive expectation is not borne out by actual calculations. It was shown by Kapusta (1979) that, already at $O(g^4)$, there are graphs which are infrared divergent, and furthermore at higher orders there are graphs which are increasingly infra red divergent. It is well-known in many-body theory that, one can take the most divergent (leading) term in each order, and take the infinite sum of them to obtain a finite result. This was first done for the electron gas by Gell-Mann and Bruckner and is known as the plasmon effect. Mathematically it is a rather simple idea. Consider terms such as $g^2/x, g^4/x^2, g^6/x^3$

etc. Each term is divergent at x=0 and increasingly so with the power of x in the denominator. However, one can sum the infinite series, and obtain,

$$S_\infty = [\frac{g^2}{x} + \frac{g^4}{x^2} + \frac{g^6}{x^3} + \ldots] = \frac{g^2}{x-g^2} \qquad (3.11)$$

Note that S_∞ is not divergent at x=0, although each term in the series is.

In view of this, one carries out the summation of the leading terms in the diagrams shown in Fig. 3.2 to obtain the next order result for Ω. The calculation (Kapusta 1979) gives

$$\Omega = \Omega_{(2)} - \frac{2g^3 T^4}{3\pi} \cdot (\frac{4}{3})^{3/2} + O(g^4) + \ldots \qquad (3.12)$$

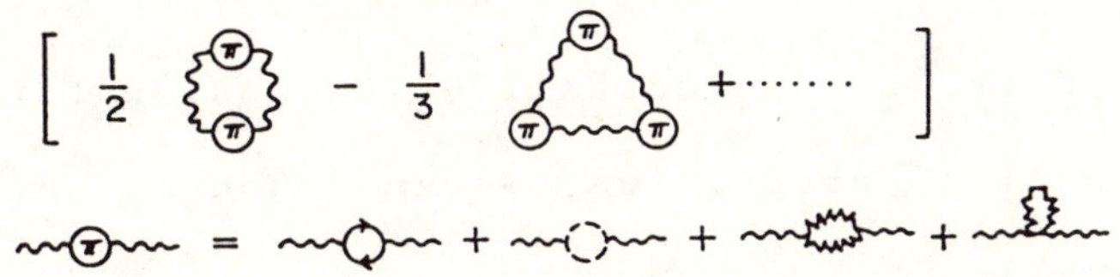

Fig. 3.2. The leading terms (divergent ones) whose summation gives the plasmon contribution to Ω (from ref. 19).

We see that, the correction to $O(g^2)$ term is $O(g^3)$, and it is a non-perturbative contribution (infinite sum of divergent terms which is finite). Consequently, although g is small there are non-perturbative effects which creep in due to the infrared divergence of the theory at high T. In the evaluation of Ω to the $O(g^3)$, shown in Eq. (3.12), this problem is manageable. Going beyond this order a more serious difficulty turns up as pointed out by Linde (1980). The source of the difficulty is again the infrared divergence of the theory. Linde (1980) showed that,at all temperatures a pure Yang Mills theory suffers from infrared divergences, if the gluons remain massless. By considering diagrams of $O(g^{2N})$, containing N four gluon vertices (leading terms only) he showed that the contribution Ω_N to the thermodynamic potential is

$$\Omega_N \sim g^6 T^4 \left(\frac{g^2 T}{m(T)}\right)^{N-3} \qquad (3.13)$$

where m(T) is an infrared cutoff. It is clear that if m(T) = 0 there would be divergences at orders g^8 and higher. In addition, if m(T) ≠ 0 but m(T) < $g^2 T$, then Eq. (3.13) implies that even though Ω_N is finite it will be larger than all Ω_p (p < N). As a result, perturbation theory will break down. Linde concludes therefore that only a cutoff m(T) >> $g^2 T$ will ensure the validity of perturbative expansion for calculating thermodynamic properties, even though the coupling constant is small.

3.3. Thermodynamic Properties from Lattice Calculations

Many calculations with different lattice sizes (N_τ)in the "time"-direction have been done to determine the equilibrium thermal properties of the pure SU(3)gauge theory. The results show scaling behaviour for $N_\tau \gtrsim 10$ and predict a first order transition. The transition temperature T_c and the latent heat $\Delta\epsilon$ have been determined with good precision in units of the string tension parameter σ. The results are (Karsch 1988).

$$T_c/\sqrt{\sigma} = 0.58 \pm 0.04$$
$$\Delta\varepsilon/(T_c)^4 = 3.44 \pm 0.02 \qquad (3.14)$$

Currently many groups are engaged in obtaining the thermal (and other) properties of full QCD--i.e. including the fermionic degrees of freedom. The results seem to depend on the mass of the quarks and the number of quark flavours but there is as yet no general agreement between various groups about the nature of the transition.
As regards the energy density it is very well described by an expression of the type (Karsch 1988)

$$\varepsilon/T^4 = a_0 + a_2 g^2 + 0\,(|g^3|) \qquad (3.15)$$

where the parameters a_0 and a_2 are deduced from calculations on the finite lattice. The results for pure SU(3) gauge theory above T_c are in very good agreement with the ideal gas values corrected to $O(g^2)$. The pressure P on the other hand exhibits significant departures from the ideal gas relation $P = 1/3\varepsilon$. The lattice results can be parameterized by adding a term linear in temperature T to the ideal gas expression. For the pure gauge theory one has, with considerable uncertainty in data

$$P = \frac{\pi^2}{45}(N^2-1)T^4 - cT \qquad (3.16)$$

with $c \approx 0.2\,(N^2-1)T^3$. The inclusion of quarks seems to lead to similar results for ε and P but again more calculations are required to have firm conclusions.

3.4. Linear Response Theory - Study of Colletive Plasma Properties

The collective behaviour of a many-body system is most easily examined within the framework of linear response theory (Fetter and Walecka 1971, Kajantie and Kapusta 1985, Heinz 1986, Heinz et al 1987). Basically, the linear response function $\tilde{\chi}_{ab}(x-x)$ relates the induced current in the medium δj_a^μ to a weak externally applied glue field $A_{\nu E}^b(x)$.

$$\delta j_a^\mu(x) = \int d^4x' \, \tilde{\chi}_{ab}^{\mu\nu}(x-x') A_{\nu E}^b(x')$$

Here μ, ν denote the Lorentz indices and a, b are colour indices. The dielectric polarizability $\chi_{ab}^{\mu\nu}(k_0, \vec{k})$ is the Fourier transform of the response function and is related to the polarization tensor $\pi_{ab}^{\mu\nu}$ by the fluctuation-dissipation theorem (Heinz 1986)

$$\text{Re } \chi_{ab}^{\mu\nu}(k_0, \vec{k}) = -\text{Re } \pi_{ab}^{\mu\nu}(k_0, \vec{k})$$

$$\text{Im } \chi_{ab}^{\mu\nu}(k_0, \vec{k}) = -\tanh(\frac{k_0}{2T}) \, \text{Im } \pi_{ab}^{\mu\nu}(k_0, \vec{k}) \qquad (3.17)$$

These are exact relations. The polarization tensor is easily evaluated in perturbation theory using Feynamn rules. The longitudinal (L) and the transverse (T) collective modes of the system are determined by solving the equations

$$k_0^2 - \vec{k}^2 - \pi_L(k_0, \vec{k}) = 0 \quad \text{and} \quad k_0^2 - \vec{k}^2 - \pi_T(k_0, \vec{k}) = 0 \qquad (3.18)$$

for k_0. The quantities π_L and π_T are obtained from $\pi_{\mu\nu}$ by applying the appropriate longitudinal and transverse projection operators. The solutions of Eq.(3.18)

$$k_0 = \omega - i\Gamma$$

give the spectrum of collective excitations of the system. The real part gives the collective oscillation frequency and the imaginary part Γ the damping rate of the collective mode. In addition, the static ($k_0=0$) long wavelength ($|\vec{k}| \to 0$) limit of the polarization tensors $\pi_\alpha(k_0=0, |\vec{k}|\to 0)$ ($\alpha=L,T$) provide information about the screening of the colour electric and magnetic fields. More precisely, the Debye screening length λ is given by

$$\lambda_E^2 = (\pi_L(k_0=0, \vec{k}\to 0))^{-1}$$

$$\lambda_M^2 = (\pi_T(k_0=0, \vec{k}\to 0))^{-1}$$

(3.19)

If λ_E and λ_M are finite then the colour fields are screened, and if they are infinite the fields remain unscreened.

3.4.1. Screening Properties

In the lowest order (1-loop, $O(g^2)$) perturbative calculation one obtains,

$$\lambda_E^{-2} \equiv m_E^2 = \frac{g^2 T^2}{3}(N + N_f/2)$$

$$\lambda_M^{-2} \equiv m_M^2 \sim \lim_{|\vec{k}|\to 0} \vec{k}^2 \ln \vec{k}^2 = 0$$

(3.20)

where N and N_f are respectively the number of colours and flavours. These results imply that the colour electric fields are screened but the colour magnetic fields are not.

Observe that the magnetic mass m_M^2 (to $O(g^2)$) has the factor $\ln \vec{k}^2$. If in higher orders there is a factor with a stronger infrared singularity (say $\frac{1}{\vec{k}^2}$ then m_M^2 may become finite and the colour magnetic fields would be screened. It seems that the infrared singularity is such as to give rise to a magnetic mass $m_M \sim O(g^2 T)$. It ought to be mentioned that the screening of transverse colour fields is the finite temperature manifestation of the confinement property of QCD at $T = 0$.

The existence of a magnetic mass $m_M \sim O(g^2 T)$ would cure the infrared divergence problem, but as in the case of the thermodynamic potential Ω (sec 3.2), it also leads to the breakdown of perturbative expansion, since the effective coupling constant $g_{eff} \sim g^2 T/m_M \sim O(1)$.

Another indication of the breakdown of perturbation theory is the evalution of Polyakov loop correlation (Nadkarni 1986). The lowest order calculation gives the same screening mass m_E as shown in Eq. (3.20). Including the higher order correction to the leading order term the Polyakov loop correlation in the infra-red limit is (Nadkarni 1986)

$$[\ln <L^\dagger(0)L(R)>]_{R \to \infty} \sim \frac{e^{-2m_E R}}{R^2} [1 - O(g^2 T R \ln R) + \ldots] \qquad (3.21)$$

which diverges even if g is small.

It is however quite reassuring to learn that lattice simulation of QCD, which is necessarily non-perturbative, gives rise to a Debye screening length λ_E of about (1/3 –

1/4) fm (Gavai et al 1988). There is as yet no reliable lattice estimate of λ_M.

3.4.2. Collective Plasma Oscillations

We first examine the longitudinal collective modes of the plasma by solving the equation

$$k_0^2 - \vec{k}^2 - \pi_L(k_0, \vec{k}) = 0 \qquad (3.22)$$

for k_0 to lowest order $O(g^2)$ in the coupling constant. One finds for the plasma oscillation frequency ω_p

$$\omega_p^2 = \frac{(N_f + 2N)}{18} g^2 T^2 \qquad (3.23)$$

Also, by explicit calculations it has been verified that one obtains the same value of ω_p in different gauges and hence the result is gauge invariant. The damping parameter Γ turns out to be less than zero in Feynman gauge. This is quite an unexpected result because it implies that, a longitudinal perturbation in the plasma would grow indicating an instability of the system (Lopez 1985, Lopez et al 1985, Parikh et al 1989). Further, it suggests that the plasma state based on the perturbative vacuum is not the state of the lowest free energy. It was also suggested that the sign of Γ is a consequence of those features of non-abelian gauge theories that give rise to asymptotic freedom.

Extreme caution however has to be exercised before one can conclude that the plasma is unstable. One of the reasons is that the sign of Γ seems to depend on the gauge. The results using linear response theory and temporal axial and Coulomb gauges give for Γ a positive value (Heinz et al 1987). On the other hand calculations done using the background field method (Hansson and Zahed 1987, Kobes and Kunstatter 1988) give a negative value for Γ. None of these are however explicitly gauge invariant calculations. Using a different approach, following the ideas of Cornwall (Cornwall 1982, Cornwall et al 1985), Nadkarni (1988,1989) carried out a manifestly gauge invariant calculation and found a negative value.

$$\Gamma = \frac{-11N}{24\pi} g^2 T \qquad (3.24)$$

Note the factor $-11N$ which provides the connection with asymptotic freedom.

A limitation of all these calculations is that they are done in the lowest order (1-loop) perturbation theory. It is possible that better calculations may alter the sign of making the plasma stable.

Unfortunately, it is not possible at the present time to determine dynamical quantities such ω_p and Γ using MC simulation of QCD on the lattice. We therefore consider classical non-perturbative studies of the plasma in the next chapter.

3.5. Conclusion

It ought to be clear from this review of the evaluation of QCD plasma properties that, although g may be small, perturbation theory is quite inadequate for the purpose, largely due to infrared problems of QCD.

4. Classical Yang-Mills Plasma

In this chapter, we discuss a <u>classical non-perturbative</u> approach to study the collective time dependent properties of the plasma. Classical results have to be viewed as giving the leading term when expanding in powers of $\hbar$. Hence one expects that they provide a reasonable model for the non-abelian non-perturbative features of the quark-gluon plasma.

Normal electromagnetic plasma (electron-ion or electron-positron) has been extensively studied using classical methods. This approach is followed here after making necessary extensions to account for the specific non-abelian features. Clearly, the most important difference between an EM and a YM plasma is that in the latter the particles exchange their colour charges by interacting with colour fields which themselves carry colour charges. As a consequence, the equations of motion for the particles and the gauge potentials are matrix equations in colour space.

The equation of motion of a coloured spin 1/2 particle in an <u>external</u> colour potential A_μ^a can be written as (Wong 1970)

$$i\hbar \frac{\partial \Psi}{\partial t} = H\Psi \qquad (4.1)$$

where

$$H = \gamma^\mu (p_\mu - gA_\mu^a I^a)$$

and I^a (a=1,2,3) are the generators of SU(2). For simplicity we shall consider the colour group SU(2) instead of SU(3). With this H in the Heisenberg representation the usual equations of motion are

$$\frac{dx_i}{dt} = i[H,x_i], \quad \frac{dp_i}{dt} = i[H,p_i] \tag{4.2}$$

As already mentioned, the matrix operators $\vec{I}$ describe the colour dynamics of the system, and hence in the Heisenberg representation one has an additional equation for the dynamical variable $\vec{I}$

$$\frac{d\vec{I}}{dt} = i[H,\vec{I}] \tag{4.3}$$

The classical limit of these equation is obtained by writing the equations for the expectation values. We then obtain

$$\frac{d\vec{p}}{dt} = gI_a(\vec{E}_a + \vec{v} \times \vec{B}_a) \tag{4.4}$$

$$\frac{dI_a}{dt} = -g\varepsilon_{abc}(A_b^0 - \vec{v} \cdot \vec{A}_b)I_c \tag{4.5}$$

In Eq.(4.4) $\vec{E}_a$ and $\vec{B}_a$ are the colour electric and magnetic fields respectively. ε_{abc} are the completely antisymmetric Levi-Civita tensors which are also the structure constants for the group SU(2). Eq. (4.4) is the non-abelian extension of the Lorentz force equation of electrodynamics. Eq. (4.5) is essentially the new ingradient for describing the colour dynamics of quarks.

Besides the particle equations (Eqs. (4.4)-(4.5)) one also has the YM field equations

$$\partial_\mu F_a^{\mu\nu} + g\varepsilon_{abc} A_{\mu b} F_c^{\mu\nu} = j_a^\nu \qquad (4.6)$$

$$F_a^{\mu\nu} = \partial^\mu A_a^\nu - \partial^\nu A_a^\mu + g\varepsilon_{abc} A_b^\mu A_c^\nu \qquad (4.7)$$

Here j_a^ν is the current due to the colour charged particles and is given by (Wong 1970)

$$j_a^\nu(x) = g\int I_a(t)\dot{\xi}^\nu \delta^4(x-\xi(t))dt \qquad (4.8)$$

where $\xi(t)$ is the world line for the single particle.

In order to describe the dynamics of the plasma involving many colour charged particles one may use either the kinetic theory or fluid theory. The latter is simpler and we shall discuss only that approach although the full quantal gauge covariant transport equations for QCD have been derived by Elze et al (1987). The fluid picture for the quarks involves three dynamic variables which are the number density $n(\vec{x}, t)$ the velocity field $\vec{v}(\vec{x}t)$ and the colour charge $I_a(\vec{x}, t)$. The fluid equations consistent with Wongs' (1970) single particle equations are (Kajantie and Montonen 1981, Heinz 1985)

$$\frac{\partial \eta_A}{\partial t} + \vec{\nabla}\cdot(\eta_A \vec{v}_A) = 0 \qquad (4.9)$$

$$\frac{\partial \vec{v}_A}{\partial t} + (\vec{v}_A \cdot \nabla)\vec{v}_A = \frac{g}{m_A} I_{Aa} [\vec{E}_a + \vec{v}_A \times \vec{B}_a] \qquad (4.10)$$

and

$$(\frac{\partial}{\partial t} + \vec{v}_A \cdot \vec{\nabla}) I_{Aa} = -g\varepsilon_{abc}[A_b^0 - \vec{v}_A \cdot \vec{A}_b] I_{Ac} \qquad (4.11)$$

The subscript A in the above equations refers to a given

species of particles. For overall colour neutrality one needs at least two species of particles. Equations (4.9) and (4.10) are the familiar continuity and force equations whereas Eq(4.11) represents the colour dynamics. In terms of these fluid variables the particle charge and current densities are given by

$$j_a^o = g \sum_A \eta_A \, I_{Aa} \qquad (4.12)$$

$$\vec{j}_a = g \sum_A \eta_A \, \vec{v}_A \, I_{Aa} \qquad (4.13)$$

The objective now is to solve the field equations (4.6) and (4.7) self consistently with the fluid equations (4.9)-(4.11) where the four current is given by Eqs.(4.12)-(4.13). These are coupled non-linear partial differential equations which are very difficult to solve in their full generality. We therefore look for some special solutions.

The longitudinal collective plasma oscillations within this framework were studied by Bhatt et al (1989) after making some simplifying assumptions. The crucial point about these approximations was that the full non-abelian non-linear dynamics was retained and neither the coupling constant g nor the wave amplitude was assumed to be small.

The basic assumption (Bhatt et al 1989) was that the gauge potentials A^μ were a function of a single variable $\xi = z + \beta t$ where β is a parameter. Also, purely longitudinal solutions were looked for so that the coupling to transverse modes was neglected. This decoupling of longitudinal modes from the transverse ones is valid when the scale length for the latter is much longer than that for the former. Further the mass of one of the species was taken to be much larger than the that for the other specie. Finally, the non-linearities arising from the fluid equations were neglected. We also choose to work in the $A^o = 0$ gauge. It is worth stressing again that in the process all the non-abelian terms have been kept.

With these approximations the fluid equations could be solved so that the variables η_A, $\vec{v}_A$ and I_A could be expressed in terms of the equilibrium values (having subscript 0) for these quantities and the field amplitude A_a^3. Using the notation $a_a = A_a^3$ we obtain the equations

$$a_a'' = -\frac{g^2 \eta_0}{m_1 \beta^2} I_{1ao}(I_{1bo} a_b) - \frac{g^2}{m_1 \beta}(\varepsilon_{abc} a_b a_c')(I_{1do} a_d) \quad (4.14)$$

$$(a=1,2,3)$$

where a sum over repeated indices is implied.

Next, we choose for simplicity $I_{110} = I_{120} = I_{130} = I_0$ and introduce new scaled variables $a_a^* = a_0 a_a$, $\eta = \frac{\omega_p}{\beta}\xi$ where $\omega_p = \frac{g^2 \eta_0 I_0^2}{m_1}$. In terms of the linear combinations $x_1 = a_1^* + a_2^* + a_3^*$, $x_2 = \sqrt{3/2}(a_1^* - a_3^*)$ and $x_3 = \sqrt{1/2}(a_1^* - 2a_2^* + a_3^*)$. Eqs (4.14) reduce to

$$\ddot{x}_1 = -3x_1 + \frac{\varepsilon}{\sqrt{3}}(x_2 \dot{x}_3 - x_3 \dot{x}_2)x_1$$

$$\ddot{x}_2 = \frac{\varepsilon}{\sqrt{3}}(x_3 \dot{x}_1 - x_1 \dot{x}_3)x_1$$

$$\ddot{x}_3 = \frac{\varepsilon}{\sqrt{3}}(x_1 \dot{x}_2 - x_2 \dot{x}_1)x_1$$

$$(4.15)$$

where

$$\varepsilon = \frac{g^2 I_o^2 a_o^2}{m_1 \omega_p}$$

ω_p is the usual abelian plasma oscillation frequency. The non-abelian physics is contained in the non-linear terms in Eqs(4.15) with the strength parameter ε. When $\varepsilon=0$, we obtain only the abelian oscillation. For $\varepsilon \neq 0$ the Eqs.(4.15) have been numerically integrated. The results are shown in Figs. 4.1 - 4.4 (Bhatt et al 1989).

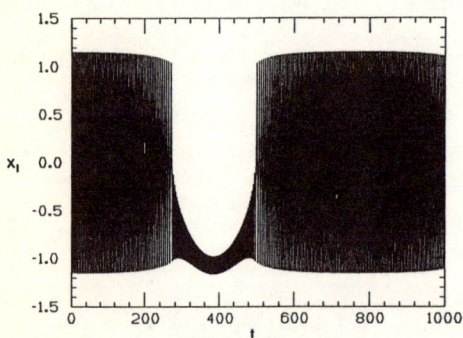

Fig. 4.1. Oscillations of the field variable x_1 for $\frac{\varepsilon}{\sqrt{3}} = 0.05$ and initial conditions $x_1=x_2=x_3=0$ and $x_1'=2$, $x_2'=1$, $x_3'=3$ (from ref.1)

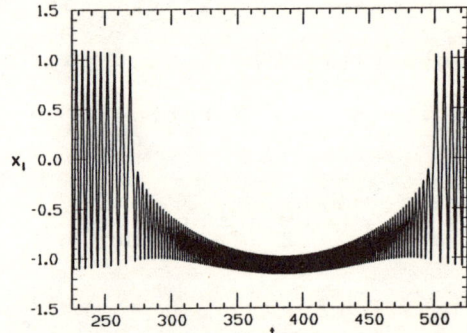

Fig. 4.2. Same as Fig. 4.1 with the scale for the variable t expanded between t=200 and t=550 (from ref. 1)

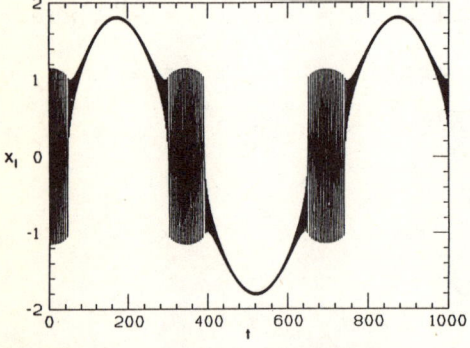

Fig. 4.3. Oscillations of x_1. $\frac{\varepsilon}{\sqrt{3}} = 0.05$ and initial conditions $x_1=x_2=x_3=0$ and $x_1'=2$, $x_2'=0.1$, $x_3' = 0.3$ (from ref. 1)

Fig. 4.4. Oscillations of x_1. $\frac{\varepsilon}{\sqrt{3}} = 0.5$ and initial conditions $x_1=x_2=x_3=0$ and $x_1'=2$, $x_2'=0.1$, $x_3'=0.3$ (from ref. 1).

We observe that for small values of ε there is the well-known abelian oscillation followed by a non-abelian mode. The frequency of the non-abelian mode is higher than that (ω_p) of the abelian mode. This is seen in Figs. 4.1 – 4.3. The non-abelian mode can be viewed as some kind of precession in colour space since Eqs.(4.15) can be cast in a form resembling Euler's equations for rigid body rotation. When the value of ε is increased by an order of magnitude, the nice periodic structure disappears and one instead observes intermittency or chaotic behaviour.

The implications of these results for the QCD plasma are far reaching. We first note that the amplitudes of oscillations are finite in all cases i.e at least at the classical level there seem to be no growing perturbations. As there is a distinct non-abelian oscillation for small values of ε, it would be of interest to explore the consequences of the two modes for the plasma produced in the laboratory. If conditions are favourable (ε large) so that the plasma is chaotic, then the energy in a given collective mode would very quickly get shared with other degrees of freedom, and push the system towards thermal equilibrium. All these questions need to be studied.

Although various simplification were made to solve the coupled field and particle equations, the results exhibit a great deal of richness due to the non-abelian nature of the system. We expect that relaxation of many of the approximations would lead to additional new features and hence it is worth pursuing such classical studies in greater depth. It is also important to examine quantum corrections to the classical results in particular those connected with the stability (see chapter 3) of the plasma.

5. Space Time Evolution of Matter in Ultra Relativistic Heavy Ion Collisions

In the previous two sections, the equilibrium properties of the QCD plasma were examined. Since one expects to study the plasma phase and its properties in nucleus-nucleus collisions, it is vital that one learns enough about the various stages of the evolution of the system, to answer questions such as:

1. At what energies is the plasma likely to be produced?

2. If it is produced, how does it evolve in time?

3. Does it reach thermodynamic equilibrium (even locally)?

4. How does the plasma hadronize?

5. What are the signatures that reveal the presence of the plasma?

Although one hopes that experiments would eventually give unambiguous answers to these questions, it is crucial to develop theoretical scenarios which provide some understanding of the phenomena as also guidelines for the experiments.

The broad picture that has emerged is based on macroscopic physics, involving ideas from kinetic theory, thermodynamics and hydrodynamics. In the present situation, where one has collisions of two finite size objects which break up, it is useful to develop a space time picture of the process.

Consider the collision of two heavy nuclei such as U+U or Pb+Pb at ultra relativistic energies in the centre of mass frame. The two nuclei approach each other with nearly the speed of light, are Lorentz contracted along their directions of motion (z-axis), and finally collide at say z=0 and time t=0. It should be pointed out that, the thickness of the nucleus does not tend to zero at higher and higher energies - in fact there is a limiting thickness which is estimated to be about 1 fm. This is because a nucleon in any frame has very low momentum constituents (wee partons) which are essentially unaffected by Lorentz transformations and hence give a minimum length scale (thickness) of ~1fm for the nucleus.

It was suggested by Fermi and Landau that, on colliding the nuclei would stop each other, thermalize quickly and give rise to a very high energy density in the collision region. Such a fireball type picture is probably reasonable at relatively low energies. However, at very high energies results of p-p and p-A collisions suggest a surprisingly different picture. Actually, the data on p-A collisions indicate that, the proton interacts just once with a nucleon while traversing the nucleus, thereby suggesting "transparency" of the nucleus. In nucleus-nucleus collisions at very high energies we therefore expect a similar behaviour. The partons that are created after the first collision can interact again only after a proper time τ_o which is estimated to be τ_o = 1 fm/C. This would not be possible for the fast particles as they would pass through the nuclear matter in time τ_o. This is the case for the baryonic component (quarks) of the target and the projectile nuclei. They remain largely unaffected by the collision and move away from the collision region ($z \simeq 0$) with high velocity ($\lesssim$ c). The slowly moving partons can however interact again within the nucleus and eventually produce a shower. Such a model for particle production is known as the inside outside cascade model.

A qualitative and physically very appealing pictorial "explanation" for some of these features has been suggested by Carruthers (1983) in term of quarks and pre-existing glue in a nucleon. He argues on the basis of relative strengths of the q-q, q-g and g-g interactions (16:36:81) that if one ignores all interactions except g-g then in high energy p-p scattering the following scenario develops: (i) the quarks do not undergo any collision, leading to "transparency" for the baryonic content of the hadronic matter, (ii) the gluonic constituents do interact and eventually thermalize supporting in a modified form the Fermi Landau mechanism, (iii) for the leading baryonic

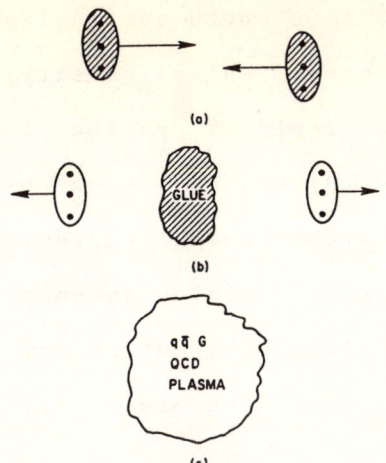

Fig. 5.1. Central collision of two protons in c.m. frame as discussed by Carruthers (ref. 3)

matter which is stripped of its "glue" in the collision and for the "gluonic" matter in the central region it would take a proper time $\tau_o \sim$ 1fm/c (with QCD scale parameter $\Lambda_{QCD} \tilde{} $ 200 MeV) to get "dressed" again. Pictorially this is represented in Fig. 5.1.

Extending these ideas to nucleus-nucleus collisions one can distinguish the central region where statistical hydrodynamical ideas may be applicable from the leading particle region where the baryonic matter is present. In the latter regions the momenta of particles are close to those of either the target or projectile nucleus. It is believed that the leading particles take away about half of the kinetic energy leaving the remaining half (in the central region) for conversion to internal energy. Although both these regions have been studied, there has been a greater emphasis on the central region, following the ideas developed by Van Hove and later by Bjorken (1983). We shall therefore discuss the evolution of this central region only.

In view of these considerations, it is worth estimating initial values of some critical parameters (energy density, mean free path) in the central rapidity region in ultrarelativistic heavy ion collisions. By taking account of the observed particle production in the central rapidity region in p-p collisions at SPS collider energies, assuming it to be the same for p-A collisions ("transparency" effect) and adding over all nucleons in the projectile, Bjorken estimated the energy density to be $\varepsilon_{initial}$ ~ 1-10 GEV/fm^3. It is believed that $\varepsilon > 2-3$ GeV/fm^3 would produce the deconfined plasma phase. Further, he assumed a mean energy/quantum ~ 400 MeV, and slab geometry, to obtain for the initial density of quanta $\rho_{initial}$ ~ 2-20/fm. This gives a mean free path λ_o ~ (1.0 - 0.1) fm. Comparing it with the transverse size (R_A) ~ 7fm of a nucleus (A ~ 200) Bjorken (1983) concluded that local thermodynamic equilibrium would be established and it would be reasonable to treat the plasma phase in the central rapidity region using relativistic hydrodynamics.

It appears therefore that for time scales of the order of 5-10 fm/c after the initial "dead" period of τ_o ~ 1 fm/c, the evolution would be goverened by hydrodynamics in the longitudinal direction and equilibrium thermodynamics in the transverse directions.

An extremely important and new element was added by Bjorken (1983) to this hydrodynamic picture. This was based on the experimental evidence that, at very high energies, the multiplicity of particles produced in p-p and p-A collisions as a function of the rapidity y has a central plateau structure. On the basis of this evidence he assumed the same for nucleus-nucleus (A-A) collisions. This implies that in all "centre of mass type" reference frames the hydrodynamical evolution should look the same -- i.e. it ought to be Lorentz invariant. Finally, for time intervals less than the transverse time-scale R_A/c, he assumed translational symmetry in the transverse direction, so that, the evolution is in longitudinal (1 space) and time directions only.

Basically then, Landau hydrodynamics is applied to the QCD plasma (with net baryon number zero) in the central rapidity region with the "boost" invariant initial conditions. The system is then described by the local energy density $\varepsilon(x)$, the local pressure $p(x)$, the local temperature $T(x)$ and the fluid 4-velocity $U_\mu(x) = \gamma(1, v_z, v_r, v_\phi = 0)$ written in a cylindrical coordinate system where $v_\phi = 0$ because of azimuthal symmetry and $U_\mu U^\mu = 1$. In the (1+1) dimensional approximation ($v_r \ll v_z$) we choose the velocity profile $v_z(z,t) = z/t$ where t is the time elapsed since the collision. This is obviously the case soon after the collision (no interactions) but it also remains nearly the same for later times in the (1+1) dimensional approximation (equilibrium is due to collisions in transvrse direction). Also, with the Lorentz invariant boundary conditions it is more appropriate to use variables which are combinations of z and t. These are taken to be the proper time τ and the space-time rapidity y (which is the same as the momentum space rapidity of the fluid)

$$\tau = \sqrt{t^2 - z^2} \tag{5.1}$$

and

$$y = \frac{1}{2} \ln \frac{t+z}{t-z} \tag{5.2}$$

Note that under Lorentz transformation τ is invariant while y shifts by a constant.

We next write down the general hydrodynamic equations. The energy momentum tensor $T_{\mu\nu}$ for a perfect fluid is,

$$T_{\mu\nu} = (\varepsilon + p) U_\mu U_\nu - g_{\mu\nu} p \tag{5.3}$$

with the conservation law,

$$\frac{\partial T_{\mu\nu}}{\partial x_\mu} = 0 \tag{5.4}$$

In addition, we have the conservation law for the baryon number, which can be written in terms of the baryon current,

$$J_{\mu b}(x) = \eta_B(x) U_\mu(x) \tag{5.5}$$

whose four divergence is zero -- i.e.

$$\partial_\mu J_b^\mu(x) = 0 \tag{5.6}$$

In the (1+1)-dimensional case, the variables will be $\varepsilon(\tau, y)$, $p(\tau, y)$ $U_\mu(\tau, y)$ with the boundary condition that at $\tau = \tau_o$ $\varepsilon(\tau_o, y) = \varepsilon_o$ = constant and

$$U_\mu(\tau_o, y) = \frac{1}{\tau_o} (t, z, o, o) \tag{5.7}$$

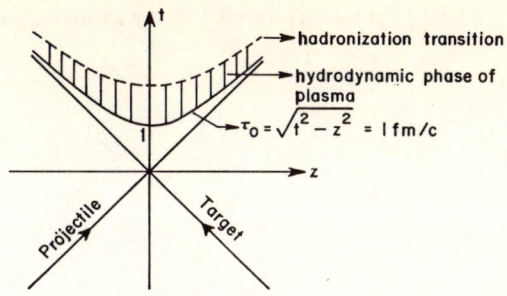

Fig. 5.2. Central collision of two nuclei, and the evolution of the system in Bjorken model

The evolution can be pictorially visualized in the 1-dimensional hydrodynamics of Bjorken as shown in Fig. 5.2.

The hydrodynamic equation (Eq. (5.4)) reduces in this simple case to

$$\frac{d\varepsilon}{d\tau} = -(\varepsilon+p)/\tau \tag{5.8}$$

These equations have to be supplemented by an equation of state for the matter and we will choose the form

$$p = c_s^2 \varepsilon \tag{5.9}$$

where c_s^2 is the speed of sound and for the ideal relativistic fluid $c_s^2 = 1/3$. By making use of this relation we can integrate Eq. (5.8) to obtain

$$\varepsilon(\tau) = \varepsilon(\tau_0) \left(\frac{\tau_0}{\tau}\right)^{c_s^2+1} \tag{5.10}$$

Also the entropy density s (from thermodynamics) is given by

$$s = \frac{1}{T}(\varepsilon+p) \tag{5.11}$$

The entropy four current $s_\mu = sU_\mu$ satisfies the continuity equation

$$\frac{\partial s_\mu}{\partial x_\mu} = 0 \tag{5.12}$$

This gives for our simple (1+1)-dimensional system the equation,

$$\frac{ds}{d\tau} = -\frac{s}{\tau} \qquad (5.13)$$

so that

$$s(\tau) = s(\tau_0) \left(\frac{\tau_0}{\tau}\right) \qquad (5.14)$$

Further it can be shown that

$$\frac{d}{d\tau}\left[\frac{dS}{dy}\right] = 0 \qquad (5.15)$$

implying thereby that the entropy/rapidity is a constant quantity. This enables one to estimate the particle multiplicity. Similarly one can also obtain for the temperature

$$T(\tau) = T(\tau_0) \left(\frac{\tau_0}{\tau}\right)^{c_s^2} \qquad (5.16)$$

We see therefore that these simple considerations make it possible to solve the hydrodynamic equations analytically and obtain the time evolution of the system. This simple behaviour will be applicable so long as one can neglect the transverse motion. The more general 3 dimensional hydrodynamics has also been studied (numerically) but we shall not discuss it. One can also consider a better equation of state and include dissipative effects.

6. Signals for Quark Gluon Plasma in Ultrarelativistic Nucleus-Nucleus Collisions

Introduction

In the previous section, we described theoretical scenarios of formation and evolution of the QGP following collisions of heavy ions at ultrarelativistic energies. Assuming that the plasma is formed in such collisions, and that it survives in this phase for a time interval of 5-10 fm/c, how does one establish from experiments, that there has been a phase transition. This is a difficult proposition because, only hadrons, leptons and photons are detected in the laboratory, and the objective is to distinguish between the possiblities

(i) nucleus + nucleus → QGP → hadrons, leptons, photons

(ii) nucleus + nucleus → excited nuclear matter → hadrons, leptons, photons

The fact that the region of formation is not macroscopic and the phase is short lived makes the task even more complicated.

In spite of these difficulties, a large number of suggestions have been made that would characterize the formation of the QGP. These include

(i)   strangeness enhancement

(ii) photon emission from QGP

(iii) dilepton emission from QGP

(iv) suppression of flavour anticorrelations

(v)   suppression of J/ψ production

and perhaps many more. The first three suggestions are essentially based on thermodynamics and evolution of the plasma but do not in any significant way depend on collective properties of the plasma. The last two suggestons, on the other hand, are directly a consequence of the screening of colour electric fields (collective behaviour) in the plasma. We briefly discuss these signals.

It is worth mentioning at the outset that almost all the suggestions, on deeper analysis, have turned out to depend on details of the evolution and hadronization of the system whose understanding is still rather primitive. In some cases (e.g. J/ψ suppression) alternative explanations have been suggested. Consequently, most of the early promise of providing a clean signal has disappeared. It is therefore extremely important to discover better ways of unambiguously detecting the formation of plasma in heavy ion collisions.

6.1. Strangeness Enhancement

This is one of the earliest proposals for detecting the formation of the plasma. The idea is that, in QGP at sufficiently high temperatures, the probability of producing strange mesons and antibaryons, would be enhanced relative to a hadron gas. This is because in the baryon rich region the production of light antiquarks ($\bar{u}$, $\bar{d}$) would be suppressed due to the chemical potential μ.

$$n(\bar{u}) = 6\int \frac{d^3k}{(2\pi)^3} \frac{1}{\exp(k+\mu)/T + 1}$$

Thus there will be a suppression of $u\bar{u}$ and $d\bar{d}$ pairs but not of $s\bar{s}$ pairs provided $\mu > m_s$ = mass of strange quark. Hence it was suggested that if the QGP is formed, then on hadronization one would observe enhanced K/π and $\bar{\Lambda}/\bar{N}$ ratio.

These naive considerations were further explored taking account of entropy of the system. It was then argued that a more appropriate measure for the strangeness of a phase is n_s/S where n_s is the density of strange particles and S is the entropy. If this is evaluated along the phase coexistence surface it is found that

$$(\frac{n_s}{s})_{QGP} > (\frac{n_s}{s})_{hadron\ gas} \quad (large\ S/A) \tag{6.1}$$

for large S/A values (similar to central region), whereas in the fragmentation regions, where the baryon density is high

$$(\frac{n_s}{s})_{QGP} < (\frac{n_s}{s})_{hadron\ gas} \quad (small\ S/A) \tag{6.2}$$

This is shown in Fig. (6.1) due to Heinz (1988). We observe that the enhancement in the central rapidity regions (large S/A) is about 20%, which in view of the various uncertainties in the calculations imply that the ratio n_K/n_π cannot be taken as a serious candidate for signalling QGP.

All the same, the density ratio $n_{\bar{s}}/n_{\bar{q}}$ of strange to light antiquarks shows a large value for QGP relative to

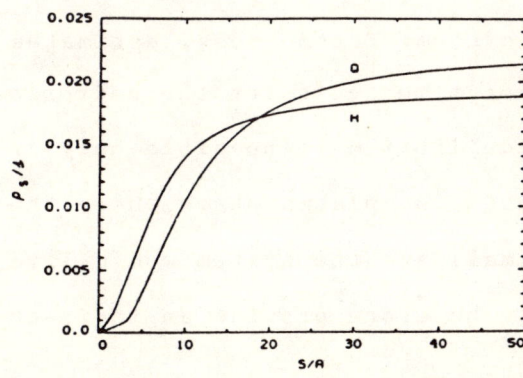

Fig. 6.1. Strangness contents in a hadron gas (H) and a quark-gluon plasma at the critical surface as a function of S/A (from ref. 15)

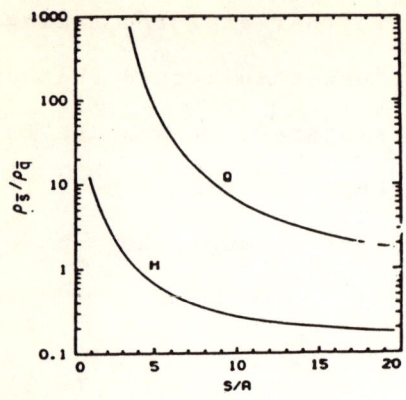

Fig. 6.2. The ratio of strange to light antiquarks in the hadronic (H) and quark-gluon plasma phase (Q) along the phase transition, as a function of S/A (from ref. 15)

the hadron gas. This is shown in Fig. 6.2 (Heinz 1988) as a function of S/A and is in concurrence with the naive expectation.

It may seem therefore that an enhancement in $n_{\bar{\Lambda}}/n_{\bar{p}}$ may provide a reliable signature for the QGP. Unfortunately, a more careful analysis shows that, this effect also loses its dramatic character so that the ratio $n_{\bar{\Lambda}}/n_{\bar{p}}$ will be of the same order in QGP and chemically equilibrated hadron gas. The reason for this is the mechanism by which gluons disappear from QGP while converting their entropy into $q\bar{q}$ and $s\bar{s}$ pairs.

The only bright aspect, of observing $n_{\bar{\Lambda}}/n_{\bar{p}}$ ratio at the equilibrium value of hadronic gas, is that it will signal chemical equilibration of the system. Furthermore, estimates indicate that the equilibration time scale for the hadronic gas is very large. Therefore, equilibrium is possible only if the system has passed through the plasma phase, where the equilibration time scale is small and the system would live sufficiently long. This may therefore provide an indirect signal for the formation of QGP.

6.2. Photons and Dileptons from QGP

Photons and dileptons do not interact strongly. It was therefore argued that they may be able to provide signals for the early stages (following the collision) of the evolution of the system -- i.e. before hadronization takes place.

It is important in such studies to find out the nature of emissions at different stages in the evolution of the system. Considering the scenario for the evolution of the system there is always (i) direct emission, (ii) emission in the pre-equilibrium stage (iii) emission from the equilibrated plasma, and (iv) emission from the hadronic products of the plasma. In view of this it is crucial to determine domains where one or the other mechanism is dominant. Clearly, for such an analysis one has to know a great deal about the evolution at each stage.

As an illustration of the kind of theoretical calculation that is useful, we show some results of Kajantie (1987) for the dilepton emission rate as a function of the dilepton mass. For this range of mass values we observe that the thermal emission is dominant.

There is more recently a suggestion (Sinha 1988) that the ratio of $(\gamma\gamma/\mu^+\mu^-)$ emissions may also provide a signal for the formation of QGP. However, it appears that suitable windows would have to be selected to obtain an unambiguous signal of the plasma.

6.3. Suppression of Flavour Anticorrelations

If QCD plasma is formed in relativistic nuclear collisions then, just as for the normal (electron-ion) plasma, the electric fields will be screened. In thermal equilibrium, perturbative QCD results suggest that the Debye length $\lambda_D \sim O((gT)^{-1})$ and give a value of $\lambda_D \simeq 0.4$ fm at $T_C = 200$ MeV with 2 flavours of quarks. Lattice QCD calculations give a smaller value, $\lambda_D \sim 0.25 - 0.3$ fm. The screening property will influence the hadronization of the plasma in a significant way because it implies a basic difference between p+p and A+A collisions. We note that, in the former case, one observes jets, due to the creation of strong colour electric field in the hard scattering of partons. On the other hand, in the latter case, due to the screening, the colour electric fields in the plasma would be weak, and hence one would not expect jet like processes.

It was predicted by Field and Feynman (1978) that, in jets, there would be charge (and more generally flavour) anticorrelations between mesons of similar rapidity. This is most easily viewed in the flux tube model which requires strong colour electric fields. This is shown in Fig. 6.4. In nuclear collisions, with plasma formation the jet production will be suppressed, which in turn gives rise to suppression of charge (and flavour) anticorrelation for the plasma (Lopez et al 1984).

For a quantitative estimate of the suppression, one ought to determine for nucleus-nucleus collisions, the

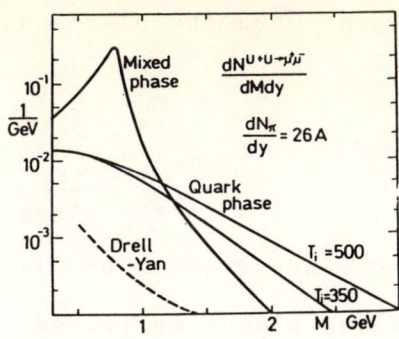

Fig. 6.3. Mass distribution of dileptons at y=0 (from ref. 18)

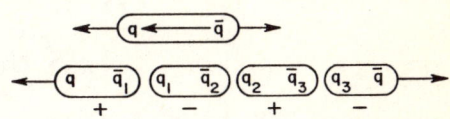

Fig. 6.4. $q\bar{q}$ creation from vacuum in strong colour electric field leading to charge anticorrelation

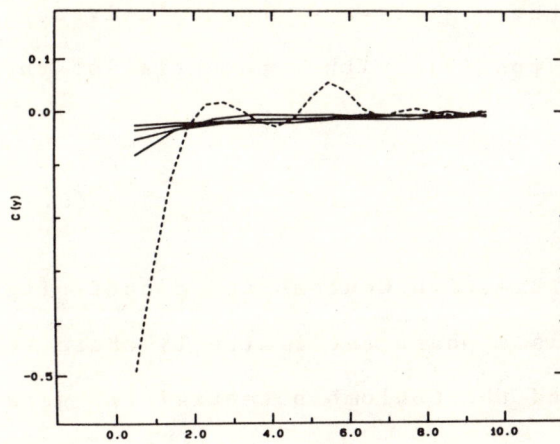

Fig. 6.5. Charge anticorrelation in the plasma (solid lines) as a function of rapidity y (from ref. 10)

charge correlation coefficient for pions of similar rapidity in the cases (a) when the plasma is not produced, and (b) when it is produced. The former can be obtained using one of the computer based event generators - in a sense, this is the background. For the latter, one would have to rely on some kind of kinetic and hydrodynamic evolution equations. Such detailed studies have not yet been carried out, but there is a simple statistical and hydrodynamical evaluation (Gupta and Parikh 1989) of the charge correlation $c(y)$ which is shown in Fig. 6.5. It shows that the charge correlations are washed out when the plasma is formed.

6.4. J/ψ Suppression in QCD Plasma

The screening of colour electric fields in the plasma leads to another signal. This is the suppression of J/ψ production (Matsui and Satz 1986) if the plasma is formed. A heavy meson (such as J/ψ) would be produced in hard scattering process during nuclear collisions. It is a $3S_1$, charm-anticharm ($c\bar{c}$) bound state. A non-relativistic treatment of $c\bar{c}$ spectra is possible with a potential of the form

$$V(r) = -\frac{\alpha}{r} + \sigma r \qquad (6.3)$$

which is a sum of an attractive Coulomb and a confining linear potential. In the plasma phase, (at finite T) there is no confining potential and the Coulomb potential $1/r$ gets screened -- i.e.

$$V(r) = -\frac{\alpha'}{r} e^{-r/\lambda_D} \qquad (6.4)$$

The crucial point is that, if the "Bohr radius" of the $c\bar{c}$ pair in the plasma is larger than the screening length λ_D, then the J/ψ meson will disintegrate in the plasma -- i.e. the $c\bar{c}$ bound state will not exist. More precisely, it is found that the condition for dissolution is

$$\lambda_D < 0.84 \, r_B \qquad (6.4)$$

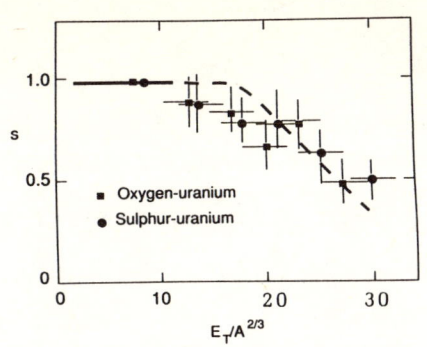
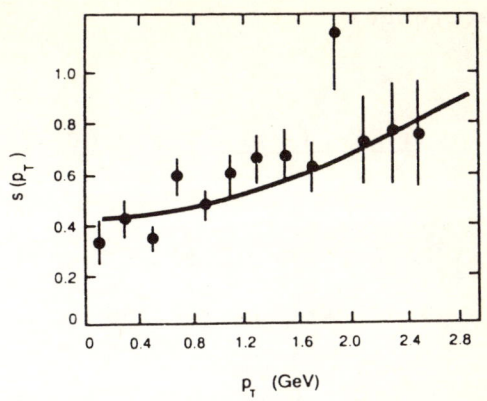

Fig. 6.6. Survival probability for J/ψ versus $E_T/A^{2/3}$ for $P_T=0$ (from ref. 21)

Fig. 6.7. Survival probability for J/ψ versus P_T for events with $(E_T/A^{2/3}) \geq 14.2$ GeV divided by those with $(E_T/A^{2/3}) \leq 8$ GeV (from ref. 21)

and since λ_D depends on temperature the dissolution temprature is estimated to be $T_{dis} \sim 210$ MeV.

Quantitative calculations including the plasma evolution have been performed and compared with data (see Figs. 6.6 and 6.7) (Karsch 1989). The data exhibits for high E_T and low p_T the suppression which was predicted theoretically and the model calculations appear to fit the data. There are, however, alternative explanations for the suppression of J/ψ production in heavy ion collisions which do not require the formation of the plasma. Unfortunately both types of models have shortcomings and hence it is not yet clear if the observed suppression is a signal for the plasma formation.

References

Quantum Field Theory and QCD

1. Ramond P., Field Theory, A Modern Primer, Benjamin, Reading (Mass),1981.

2. Quigg C., Gauge Theories of the Strong, Weak and Electromagnetic Interaction, Benjamin/Cummings,Reading (Mass) 1983.

Quark Gluon Plasma

Reviews

1. Shuryak E.V.,1980,Phys. Rep. $\underline{61}$,71.

2. Gross D.J.,Pisarski R.D. and Yaffe L.G.,1981,Rev. Mod. Phys. $\underline{53}$,43.

3. Cleymans J., Gavai R.V. and Suhonen E.,1986,Phys. Rep. $\underline{130}$,217.

Conference Proceedings

1. Quark Matter Formation and Heavy Ion Collisions,M. Jacob and H. Satz (editors),World Scientific,Singapore (1982).

2. Quark Matter '83,T.W. Ludlam and M.E. Wegner (editors), Nuclear Physics $\underline{A418}$ (1984).

3. Quark Matter '84, K. Kajantie (editor), Lecture Notes in Physics 221, Springer Verlag (Berlin) 1985.

4. Quark Matter '86, L.S. Schroeder and M. Gyulassy (editors), Nuclear Physics A461, 1987.

5. Quark Matter '87, H. Satz, H.J. Specht and R. Stock (editors) Z. Phys. C 38, 1988.

6. Physics and Astrophysics of Quark Gluon Plasma, B. Sinha and S. Raha (editors), World Scientific (Singapore), 1988.

References to Specific Papers in the Text

1. Bhatt J.A., Kaw P.K. and Parikh J.C. (1989) Phys. Rev. D39, 646.

2. Bjorken J.D. (1983) Phys. Rev. D27, 140.

3. Carruthers P. and Minh Duong-Van (1983) Phys. Rev. D28, 130.

4. Celik T., Engels J. and Satz H. (1983) Phys. Lett. B129, 323.

5. Cornwall J.M. (1982) Phys. Rev. D26, 1453.

6. Elze H.T., Gyulassy M. and Vasak D. (1986) Phys. Lett. B177, 402; Nuclear Physics B276, 706.

7. Fetter A.L. and Walecka J.D. (1971) "Quantum Theory of Many-Particle Systems",McGraw-Hill (New York).

8. Field R.D. and Feynman R.P. (1978) Nuclear Physics B136,1.

9. Gavai R.V., Lev M., Petersson B. and Satz H. (1988) Phys. Lett. B203,295.

10. Gupta S. and Parikh J.C. (1989) Phys. Lett. B219,354.

11. Hansson T.H. and Zahed I. (1987) Phys. Rev. Lett. 58, 2397;Nuclear Phys. B292,725.

12. Heinz U. (1985) Ann. Phys. 161,48.

13. Heinz U. (1986) Ann. Phys. 168,148.

14. Heinz U.,Kajantie K. and Toimela T. (1987),Phys. Lett. B183,96;Ann. Phys. 176,218.

15. Heinz U. (1988) in "Physics and Astrophysics of Quark-Gluon Plasma", B. Sinha and S. Raha (editors), World Scientific (Singapore),pp. 463-474 (1988).

16. Kajantie K. and Montonen C. (1981),Phys. Scr. 22,555.

17. Kajantie K. and Kapusta J. (1985) Ann. Phys. 160,477.

18. Kajantie K. (1987) in "Quark Matter 1986", L.S. Schroeder and M. Gyulassy (editors),Nuclear Physics A461,pp. 225C-238C.

19. Kapusta J.I. (1979) Nuclear Physics B418,461.

20. Karsch F. (1988) in "Quark Matter 1987",H. Satz,H.J. Specht and R. Stock (editors),Z. Phys. C 38, pp.147-155.

21. Karsch F. (1989) Particle World 1,24.

22. Kobes R. and Kunstatter G. (1988) Phys. Rev. Lett. 61, 392.

23. Linde A.D. (1980) Phys. Lett. B96,289.

24. Lopez J.A., Parikh J.C. and Siemens P.J. (1984),Phys. Rev. Lett. 53,1216.

25. Lopez J.A. (1985), Ph.D. dissertation, Texas A&M University.

26. Lopez J.A., Parikh J.C. and Siemens P.J. (1985),Texas A&M University preprint.

27. Matsui T. and Satz H. (1986) Phys. Lett. B178,416.

28. Nadkarni S. (1986),Phys. Rev. D33,3738;Phys. Rev. D34, 3904.

29. Nadkarni S. (1988) Phys. Rev. Lett. 61,396.

30. Nadkarni S. (1989) Physica A158,226.

31. Parikh J.C.,Siemens P.J. and Lopez J.A. (1989),Pramana (Jnl. of Physics) 32,555.

32. Sinha B. (1988) in "Physics and Astrophysics of Quark-Gluon Plasma", B. Sinha and S. Raha (editors), World Scientific (Singapore), pp. 328-340 (1988).

33. Wong S.K. (1970), Nuovo Cimento 65A, 689.

Quarks and Gluons in Hadrons and Nuclei

F.E. Close

Oak Ridge National Laboratory*, Oak Ridge, TN 37831, USA
and University of Tennessee, Knoxville, TN 37996, USA

These lectures discuss (1) the particle-nuclear interface — a general introduction to the ideas and application of colored quarks in nuclear physics, (2) color, the Pauli principle, and spin flavor correlations — this lecture shows how the magnetic moments of hadrons relate to the underlying color degree of freedom, and (3) the proton's spin — a quark model perspective. This lecture reviews recent excitement which has led some to claim that in deep inelastic polarized lepton scattering very little of the spin of a polarized proton is due to its quarks. Lecture (4) discusses the distribution functions of quarks and gluons in nucleons and nuclei, and how knowledge of these is necessary before some quark-gluon plasma searches can be analyzed.

1. THE PARTICLE-NUCLEAR INTERFACE

Once upon a time nuclear physics was the study of nucleons and pions vibrating and oscillating at the center of the atom; particle physics was the study of nucleons and pions interacting and producing resonances. From the latter, people gradually realized that hadrons are built from quarks; the fundamental rules governing their interactions were deduced ("QCD" — quantum chromodynamics) and the similarities with QED suggested the possibility that all of the natural forces can be described in a grand unified theory. Today, particle physics deals with questions ranging from the origins of matter in the first microseconds of the universe, whose experimental investigation requires the energies of the SSC, down to the quark structure of protons and nuclei that can be studied at relatively low energies.

Nuclear physics theory has taken QCD on board. There are still detailed studies going on in nuclear excitations and there are important overlaps with nuclear astrophysics. The field is very rich. The attempt to understand nuclear phenomena at the quark level causes many nuclear physicists to be concerned with the same sort of problems as their colleagues in particle physics.

During the last three years, there has been a blossoming in this overlap area which I would like to call "hadron physics". I would like to make the following definitions.

---

*Operated by Martin Marietta Energy Systems, Inc. under Contract DE-AC05-84OR21400 with the U.S.D.O.E.

Nuclear Physics deals with the collective properties of the many-body nucleus consisting of nucleons. Hadron physics deals with the structure and interaction dynamics of those hadrons. Particle physics uses these particles as tools with which to elucidate deeper truths, seeking the origins of matter, the nature of mass, and the several other parameters which presently have to be invoked ad hoc (the weak mixing angles, fermion masses, etc.).

With tongue in cheek, one might contrast the two extremes. In extreme high-energy particle physics, there is infinite theorizing but almost no data; in the nuclear structure field, we have copious data but no truly fundamentally useful theory. In hadron physics we have much data and the hope of confronting them with the fundamental QCD theory of interacting quarks and gluons. This is stimulating but also difficult. A major problem is confinement of colored quarks and gluons within the nucleons — this is simulated on computers where space-time is described as a discrete lattice, but its origins analytically are still rather poorly understood.

This confinement phenomenon is also the catch-22 of "deriving" nuclear physics from QCD. As Bob Jaffe once remarked, "Looking for evidence of quarks in nuclei is like looking for the mafia in Sicily: everyone knows they are there, but it's hard to find the evidence." The quarks are confined in nucleons, and so any successful description of the nuclear structure must reduce to that of quarks clustered inside individual nucleons. Thus we have first to understand the proton and neutron — hence the interest in hadron physics. There is the interesting possibility that the interactions and overlaps among closely packed nucleons in large or dense nuclei, or in high-energy collisions of heavy ions, may disturb the distributions of quarks — their spatial or momentum distributions — relative to their behavior when confined in isolated free nucleons. There are indeed hints of such behavior, e.g., the "EMC effect" where quarks in iron have a slightly different momentum distribution relative to that in the deuteron. Their mean momenta are reduced in iron and other heavy nuclei suggesting greater spatial freedom. Is this "liberation" a "cold" precursor to "hot" deconfinement? Are nucleons in nuclei physically "enlarged", or is this a manifestation of quarks exchanged between nucleons, tying the nucleus together and having more spatial mobility than when in a single free nucleon? Future experiments may help to answer these questions — questions raised, in large degree, by the underlying quark theory.

I would like now to draw some analogies between QED (electrical charges, atoms, and molecules) on the one hand, and on the other, QCD (color, hadrons, and nuclei). The similarity is such that one could rewrite Bjorken and Drell's QED text by inserting a traceless 3×3 matrix (λ of SU(3)) at the fermion gauge boson vertices and let QED become QCD (with α replaced by $\alpha_s \simeq 1/10$). However, the gluons themselves have color and so mutually interact via the color forces (contrast the photon of QED which transmits but does not directly "feel" the electromagnetic force). These new intergluon interactions give rise to vertices involving three or four gluons at a point, and so a text on QCD requires more than just a coloring of Bjorken and Drell.

Now let's make a matrix to summarize how systems variously react to the forces.

Notice that the gluon and photon are in different slots. This small difference gives rise to the different long-range phenomena in QCD compared to QED (e.g. confinement versus ionization).

QED	QCD
electric charge	3 colors
attraction of opposites	attraction of unlike colors
electrically zero atoms	colorless hadrons
radiation photon	radiation gluons
magnetic effects	chromomagnetic effects
hyperfine splitting $^3S_1 - {}^1S_0$	color hyperfine $m_\rho - m_\pi$, $m_\Delta - m_N$
Fermi-Breit in hydrogen	Fermi-Breit splittings in hadron spectroscopy

		QED	QCD
Feel the force	Carry the charge	e^- Z^+ Na^+ Cl^-	Quarks Gluons
	Contain the charge	Atoms Molecules	Hadrons Nuclei
Do not feel the force Do not contain the charge		Neutrinos Photon	Leptons (ν, e)

Within atoms and hadrons one finds analogues. The Coulomb potential of hydrogen has an analogue in quark systems: as the quarks' relative separations $r \to 0$, $V(r) \sim 1/r$, but at large r, $V(r) \sim r$, presumably due to the detailed self-interactions among the gluons that are transmitting the force. In the ground state of hydrogen the magnetic interaction ("one photon exchange") splits the $^3S_1$ and $^1S_0$ levels. In the ground-state quark conglomerates, the high-J combinations have enhanced masses relative to their low-J counterparts due to "one gluon exchange"; thus the Δ(1230) resonance, with J = 3/2, has greater mass than the J = 1/2 nucleon.

Now move up a layer in complexity to the world of molecules (QED) and nuclei (QCD).

At the risk of being accused of oversimplification by the atomic experts, I will divide the interatomic forces into three broad classes, then make analogy in the QCD world with interhadronic forces at the level of the quarks and gluons.

	Covalent	van der Waals	Ionic
Atoms Molecules	e^- exchange	"two photon"	Na^+Cl^-
Hadrons Nuclei	quark exchange	"two gluon"	no analogue if color confined in neutral clusters

If this was the whole story, then nuclear physics from QCD would be a rerun of molecules from QED. However, confinement of color breaks the simple analogy. The quark exchange at large distances ($\gtrsim$ 1 fm) is contained within the confined packages, dominantly pions. The confinement of gluons in glueballs also breaks the analogy with van der Waals' forces. The hope that QCD would predict observable color van der Waals' forces in nuclei is most probably flawed, as the gluons will be confined within colorless glueballs. Computer simulations of QCD suggest that the lightest glueballs have masses in excess of 1 GeV and so transmit forces over much less than a nucleon radius. Thus their presence is hidden in nuclear physics.

It is an open question whether analogues of ionic forces occur in dense or hot nuclear systems; whether multiquark clusters occur within nuclei; whether quark-gluon plasma may form in hot-dense systems.

If color attractions among quarks are the source of internucleon forces, then there could exist analogous clusters of mesons — "meson molecules". The instability of most mesons prevents formation of these systems, but π, K, η are stable on the time scales of the strong interactions and may have the chance to bind. Indeed Weinstein and Isgur find that such attractions occur in S-wave. The ππ system has a strong enhancement above $2m_\pi$ which may be manifested in the $\psi \to \omega\pi\pi$ dipion spectrum. The KK system binds forming nearly degenerate I = 0, 1 systems 10 MeV below $2m_K$. The S^*(975 MeV) and δ(980 MeV), scalar "mesons", thus appear to be meson molecules; meson analogues of the I=0 deuteron (whose I=1 partner is above $2m_N$).

The color attractions among quarks and gluons lead to the prediction of glueballs and hybrid hadrons — the latter where gluons play a dynamical role, attracted to quarks to form hybrid mesons and baryons.

The problem in predicting the masses of these states is that we have to simulate the effects of confinement. Perhaps the simplest way of doing this is to suppose that the constituent quarks or gluons are free until they hit an infinitely high wall. This is the essence of cavity or bag models. Confine a massless $J=1/2$ quark in a radius, R, and it gains an energy that scales as $1/R$. This energy becomes of the order of 350 MeV if R is of the order of the proton radius, hence the proton mass may be modeled. For gluons, one solves the eigenvalue equations for $J = 1$ rather than $J = 1/2$ confined fields. There are electric or magnetic modes (actually TE and TM in the language of classical electrodynamics) with different eigenvalues. If R is the same as for quark systems, the typical confined-mass-scale is some 500 MeV per TE mode and 750 MeV per TM mode. Thus follows the prediction that the lightest systems consisting of at least two confined gluons weigh in at O(1 GeV) and that the lightest hybrid baryons weigh in at O(1.5 GeV). A problem is that as soon as the hyperfine shifts in energy are taken into account (this involves one first calculating the propagators of confined quarks and gluons), the lowest spin-J systems are pulled down significantly in mass. The lightest hybrid baryon might thus appear to have a mass near that of the proton which suggests either a profound rethink of baryon spectroscopy or that we have unearthed a naievity.

I suspect it is the latter. No one yet has convincingly set up a study of loop effects with renormalization within a cavity. These loop diagrams enter at the same order in perturbation theory to which the hyperfine shifts have been calculated and may alter the naive "effective" energies per confined gluon. In the case of quarks, their effects were subsumed in the MIT bag by an input mass parameter for the quark; this mass fitted to the overall mass scale of the spectroscopy. In the gluonic sector we have no mass scale to set the scale, and until we make sense of the (infinite!) self-energy diagrams, we cannot predict the absolute scale. So the mass separations among the various states may be reliable, but the absolute mass scale is beyond present analysis. To predict the masses of glueballs and hybrid hadrons, we have to resort to computer simulations — lattice QCD. This has proved to be a harder task than was originally thought.

The eventual discovery of the gluonic spectroscopy may give important insights into the nature of confinement of gluons. If lattice calculations, including quarks <u>and</u> gluons (to date, people work in the "quenched" approximation, which roughly translated means "ignore the quarks") merely point out masses of states that correspond to the particle data tables, we will confirm QCD but may still require much study to elucidate the analytic dynamics of confinement. The main outcome of such a success may be the advances that will have come in the art of computation and design of machines. Thus the significant questions posed by hadron physics are having a spinoff in the intellectual stimulation they provide to computational science and, in turn, the subsequent ability to encode problems in field theory, condensed matter, and other areas of science. I am reminded of the title of Tony Hey's talk at a recent meeting of the British Association for the Advancement of Science, and it provides an apt one-line summary of the multi-disciplinary efforts flowing from computation at the nuclear-particle interface. It was: "Quarks, Supercomputers, and Oil Prospecting".

2. COLOR, THE PAULI PRINCIPLE, AND SPIN-FLAVOR CORRELATIONS

2.1. Color

If quarks possess a property called color, any quark being able to carry any one of three colors (say red, yellow, blue), then the Ω^- (and any baryon) can be built from distinguishable quarks:

$$\Omega^- \left(s_R^\uparrow \; s_Y^\uparrow \; s_B^\uparrow \right).$$

If quarks carry color but leptons do not, then it is natural to speculate that color may be the property that is the source of the strong interquark forces — absent for leptons.

Electric charges obey the rule "like repel, unlike attract" and cluster to net uncharged systems. Colors obey a similar rule: "like colors repel, unlike (can) attract". If the three colors form the basis of an SU(3) group, then they cluster to form "white" systems — viz. the singlets of SU(3). Given a random soup of colored quarks, the attractions gather them into white clusters, at which point the color forces are saturated. The residual forces among these clusters are the nuclear forces whose origin will be mentioned later.

If quark (Q) and antiquark ($\bar{Q}$) are the $\underline{3}$ and $\underline{\bar{3}}$ of color SU(3), then combining up to three together gives SU(3) multiplets of dimensions as follows (see Ref. 3):

$$QQ = \underline{3} \times \underline{3} = \underline{6} + \underline{\bar{3}}$$

$$Q\bar{Q} = \underline{3} \times \underline{\bar{3}} = \underline{8} + \underline{1}$$

The $Q\bar{Q}$ contains a singlet — the physical mesons

$$QQ\bar{Q} = \underline{15} + \underline{6} + \underline{3} + \underline{3}$$
$$QQQ = \underline{10} + \underline{8} + \underline{8} + \underline{1}.$$

Note the singlet in QQQ — the physical baryons.

For clusters of three or less, only $Q\bar{Q}$ and QQQ contain color singlets and, moreover, these are the only states realized physically. Thus are we led to hypothesize that only color singlets can exist free in the laboratory; in particular, the quarks will not exist as free particles.

2.2. Symmetries and Correlations in Baryons

To have three quarks in color singlet:

$$1 \equiv \frac{1}{\sqrt{6}} [(RB-BR)Y + (YR-RY)B + (BY-YB)R] \tag{1}$$

any pair is in the $\bar{3}$ and is antisymmetric. Note that $3 \times 3 = 6 + \bar{3}$. These are are explicitly

$\bar{3}_{anti}$	6_{sym}
RB-BR	RB+BR
RY-YR	RY+YR
BY-YB	BY+YB
	RR
	BB
	YY

(2)

Note well: <u>Any Pair is Color Antisymmetric</u>

The Pauli principle requires total antisymmetry and therefore any pair must be:

<u>Symmetric in all else</u>

("else" means "apart from color").

This is an important difference from nuclear clusters where the nucleons have no color (hence are trivially <u>symmetric</u> in color!). Hence for nucleons Pauli says

<u>Nucleons are Antisymmetric in Pairs</u> (3)

and for quarks

<u>Quarks are Symmetric in Pairs</u> (4)

If we forget about color (color has taken care of the antisymmetry and won't affect us again), then

(i) Two quarks can couple their spins as follows

$$\begin{cases} S = 1: & \text{symmetric} \\ S = 0: & \text{antisymmetric} \end{cases} \tag{5}$$

(ii) Two u,d quarks similarly form isospin states

$$\left\{\begin{array}{l} I = 1: \text{ symmetric} \\ I = 0: \text{ antisymmetric} \end{array}\right\} \qquad (6)$$

(iii) In the ground state L = 0 for all quarks; hence the orbital state is trivially symmetric. Thus for pairs in L = 0, we have

$$\left\{\begin{array}{l} S = 1 \text{ and } I = 1 \text{ correlate} \\ S = 0 \text{ and } I = 0 \text{ correlate} \end{array}\right\}. \qquad (7)$$

Thus the Σ^0 and Λ^0 which are distinguished by their u,d being I = 1 or 0 respectively also have the u,d pair in spin = 1 or 0 respectively:

$$\left\{\begin{array}{l} \Sigma^0(u,d)_{I=1}s \leftrightarrow (u,d)_{S=1}s \\ \Lambda^0(u,d)_{I=0}s \leftrightarrow (u,d)_{S=0}s \end{array}\right\}. \qquad (8)$$

Thus, the spin of the Λ^0 is carried entirely by the strange quark.

This is the source of the $\Sigma-\Lambda$ mass difference. The $\vec{S}\cdot\vec{S}$ interaction acts between all possible pairs; thus

$$\Sigma^0 \ [(u,d)_1 s]: \quad \langle\vec{S}\cdot\vec{S}\rangle_1 + \langle\vec{S}\cdot\vec{S}\rangle_{s,1} \qquad (9)$$

$$\Lambda^0 \ [(u,d)_0 s]: \quad \langle\vec{S}\cdot\vec{S}\rangle_0 \qquad (10)$$

(note $\langle\vec{S}\cdot\vec{S}\rangle$ between a spinless diquark and anything vanishes; hence the absence of $\langle S \cdot S \rangle_{s,0}$).

Now

$$\langle\vec{S}\cdot\vec{S}\rangle_0 = -3 \langle\vec{S}\cdot\vec{S}\rangle_1, \qquad (11)$$

(see p. 91 of Ref. 3). Further, if $m_s = m_{u,d}$, the Σ and Λ become mass degenerate, and so in this limit

$$\langle\vec{S}\cdot\vec{S}\rangle_{s,1} = -4 \langle\vec{S}\cdot\vec{S}\rangle_1. \qquad (12)$$

For unequal masses of u and s, the magnetic interaction scales as the inverse mass. Hence finally

$$\Sigma^0 \sim \langle\vec{S}\cdot\vec{S}\rangle_1 \left\{1 - 4 \frac{m_u}{m_s}\right\} \qquad (13)$$

$$\Lambda^0 \sim \langle\vec{S}\cdot\vec{S}\rangle_0 \{-3\ \}. \qquad (14)$$

Then with $m_s > m_u$, we find $m_\Sigma > m_\Lambda$ as observed. Increasing m_s/m_u enhances the effect (e.g., for the charmed analogues $\Sigma_c[(u,d)c]$ and $\Lambda_c[(u,d)c]$ the splitting will be larger — again observed).

2.3. Color, the Pauli Principle, and Magnetic Moments

The electrical charge of a baryon is the sum of its constituent quark charges. The magnetic moment is an intimate probe of the correlations between the charges and spins of the constituents. Being wise, today we can say that the neutron magnetic moment was the first clue that the nucleons are not elementary particles. Conversely the fact that quarks appear to have g ≃ 2 suggests that they <u>are</u> elementary (or that new dynamics is at work if composite).

A very beautiful demonstration of symmetry at work is the magnetic moment of two similar sets of systems of three, viz.

$$\left\{ \begin{array}{c} N \; ; \; P \\ ddu; \; uud \end{array} \right\} \quad \mu_P/\mu_N = -3/2$$

and the nuclei

$$\left\{ \begin{array}{c} H^3 \; ; \; He^3 \\ NNP; \; PPN \end{array} \right\} \quad \mu_{He}/\mu_H = -2/3.$$

The Pauli principle for nucleons requires He^4 to have <u>no</u> magnetic moment:

$$\mu[He^4; \; P^\uparrow P^\downarrow N^\uparrow N^\downarrow] = 0.$$

Then

$$He^3 \equiv He^4 - N$$
$$H^3 \equiv He^4 - P$$

and so

$$\frac{\mu_{He^3}}{\mu_{H^3}} = \frac{\mu_N}{\mu_P}$$

To get at this result in a way that will bring best comparison with the nucleon three-quark example, let's study the He^3 directly.

He^3 = ppn: pp are flavor symmetric; hence, spin antisymmetric; i.e., S = 0.

Thus

$$[He^3]^\uparrow \equiv (pp)_0 \; n^\uparrow \tag{15}$$

and so the pp do not contribute to its magnetic moment. The magnetic moment (up to mass scale factors) is

$$\mu_{He^3} = 0 + \mu_N. \tag{16}$$

Similarly,

$$\mu_{H^3} = 0 + \mu_P. \tag{17}$$

Now let's study the nucleons in an analogous manner.

The proton contains u,u flavor symmetric and <u>color antisymmetric</u>; thus the spin of the "like" pair is symmetric (S = 1) in contrast to the nuclear example where this pair had S = 0. Thus coupling spin 1 and spin 1/2 together, the Clebsches yield

$$p^\uparrow = \frac{1}{\sqrt{3}} (u,u)_0 d^\uparrow + \sqrt{\frac{2}{3}} (u,u)_1 d^\downarrow \tag{18}$$

(contrast Eq. (15)), and (up to mass factors)

$$\mu_p = \frac{1}{3}(0+d) + \frac{2}{3}(2u-d). \tag{19}$$

Suppose that $\mu_{u,d} \propto e_{u,d}$, then

so

$$\mu_u = -2\,\mu_d \tag{20}$$

$$\frac{\mu_P}{\mu_N} = \frac{4u-d}{4d-u} = -\frac{3}{2} \tag{21}$$

(the neutron follows from proton by replacing u ↔ d).

I cannot overstress the crucial, hidden role that color played here in getting the flavor-spin correlation right.

We can extend this discussion to the full baryon octet. Six of these states contain two identical quark flavors (which by symmetry necessitates that this pair have total spin S = 1):

$$\left.\begin{array}{l} [P(uu)_1 d] \\ [N(dd)_1 u] \\ [\Sigma^+(uu)_1 s] \\ [\Sigma^-(dd)_1 s] \\ [\Xi^0(ss)_1 u] \\ [\Xi^-(ss)_1 d] \end{array}\right\} \tag{22}$$

The remaining pair are Σ^0 and Λ^0, both u,d's. In the former, the (ud) have I = 1 and hence S = 1. For the Λ^0, on the other hand, the (ud) have I = 0 and hence S = 0.

Thus the Λ^0 $[(u,d)_0 s]$ is analogous to the He³ nuclear example. The magnetic moment is carried entirely by the third quark, namely s. The data yield[8]

$$\mu_\Lambda \rightarrow \mu_s \simeq \frac{3}{5}\mu_d. \tag{23}$$

The strange and down quarks have the same charge (-1/3) and so the datum fits with

$$m_d = \frac{3}{5} m_s \tag{24}$$

as already noted.

If we approximate $m_u = \frac{1}{3} m_p$ (thus the proton would have g = 3), then we can do a quick computation of baryon magnetic moments where the individual contributions to the g factors are

$$\left.\begin{array}{l} u = 2 \\ d = -1 \quad (\text{ratio of } e_d/e_u) \\ s = -3/5 \quad (\text{ratio of } m_d/m_s) \end{array}\right\} \quad (25)$$

From the general spin structure of Eq. (18)

$$B^\uparrow = \frac{1}{\sqrt{3}} (q_1 q_2)_\rightarrow q_3^\uparrow + \sqrt{\frac{2}{3}} (q_1 q_2)_\uparrow q_3^\downarrow$$

we have

$$\mu = \frac{2}{3} (q_1 + q_2) - \frac{1}{3} q_3 \quad (26)$$

into which the (25) are to be substituted as required. The resulting pattern is as follows

		Prediction	Data	
P	$[(uu)_1 d]$	$\frac{1}{3}(4u-d) = 3$	2.79	(27)
N	$[(dd)_1 u]$	$\frac{1}{3}(4d-u) = -2$	-1.9	(28)
Σ^+	$[(uu)_1 s]$	$\frac{1}{3}(4u-s) = 2.8$	2.33 ± 0.13	(29)
Σ^-	$[(dd)_1 s]$	$\frac{1}{3}(4d-s) = -1.1$	-1.41 ± 0.25	(30)
Ξ^0	$[(ss)_1 u]$	$\frac{1}{3}(4s-u) = -1.5$	-1.25 ± 0.02	(31)
Ξ^-	$[(ss)_1 d]$	$\frac{1}{3}(4s-d) = -0.5$	-0.75 ± 0.06	(32)

The trend is exceptionally well described. There are undeniably 20% effects not fully accounted for.

Encouraged by this success, we might look further at this problem since, after all, there are exchange effects in nuclei that cause 20% deviations from the naive additive approach analogous to that which we have used for quarks.

We can form contributions of u and d quarks from (27) and (28) for nucleon, (29,30) for Σ, (31,32) for Ξ, and the data yield

$(u-d)_N = 2.9$

$(u-d)_\Sigma = 1.7 \pm 0.15$

$(u-d)_\Xi = 0.9 \pm 0.12$.

As we go to systems with more strange quarks, the u,d quarks act as if their effective mass increases (by a factor of three??). There is a systematic trend but extremely dramatic.

Now let's study the strange quark. We can do this by supposing that the environmental dependence for u and d flavors is the same. Then

$(s)_\Xi \equiv - (\Xi^0 + 2\Xi^-) = -0.69 \pm 0.03$

$(s)_\Sigma \equiv (\Sigma^+ + 2\Sigma^-) = -0.5 \pm 0.5$

$(s)_\Lambda = -0.6$

So the strange quark gives its "canonical" contribution to baryons containing either one or two strange quarks.

3. THE PROTON'S SPIN: A QUARK MODEL PERSPECTIVE

Inelastic lepton scattering from nucleons at high momentum transfer measures the number densities of charged constituents, $q(x)$, $\bar{q}(x)$, as a function of the Bjorken variable x (essentially the ratio of the constituent and target longitudinal momenta in an infinite momentum frame). There is a weak dependence of these distributions on the momentum transfer, Q^2, but I shall suppress this in much of what follows.

If the beam and target are polarized, one can extract the helicity-dependent distributions for quarks or antiquarks polarized parallel $(q^\uparrow(x))$ or antiparallel $(q^\downarrow(x))$ to the target polarization. I shall define $\Delta q(x) \equiv q^\uparrow(x) - q^\downarrow(x)$; $q(x) \equiv q^\uparrow(x) + q^\downarrow(x)$, and similarly for antiquarks, $\bar{q}$.

Data are presented in two ways.[1-3] One is in terms of the polarization asymmetry

$$A(x) = \sum_i e_i^2 (\Delta q_i(x) + \Delta\bar{q}_i(x)) / \sum_i e_i^2 (q_i(x) + \bar{q}_i(x)), \qquad (33)$$

(note that $-1 < A < +1$). The other involves the polarized structure function

$$g_1(x) = \frac{1}{2} \sum_i e_i^2 (\Delta q_i(x) + \Delta\bar{q}_i(x)), \qquad (34)$$

thus

$$g_1(x) \equiv A(x) F_1(x). \qquad (35)$$

In advance of the data, the expectations were that

(i) At $x > 0.2$ where valence quarks dominate, $A(x)$ should be large and positive.[3,4] This follows from intuition developed for constituent valence quarks in baryon spectroscopy where the Pauli principle requires $\Delta u > 0$, $\Delta d < 0$. As the charge-squared weighting of Δu is four times that of Δd in protons, so $A^p(x > 0.2) > 0$. Data confirm this brilliantly. For a neutron target, it is Δd that is weighted 4:1 relative to Δu, hence these tend to cancel and one predicts[4] a small (zero?) asymmetry on the neutron.

(ii) Form $g_1(x)$, which directly shows the charge weighted helicity-dependent distributions and integrate over all x.[5,6] If it were not for the charge weightings, this would measure the net $\Delta q + \Delta \bar{q}$ ($\Delta q \equiv \int_0^1 dx \Delta q(x)$ = net quark polarization). Explicitly, in the quark parton model

$$I^p \equiv \int dx\, g_1^p(x) = \frac{1}{2}\left\{\frac{3}{9}\Delta u + \frac{1}{9}(\Delta u + \Delta d + \Delta s)\right\} + (\Delta q_i \to \Delta \bar{q}_i). \tag{36}$$

The surprise[2] is that I^p(EMC) $\simeq 0.12$ and is almost saturated by[7] $\Delta u(\simeq 0.75)$ leaving

$$\sum_i (\Delta q + \Delta \bar{q})_i \simeq 0, \tag{37}$$

hence, the much-advertised claim that maybe "none of the proton's spin polarization is carried by quarks". This is a misinterpretation of Eq. (37). The valence quarks are highly polarized (point (i) above); thus, the interpretation of Eq. (37) is that something cancels or hides it. Candidates include a highly polarized sea spinning opposite to the valence quarks, orbital angular momentum, or gluon polarization.[8-10]

One can cancel out some charge weighting effects by looking at the difference of proton and neutron for which

$$I^p - I^n = \frac{1}{6}(\Delta u - \Delta d) \equiv \frac{1}{6}\left|\frac{g_A}{g_V}\right|, \tag{38}$$

which is Bjorken's sum rule.[5] The various g_A in the baryon octet give information on the <u>differences</u> of Δu, Δd, and Δs which are summarized by a measured parameter known as F/D. To extract the <u>sum</u>, Δq, we need the proton integral (Eq. (36)) or information on neutral current form factors

$$\tilde{g}_A(\nu p \to \nu p) = \Delta u - \Delta d - \Delta s. \tag{39}$$

I shall discuss this at the end of the talk. Preceding that, I shall discuss the question of Δs, since the measured F/D and the measured I^p can be combined to extract a value for Δs. This appears to be substantial; EMC claiming that

$$\Delta s = -0.23 \pm 0.08. \tag{40}$$

Implications and criticisms of this startling result will occupy the latter half of this talk. First, I will discuss what we know about the (constituent) quark polarization from static properties of the nucleon (magnetic moments, g_A/g_V) and review the extent to which the new insights do or do not require revision of this simple picture.

3.1. <u>Spin Polarization of Valence (Constituent) Quarks</u>

In the constituent quark model where $L_z = 0$ the charges and the magnetic moments of neutron and proton place the following constraints on the probabilities for finding the flavors and spin correlations of "valence" quarks,

$$u_v = 2d_v \frac{\mu_n}{\mu_p} = -\frac{2}{3} \rightarrow \Delta u_v = -4\Delta d_v. \tag{41}$$

The 56, $L_z = 0$ wave function of the nonrelativistic quark model (NRQM) satisfies (41) but it is by no means unique. A hybrid state, where a gluon ($J_z = \pm 1$) is partnered by qqq in 70 (required by the Pauli principle for qqq in color 8) satisfies Eq. (41) for the coherent combination[11] $g(^28+^48)$ where the superscripts refer to the 2S+1 of the net spin of the qqq system. The "valence quarks" here are significantly depolarized relative to 56. One can also have a significant polarized sea without destroying the magnetic moment relations. This is because

$$\frac{\mu_n}{\mu_p} = \frac{2\Delta d - \Delta u + (-2\Delta\bar{u} + \Delta\bar{d} + R\Delta\bar{s})}{2\Delta u - \Delta d + (-2\Delta\bar{u} + \Delta\bar{d} + R\Delta\bar{s})}, \tag{42}$$

where $R = m_d/m_s \simeq 3/5$. The electrical neutrality of the sea tends to shield its contribution. A detailed fit is made in Ref. 12.

The $(g_A g_V)$ for the octet of baryons also relate to the spin polarized probabilities such as

$$\left(\frac{g_A}{g_V}\right)_{np} = \Delta u_v - \Delta d_v \rightarrow -5\Delta d_v, \tag{43}$$

where we used Eq. (41). Thus immediately

$$\Delta d_v = -0.25; \quad \Delta u_v = 1. \tag{44}$$

In the 56 NRQM one would have[3]

$$\Delta d_v = -1/3; \quad \Delta u_v = 4/3; \quad \Delta u_v + \Delta d_v = 1, \tag{45}$$

and the entire spin polarization comes from the quarks. However, from Eq. (44), we see that

$$\Delta u_v + \Delta d_v \simeq 3/4, \tag{46}$$

and so, in advance of the EMC data, only naive "quarkists" would have expected 100% for Δq_v. Anyone who worked with four-component spinors, of which the MIT bag is a specific model example, knew that the "orbital dilution" in the lower components played an essential role.[13] In fact, the Δq_v expectation is even less than Eq. (46). When one makes a best fit to all of the baryon octet g_A/g_V, one finds

$$\Delta q_v (\equiv 3F-D) = 0.55 \pm 0.10. \tag{47}$$

Note the appearance of F and D which summarizes the g_A/g_V. This parameter will appear later. Note that many analyses of the polarization data use[2,6,10,15] F/D = 0.63 (Ref. 16). However, this value fitted a value of the neutron lifetime that we now know to have been incorrect.[17,18] The correct current value[19-25] is lower than 0.63 and is dependent upon assumptions about SU(3) flavor breaking.

The earliest predictions for the deep inelastic polarization asymmetry in the valence-dominated region assumed that all Δq and q (valence) have the same x dependence. Thus (see Refs. 3 and 4 for origins of these formulae)

$$A^n(x) \simeq 4\Delta d + \Delta u \to 0,$$

(the zero following immediately from Eq. (41)) and

$$A^p(x) = \frac{5}{3}(-\Delta d) \to \frac{1}{3}(g_A/g_V).$$

The prediction that $A^p > 0$ is non-trivial as <u>a priori</u> it could be anywhere in the range $-1 \leq A \leq +1$. The presence of a $q\bar{q}$ sea as $x \to 0$ was expected to cause $A(x \to 0) \to 0$. The other qualitative expectation[26,27] was that $A(x \to 1) \to 1$ as follows.

The valence picture above implicitly assumed that $u_v(x) = 2d_v(x)$ for all x. However, unpolarized data show this to be untrue in that it would require that

$$\frac{F_1^n(x)}{F_1^p(x)} = 2/3.$$

In practice, this ratio drops as $x \to 1$, suggesting that the $u(x \to 1) \gg d(x \to 1)$, a phenomenon which follows from spin dependence via single gluon exchange. Chromomagnetic hyperfine energy shifts split the Δ-N masses and elevate $u(x \to 1)$ over $d(x \to 1)$. They also cause $u^\uparrow(x \to 1)$ to dominate over $u^\downarrow(x \to 1)$, which the consequence that $A^{p,n}(x \to 1) \to 1$. Thus, a qualitative expectation for A^p emerged:

$$A^p(x \to 0) \to 0; \quad A^p(x \simeq 1/3) \simeq 1/3 \left|\frac{g_A}{g_V}\right|; \quad A^p(x \to 1) \to 1.$$

These predictions turned out to be remarkably well verified and even agree with the latest EMC data.

Recently Close and Thomas[28] showed that, within the framework of the MIT bag model, one could relate the x-dependent distortion of the valence distributions to the measured chromomagnetic energy shift in the Δ-N masses. All of this suggests that the valence quark polarizations measured in polarized deep inelastic scattering are similar to the polarizations of the constituent quarks manifested in low-energy spectroscopy. This is an important constraint on model builders. The memory of the constituent quark spins is not lost as one proceeds to the deep inelastic: the <u>valence quarks are highly polarized</u>.

If, as is being claimed, the quarks and antiquarks contribute (within errors) nothing to the net spin polarization of the proton, then we must conclude that something is canceling the contribution of the valence quarks. Candidates include orbital angular momentum polarized gluons or a negatively polarized sea.

We already noted that in the constituent limit it is over naive to ignore orbital angular momentum. The presence of polarized gluons may be probed by studying the polarization dependence of direct photon production or spin dependence of heavy flavor production; a polarized sea may affect the inclusive production of hadrons[33,34] and fast $K^-(s\bar{u})$ production may be a tag for scattering from the sea.[34]

Dziembowski et al.[35] have studied the relation between constituent quarks and partons. They view the constituent quarks as being a conglomerate of partons-quarks, antiquarks, and gluons, thus

$$q_i^{\lambda_i}(x,Q^2) = \sum_{v,\lambda_v} \int_x^1 \frac{dy}{y} G_{v/N}^{\lambda_v}(y) q_{i/v}^{\lambda_i \lambda_v}\left(\frac{x}{y}, Q^2\right),$$

where the λ are helicity labels. The constituent quark distributions $G_{v/N}(y)$ reflect the dynamics that binds the quarks to form hadrons, and are determined by a light cone nucleon wavefunction. The constituent quark structure functions $q(x/y, Q^2)$ are adapted from Altarelli et al.[36] together with Carlitz and Kaur's ansatz[37] for the spin of soft valence partons within a polarized constituent quark.

This picture of partons convoluted within constituents generates some effective $L_z \neq 0$ but not enough to account for the spin deficit claimed by EMC. The data seem to fall below the model systematically for $x \lesssim 0.1$. If these small x data survive further experiments, then it seems that polarization of the sea (not included in Ref. 35) must be allowed for. This naturally leads to the question of whether there are polarized strange quarks in the proton.

3.2. Polarized Strange Quarks?

One exciting possibility is that the EMC data imply a large polarization of strange quarks and/or antiquarks within the proton. If true, this could have significant consequences. In particular, it could modify earlier analyses of electroweak parity violation in deuterium where Campbell et al. argue,[15] the polarized strange quarks could give contributions that dominate over electroweak radiative corrections. An extreme claim has appeared in the literature that the large value for Δs is in conflict with perturbative QCD. If true, this would be devastating. This claim comes about, in part, because an incorrect value of F/D has been used in the analyses. It is this parameter, and its implications for Δs, that I will now discuss.

Given the integral, I_p, of the polarized structure function $g_1^P(x,Q^2)$, one extracts Δs (including new QCD corrections)

$$I_p \equiv \int dx g_1^P(x,Q^2) = \frac{1}{18}\left(\frac{g_A}{g_V}\right)\left[\frac{9f-1}{f+1} - \frac{\alpha_s(Q^2)}{\pi}\frac{3f+1}{f+1}\right] + \frac{\Delta s}{3}, \quad (48)$$

where $f \equiv F/D$ with $\alpha_s(Q^2) = 0.27$, $g_A/g_V = 1.254 \pm 0.006$ and $I_p = 0.126 \pm 0.022$. A feeling for the sensitivity of Δs to f can be gauged from the approximate relation

$$\Delta s \simeq (f-0.40) \pm 0.07. \quad (49)$$

The widely used value, following the much-quoted fit of Ref. 16 has been

$$F/D = 0.63 \pm 0.02 \rightarrow \Delta s = -0.23 \pm 0.09. \tag{50}$$

If the sea is flavor-independent, then Eq. (50) summarizes the widely accepted interpretation of the EMC polarized structure function data where a significant negative polarization of the sea cancels out the positive polarization of the valence quarks.

This value was based on the original value for I_p quoted by EMC,[2] namely I_p = 0.116 ± 0.022. However, the revised value,[29] I_p = 0.126 ± 0.022, reduces the magnitude of Δs by 0.03, and so Δs = -0.20 ± 0.09 should replace Eq. (50).

However, it does not seem to be widely appreciated that the F/D of Ref. 16 was much constrained by an outdated value of the neutron lifetime, and that Ref. 16 chose "to omit from (their) fit the neutron decay correlation (which yields) g_A = 1.258 ± 0.009, which differs significantly from the result 1.239 ± 0.009 required by the neutron lifetime measurements". The value accepted as correct today[18] differs by some 3σ from the old value, and this, together with other data on hyperon beta decays,[16,18,19] shows that F/D is much smaller than the old value. Flavor symmetry breaking causes a spread in values of F/D, depending on which partial set of data one uses; indeed, the symmetry breaking even calls into question the utility of the F/D parameter,[20] and so Refs. 17 and 21 set up their analyses without direct reference to F/D. Translating their work into F/D, one finds that the value subsumed in Ref. 17 is F/D = 0.56 consistent with that implicit in Ref. 21 and, within errors, with the fitted value in Ref. 22. Reference 23 obtained an even smaller value of F/D = 0.545 ± 0.02. Recent improvements in the Σn beta decay data, in particular, may raise F/D to 0.58 (Ref. 24), but nowhere as high as the 0.63 used previously.

The magnitudes for Δs implied by these values for F/D are

$$F/D = 0.548 \pm 0.01 \rightarrow \Delta s = -0.12 \pm 0.06 \text{ Ref. 23} \tag{51}$$

$$F/D = 0.58 \pm 0.01 \rightarrow \Delta s = -0.15 \pm 0.08 \text{ Ref. 24} \tag{52}$$

Thus we see that the magnitude of the (negative) strange polarization may be only half as big as that previously assumed. The QCD-corrected value of the Ellis-Jaffe sum rule falls from 0.19 (the cited value when F/D = 0.63) to 0.17 if F/D = 0.56, thereby reducing the statistical significance of the much-advertised failure of this sum rule.

What independent information exists on Δs? Elastic neutrino-proton scattering can, in principle, probe this quantity,[32] and a fit to these data give

$$\Delta s = -0.15 \pm 0.09.$$

Note that this agrees with the revised value in the present paper arising from the smaller F/D and the revised EMC integral (Ref. 29).

One should also be aware that the neutrino experiment is also consistent with Δs = 0 which, in advance of the controversial EMC experiment, was the expectation.

Flavor-changing weak interactions, such as neutron beta decay, can yield

$$\frac{g_A}{g_V} \simeq 1.25 = \Delta u - \Delta d,$$

while the zero momentum limit of $\nu p \to \nu p$ can probe

$$\tilde{g}_A(0) = \Delta u - \Delta d - \Delta s \left(\frac{g_A}{g_V}\right)\left(1 - \frac{\Delta s}{1.25}\right),$$

and so a difference between $\tilde{g}_A(0)$ and g_A/g_V can, after radiative corrections, reveal nonzero Δs. (Our Δs ≡ 1.25η of Ref. 32.)

A practical problem is that $\nu p \to \nu p$ is detected by proton recoil and so an extrapolation to $\vec{q} = 0$ is needed. One fits the $q^2 \neq 0$ data with a form factor, in essence

$$\frac{1 - \Delta s/1.25}{(1 + Q^2/M_A^2)^2},$$

where M_A is a mass scale to be fitted. Other experiments have determined this to have the value $M_A = 1.032 \pm 0.036$ GeV. If one fixes M_A to equal the world average, then Δs = -0.15 ± 0.09; hence the claim to support the nonzero strange polarization. However, Ref. 32 also makes another, less well-advertised, fit. They constrain Δs = 0 and find that in this case Δs = 0; $M_A = 1.06 \pm 0.05$ GeV. Thus, one sees that Δs = 0 yields M_A consistent with the world average and hence is equally acceptable as a solution. The crucial statement in Ref. 32 is that "M_A and η(Δs) are strongly correlated". Thus, Ref. 32 does not require Δs < 0 and thereby does not necessarily lend support to those who desire Δs ≠ 0. Thus the question of the magnitude of the (strange) sea polarization is open. It is likely to be significantly nearer to zero than is being assumed in much of the current literature. Some of the inferences claimed from the EMC polarization data may need re-evaluation therefore. In particular, there need be no conflict with perturbative quantum chromodynamics.[30]

3.3. Polarized Gluons?

It has recently been realized[8] that the perturbative QCD correction to the singlet part of $g_1^p(x)$ effectively scales (to $O(\alpha_s^2)$) and may be important. This may be incorporated by replacing the Δq in Section 1 by $\tilde{\Delta}q \equiv \Delta q - \alpha_s/2\pi \, \Delta G$, where $\Delta G \equiv \int_0^1 dx \, \Delta g(x)$ and $\Delta G(x) = g_\uparrow(x) - G_\downarrow(x)$ is the polarized gluon distribution. This modifies the polarized lepton analysis, but cancels out in the expressions for (g_A/g_V) and does not enter the magnetic moment (Section 2) analysis.

One consequence is that there may be a continuity between the low-energy polarization revealed in constituent quarks (magnetic moments and spin dependence of resonance excitation) and the deep inelastic polarization.

First of all, we summarize the data on the Δq (or equivalently $\tilde{\Delta}q$) from the various (g_A/g_V).

If we assume $SU(3)_F$ symmetry in the sense that $s(\Sigma^+) \equiv d(P)$, then we may write the various g_A in terms of F, D, or Δq as follows:

g_A	F,D	$\Delta q^{(P)}$	Data
np	$F + D$	$\Delta u - \Delta d$	1.26 ± 0.005
Λp	$F + \frac{1}{3} D$	$\frac{1}{3}(2\Delta u - \Delta d - \Delta s)$	0.72 ± 0.02
$\Xi\Lambda$	$F - \frac{1}{3} D$	$\frac{1}{3}(\Delta u + \Delta d - 2\Delta s)$	0.25 ± 0.05
Σn	$F - D$	$\Delta d - \Delta s$	-0.33 ± 0.02

Thus $F/D \equiv (\Delta u - \Delta s)/(\Delta u + \Delta s - 2\Delta d)$. Extracting the individual contributions involves a correlated fit. The EMC values, corrected for F/D, become $\tilde{\Delta}u = 0.80 \pm 0.06$, $\tilde{\Delta}d = -0.45 \pm 0.06$, and $\tilde{\Delta}s = -0.15 \pm 0.06$. One possibility is that $\Delta s = 0$, so that $\tilde{\Delta}s = -\alpha/2\pi \, \Delta G$. In this case, we obtain for

$$\Delta u \equiv \tilde{\Delta} u - \tilde{\Delta} s = 0.95 \pm 0.06 \tag{53}$$

$$\Delta d \equiv \tilde{\Delta} d - \tilde{\Delta} s = -0.30 \pm 0.06 \tag{54}$$

It is interesting to note that these values are consistent with those extracted from the magnetic moments (Eq. (4)) viz

$$\Delta u_v = 1, \quad \Delta d_v = -0.25$$

The proton helicity is given by

$$\frac{1}{2} = \frac{1}{2} \Delta q + (\Delta G + L_Z) \tag{55}$$

Hence $\Delta G = 3.5$ and $L_Z = -3.35$ at $Q^2 \simeq 10$ GeV^2. As one devalues to lower Q^2, $d/dQ^2 (\Delta G + L_Z) = 0$, and the individual contributions fall. It is an open question whether the "passive" L_Z in the constituent model (i.e., the dilution of S_Z due to relativistic spinors) at low Q^2 provides a consistent picture between constituent spin polarization and "parton" polarization.

Ellis et al.[31] suggest that the modification to $\Delta q(x)$ be driven by evolution

$$\tilde{\Delta}q(x) = \Delta q(x) - \int_x^1 \frac{dy}{y} \Delta G(y) \sigma(x/y)$$

where $\sigma(Z)$ is the cross section for $\gamma^* g \to q\bar{q}$. If so, then $g_1^p(x \to 0) < 0$, the crossover from positive to negative moving to smaller x values as Q^2 increases.

It is tantalizing that such a picture may already be manifested at low Q^2 in the resonance region. It is well known that the prominent $D_{13}(1520)$ and $F_{15}(1690)$

resonances are excited dominantly in $\sigma_{3/2}$ when $Q^2 = 0$, but in $\sigma_{1/2}$ for $Q^2 \neq 0$. The change in helicity structure,[38] or change in sign of $g_1^P(Q^2)$, occurs at $Q^2 \simeq 0.4$ GeV^2 for D_{13} and $Q^2 \simeq 0.7$ GeV^2 for F_{15}. It is amusing that these correspond to $x \simeq 0.2$, and so Bloom-Gilman duality may approximately hold true even for polarized leptoproduction, with a Q^2 dependence to the x_c where $g_1^P(x_c) = 0$. The first resonance $P_{33}(1236)$ sits on top of an S-wave background; the relative Q^2 dependences are not well known. However, perturbative QCD applied to resonance excitation suggests that this excitation may also change its character with Q^2 such that its contribution to $g_1^P(x)$ changes sign at $x \to 0$. It will be interesting at CEBAF to verify if the resonance region indeed matches onto the deep inelastic, and at high energy labs to verify whether $g_1^P(z \to 0) < 0$.

4. PARTON DISTRIBUTIONS IN NUCLEI: QUAGMA OR QUAGMIRE?

In this lecture I review the emerging information on the way quark, antiquark, and gluon distributions are modified in nuclei relative to free nucleons. I place particular emphasis on Drell-Yan and ψ production on nuclei and caution against premature use of these as signals for quagma in heavy-ion collisions.

If we are to identify the formation of quark-gluon plasma in heavy-ion collisions by changes in the production rates for ψ relative to Drell-Yan lepton pairs, then it is important that we first understand the "intrinsic" changes in parton distributions in nuclei relative to free nucleons. So, I will review our emerging knowledge on how quark, antiquark, and gluon distributions are modified in nuclei relative to free nucleons, and briefly summarize the emerging theoretical concensus.

4.1. Partons in Nuclei

The best known nuclear distortion is that of the EMC effect which reveals a modification of the valence quark distributions in nuclei relative to those in free nucleons.

All experiments now show broad agreement.[39,40] The rise in F_A/F_N at $x > 0.7$ is due to Fermi motion causing the structure function F_A to leak out to $x > 1$; dramatic as this appears, it occurs where $F_{A,N} \simeq 0$, and is, in fact, a very minor contributor to the overall phenomenon. Indeed, overall, the effect is a subtle 10% affair, and we don't need to rewrite the nuclear physics textbooks. As $x \to 0$, we are beginning to see evidence for shadowing, a subject on which theory is now also starting to develop.[41]

4.2. Quarks in Nuclei

In the "intermediate" region $0.2 \leq x \leq 0.6$ the ratio falls below unity as (valence) quarks lose momentum due to nuclear binding. The A dependence was successfully predicted in advance of data[42] and is rather well understood. It was first <u>predicted</u> in the context of the rescaling analysis and subsequently verified by experiments at SLAC.[39] However, the A-dependence is more general than the rescaling model[43] and will arise in any model in which (i) the EMC effect is fitted or predicted for one A value, say iron, (ii) the physics underwriting the effect has a finite range in coordinate space, and (iii) the effect is expressible as a functional of the nuclear density operator, $\rho(r)$. If $g_k(r_1,\ldots r_k,x,Q^2)$ is an A-dependent function expressing the change in F_2^A arising from the overlap of k-nucleons, $\rho^A(r)$ is the mean nuclear density, then

$$\delta F_{k=2}^A(x,Q^2) = \frac{A-1}{2}\int d^3r_1\, d^3r_2\, \rho^A(r_1)\rho^A(r_2)[f(r_{12})-1]g_2(r_1,r_2;x,Q^2)$$

where $f(r_{12})$ was assumed to be A independent. Qualitatively, the effect is driven by the chance that there is a nearby nucleon correlated with the target nucleon. Reference 4 made a geometrical model for g_2 and predicted δF_2^A for all A. (As the volume of a nucleon is only about 40% of the available volume, the two-nucleon contribution is $\leq$ 10% effect, and three-body and higher contributions can be ignored.) The fit is excellent and shows that the EMC effect is sensitive to details of nuclear structure, reflected in $\rho^A(r)$ as a result of which significant fluctuations are predicted at small A which have yet to be studied. At large A the behavior is smooth and it is safe to interpolate. Thus, one can infer the $F_A(x)$ when A = tungsten, say, and use this as input to $\pi W \to \mu\bar{\mu}$ ... analyses for example.

A common feature of models is that the degradation of the valence quarks transfers energy momentum to some other component (gluons and $q\bar{q}$ in rescaled QCD[44] or the partons in the π that are responsible for nuclear binding).[45,46] Thus, they generate an increased sea in nuclei relative to that measured in free nucleons. In turn, this implies that $F_A/F_N > 1$ as $x \to 0$. However, this predicted enhancement will probably be blacked out by shadowing (which has not been incorporated in these models so far). Mueller and Qiu[42] have begun to illuminate us about the x and Q^2 dependence of nuclear shadowing; the quantitative combination of their work with "soft π"[45,46] or rescaled QCD[42,44] remains to be completed. What impact does this have on Drell-Yan?

If $x_{1,2}$ refers to the beam and target partons, $x_F = x_1 - x_2$, and $x_1 x_2 = Q^2/s$, then the ratio of cross sections for some fixed Q^2/s is

$$\frac{\sigma^{bA}}{\sigma^{bN}} \sim \frac{\bar{q}^b(x_1)q^A(x_2) + q^b(x_1)\bar{q}^A(x_2)}{\bar{q}^b(x_1)q^N(x_2) + q^b(x_1)\bar{q}^N(x_2)}$$

where sum over flavors weighted by their squared charge is understood. In the case of π^- beams, if $x_2 > 0.2$ so that $\bar{q}^A \ll q^A$, the DY process is dominantly due to q^π annihilating with $q^{A,N}$. Thus, in this kinematic regime

$$\frac{\sigma^{\pi-A}}{\sigma^{\pi-N}} \sim \frac{u^A(x_2)}{u^N(x_2)} \sim \frac{F^A(x_2)}{F^N(x_2)}$$

This is the same ratio as measured in inelastic lepton scattering ("EMC effect") and must be obtained here too if factorization is valid. Thus, we should not be surprised by the results from NA10[47] who study

$$\frac{\sigma(\pi^- W \to \mu^+ \mu^- ...)}{\sigma(\pi^- D \to \mu^+ \mu^- ...)}$$

and by varying Q^2 and x_F can separate both the pion and target structure functions. In Ref. 47 they exhibit the resulting ratio of $q^A/q^N(x_2)$ and $\bar{q}^{\pi(A)}/\bar{q}^{\pi(N)}(x_1)$.

The latter should be unity and is within errors when one combines data from two energies, 140 GeV and 286 GeV incident π beams. (However, if one restricts attention to the lower energy sample, the situation is more messy and the pion distributions do not seem to factorize. Why this should be is unclear to me, but bear it in mind as an empirical observation for later reference.)

So the message is: do not be misled by $\sigma = A^\alpha$, $\alpha \simeq 1$ for Drell-Yan pair production on nuclei. While this may be approximately true for the total rate, there can be (and are) non-trivial effects in x and p_t^2. A depletion at large x may be compensated by an enhancement as $x \to 0$ for example. The kinematic conditions of experiments may emphasize different regions of x for the beam and target. The extent to which A-dependent effects will arise depends on $\bar{q}^A(x)$ and $g^A(x)$, about which we know almost nothing. Rescaling and pion models imply that both $\bar{q}(x,Q^2)$ and $g(x,Q^2)$ have non-trivial A dependence; moreover, shadowing effects will modify them as $x \to 0$. Empirical information is only now beginning to emerge.

4.3. Antiquarks in Nuclei

The Drell-Yan process with incident nucleons can probe q in the target if suitable kinematics are chosen, e.g., $x_1 \simeq 0.7$ and $x_2 \ll x_1$. An investigation of this in various models has been made by Bickerstaffe et al.[48] and by Berger et al.[46] As an example, in Fig. 1, I show the predictions for $\bar{q}^A/\bar{q}^N(x)$ in iron in three models compared with information gleaned from CDHS.[49] The dramatic rise in the Berger-Coester model at x > 0.3 is due to their prediction that $\bar{q}$ leak out to moderate x values in nuclei. However, it is illusory to some degree as both $\bar{q}^A$ and $\bar{q}^N$ are vanishingly small; even so, experiment E772 may be able[54] to test this. Independent of specific models, it is an interesting question whether $\bar{q}$ leak to "large" x in nuclei as this will have a bearing on the Q^2 shape of Drell-Yan pairs in nuclei which may differ from the ψ production (produced by gluons) and potentially provide a background to the plasma signal sought in heavy-ion collisions. Indeed, there are hints of a change in the ψ/Drell-Yan ratio in com-

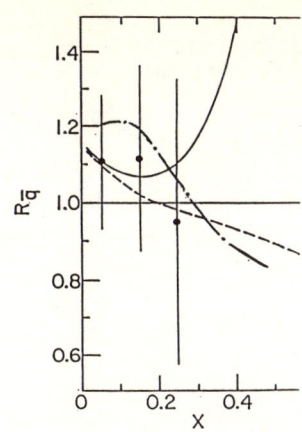

Fig. 1. Data from Ref. 49 on the ratio $\bar{q}^{Fe}/\bar{q}^{D}(x)$ from neutrino scattering. Here $\bar{q} = (\bar{u}=\bar{d}=2\bar{s})$. The solid curve illustrates predictions of pion exchange model (Refs. 46 and 49), the dot-dash is the pion model of Ref. 45, and the dashed curve is the rescaling model, Ref. 44.

paring nucleon data with that from oxygen-uranium (NA38 collaboration at CERN,[55] but close examination suggests that the shape of the Drell-Yan continuum has changed. This implies that "conventional" A dependence of $\bar{q}(x)$ may be significant, and that this is not a first hint of plasma formation.[56]

Recently WA25 and WA59 in collaboration have studied the EMC effect using ν and ν interactions in neon and deuterium. The x,y distributions allow separation of quark and antiquark distributions, and there is some indication that the sea <u>decreases</u> in going from deuterium to neon. Depending upon the model assumed for σ_L/σ_T, the fractions of sea in neon is 7 ± 6% or 18 ± 10% less than in deuterium. Presumably, these data are being dominated by nuclear shadowing, like the electromagnetic data for $x \leq 0.1$, again highlighting the need to see how shadowing modifies the curves in Fig. 1 at small x.

4.4. <u>Gluons in Nuclei</u>

Insofar as inelastic $\gamma A \rightarrow \psi + \ldots$ proceeds via photon-gluon fusion and $NA \rightarrow \psi + \ldots$ involves gluon-gluon fusion, these processes probe $g^A(x)$. There have been early claims that $g^A > g^N$ ($x \sim 0.05$), this based on the EMC data[50] for $(\gamma Fe \rightarrow \psi)/(\gamma D \rightarrow \psi \ldots)$. Expressed as A^α this gave $\alpha = (1.10 \pm 0.03 \pm 0.04)$. However, it is now less clear whether <u>coherent</u> ψ production (for which $\alpha \simeq 4/3$) has been entirely removed from the data. Indeed E691 (Sokoloff <u>et al.</u>) report[51] that (at $Q^2 = 0$)

$$\alpha_{coherent} = 1.40 \pm 0.06 \pm 0.04$$
$$\alpha_{incoherent} = 0.94 \pm 0.02 \pm 0.03$$

This suggests that gluons are shadowed in nuclei.

There are also confusing signals[52] coming from E537 who probe the gluon distribution with $\pi^- W/Be \rightarrow J/\psi \ldots$ at 125 GeV, measuring the x_F distributions and thereby enabling $g^{W/Be}$ to be measured for x_{Bj} 0.2. If $x_F = x_1 - x_2$, then for J/ψ

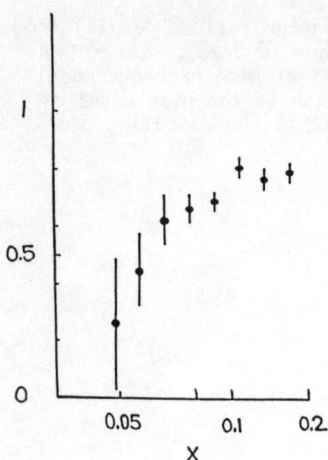

Fig. 2. Ratio of ψ production in π^-W/π^-Be from E537 (Ref. 52) which is indicative of $g^W/g^{Be}(x)$, plotted against x-Bjorken for the nuclear target (x_2).

production at 125 GeV

$$x_F \simeq \frac{1}{25x_2} - x_2$$

Thus, we can replot the data from E537 against X_2 (Fig. 2). This is equivalent to $g^W(x)/g^{Be}(x)$ only if $g^{\pi(W)} \neq g^{\pi(Be)}$ cancels out. However, we have no immediate way of knowing if this is true empirically as $Q^2(=m_\psi^2)$ is fixed. <u>Prima facie</u>, one may justifiably be worried. First, which is theoretical prejudice, if Fig. (2) is interpreted as $g^{W/Be}(x)$, it implies that gluons are significantly shadowed for x as large as 0.2. Our understanding of shadowing is still rather primitive, but such behavior would be against all current models.

The NA(10) data[47] on $\pi^- W/D \rightarrow \mu^+\mu^- \ldots$ may give hints that we are right to be wary of E537. Recall their extracted <u>quark</u> ratio which is in line with the EMC data from inelastic muon scattering — "EMC effect".[41] This is fine for the 140-GeV and 286-GeV data combined, but when one looks at the NA(10) data 140 GeV sample alone, things are less clear. The x_F distribution from NA(10) (which is a convolution of beam and target and thus "nearest" to the E537) for 140 GeV matches smoothly onto E537 at x ≃ 0.2 — and the reason is that $q^{\pi(W)} \neq q^{\pi(Be)}$ in the 140-GeV data sample. I have no idea why this should be so, but if it is true for gluons too that $g^{\pi(W)} \neq g^{\pi(Be)}$ at 125 GeV, then it raises a question about extraction $g^W(x)/g^{Be}(x)$ from E537. If we take the NA(10) data on quarks as a guide, then it is possible that the $g^{W/Be}(x)$ is, in effect, to be renormalized upwards by 20%. (More legitimately, I don't know why there is such an energy dependence, or even if it is real, but I would be happier to see the E537 experiment with 300 GeV incident beams or comparison with Υ production so as to get some lever on the $x_1 x_2$ separation directly.)

If one renormalizes the ratio in Fig. 2 upwards by 20%, then there is no shadowing at x ≃ 0.2 and, furthermore, the data then look quantitatively as expected in Mueller-Qiu theory of shadowing.[42] It is important that this problem with separation be better understood before we can conclude very much on the $g^{W/Be}(x)$ ratio. (Another reason why one might regard this renormalization as reasonable is that "infinite" shadowing leads to an $A^{2/3}$ behavior and for W/Be ratio this is 0.37. The trend of the E537 data looks set to violate this, whereas a 20% increase would bring this into line.)

Experiment E672 at Fermilab is measuring[53] $\pi^-A \rightarrow \psi$ on four nuclei at 530 GeV. I await their "high energy" extraction of $g^{A/N}(x)$. Until the conundrum of energy dependence (i.e. the non-factorization of the partons in the incident beam) is settled, I conclude that $g^{A/N}(x)$ probably falls as $x \rightarrow 0$, in qualitative agreement with the shadowing phenomenon, but the quantitative measure is unclear.

Thus, with the exception of valence quarks for $x > 0.2$, there is little or no evidence for non-trivial behavior for $\bar{q}^{A/N}$ and $g^{A/N}$. It is imperative to know these quantities much better, and to understand the anomalous energy dependence manifested by NA(10), and implicitly hinted at by E537. Until we do, then we cannot, with any confidence, use J/ψ relative to Drell-Yan production in AA collisions as a signal for quark-gluon plasma formation. Note that if the E537 experiment's dramatic suppression of nuclear glue is true, then ψ production per nucleon in heavy-ion collisions will be markedly suppressed relative to that in pp or even pA interactions. There is no reason to anticipate such drama for Drell-Yan.

References

1. G. Baum et al., Phys. Rev. Lett. 51, 1135 (1983); V. W. Hughes and J. Kuti, Ann. Rev. Nucl. Part. Sci. 33, 611 (1983).
2. J. Ashman et al. (EMC), Phys. Lett. B206, 364 (1983; V. W. Hughes et al., Phys. Lett. B212, 511 (1988).
3. F. E. Close, "Introduction to Quarks and Partons," Academic, New York (1979) Chap. 13.
4. J. Kuti and V. Weisskopf, Phys. Rev. D4, 3418 (1971).
5. J. D. Bjorken, Phys. Rev. D1, 1976 (1970).
6. J. Ellis and R. L. Jaffe, Phys. Rev. D9, 1444 (1984).
7. J. Ellis, R. A. Flores, and S. Ritz, Phys. Lett. B194, 493 (1987).
8. G. Altarelli and G. G. Ross, Phys. Lett. B212, 391 (1988); R. Carlitz, J. Collins, and A. Mueller, Phys. Lett. B214, 229 (1988).
9. L. M. Sehgal, Phys. Rev. D10, 1663 (1974).
10. S. J. Brodsky, J. Ellis, and M. Karliner, Phys. Lett. B206, 309 (1988).
11. T. Barnes and F. E. Close, Phys. Lett. 128B, 277 (1983; F. Wagner, Proc. XVI Rencontre de Moriond (1982), ed. J. Tranthanhvan.
12. C. Carlson and J. Milana, College of William & Mary report, WM-89-101. The role of gluon exchange is discussed by H. Hogassen and F. Myhrer, Phys. Rev. D37, 1950 (1988).
13. For example, p. 117 in Ref. 3.
14. This comes from combinations of g_A/g_V for np with that for Λp, Σn, and $\Xi\Lambda$; see Eq. (11) in Ref. 17.
15. B. A. Campbell, J. Ellis, and R. A. Flores, CERN-TH-5342/89.
16. M. Bourquin et al., Z. Phys. C21, 27 (1983).
17. F. E. Close and R. G. Roberts, Phys. Rev. Lett. 60, 1471 (1988).
18. M. Aguilar-Benitez et al. (Particle Data Group) Phys. Lett. B204, 1 (1988).
19. S. Hsueh et al. (E715 collaboration) Phys. Rev. D38, 2056 (1988).
20. H. J. Lipkin, Phys. Lett. 214B, 429 (1988).

21. M. Anselmino, B. Ioffe, and E. Leader, Santa Barbara ITP report (1988) unpublished.
22. D. Kaplan and A. Manohar, Nucl. Phys. B310, 527 (1988).
23. J. Donoghue, B. Holstein, and S. Klint, Phys. Rev. D35, 934 (1987).
24. A. Beretvas, private communication; Z. Dziembowski and J. Franklin, Temple University, Philadelphia report TUHE-89-11 (1989).
25. R. Jaffe and A. Manohar, MIT report, MIT-CTP-1706 (1989) (but note that this inputs outdated neutron lifetime, which artificially increases the errors).
26. F. E. Close, Phys. Lett. 43B, 422 (1973).
27. G. Farrar and D. Jackson, Phys. Rev. Lett. 35, 1416 (1975).
28. F. E. Close and A. W. Thomas, Phys. Lett. B212, 227 (1988).
29. EMC Collaboration, CERN preprint, CERN EP/89-73, June 1989.
30. F. E. Close, Phys. Rev. Letts. (in press).
31. J. Ellis, M. Karliner, and C. Sachrajda, CERN-TH-5471/89.
32. L. Ahrens et al., Phys. Rev. D35, 785 (1987).
33. M. Frankfurt et al. (private communication).
34. F. E. Close and R. Milner, Oak Ridge National Laboratory report (1989).
35. Z. Dziembowski et al., Phys. Rev. D39, 3257 (1989).
36. G. Altarelli et al., Nucl. Phys. B69, 531 (1974).
37. R. Carlitz and J. Kaur, Phys. Rev. Lett. 38, 673 (1977).
38. See, e.g., V. Burkerdt, Research Program at CEBAF (Report of 1985 study group) ed. F. Gross (CEBAF, Newport News, 1986).
39. EMC collaboration: Phys. Lett. 123B, 275 (1983); BCDMS: Phys. Lett. 163B, 282 (1985); A. Bodek et al., Phys. Rev. Lett. 50, 1431 and 51, 534 (1983); R. Arnold et al., Phys. Rev. Lett. 52, 727 (1984).
40. T. Sloan, Proceedings of Uppsala International Conference on HEP (1987).
41. A. Mueller and J. Qiu, Nucl. Phys. B268, 427 (1986); J. Qiu, Nucl. Phys. B291, 746 (1987). Early qualitative ideas are in N. N. Nikolaev and V. Zakharov, Phys. Lett. 55B, 397 (1975).
42. F. E. Close, R. L. Jaffe, R. G. Roberts, and G. G. Ross, Phys. Letts. 134B, 449 (1984).
43. R. L. Jaffe, Proc. of XI Int. Cong. PANIC, Kyoto, May 1987; MIT-CTP-1466.
44. F. E. Close, R. G. Roberts, and G. G. Ross, Phys. Letts. 129B, 346 (1983).
45. C. H. Llewellyn Smith, Phys. Lett. 128B, 107 (1983); M. Ericson and A. Thomas, Phys. Lett. 128B, 112 (1983).
46. E. Berger and F. Coester, Phys. Rev. D32, 1071 (1985); E. Berger, F. Coester, and R. Wiringa, Phys. Rev. D29, 398 (1984); E. Berger, Nucl. Phys. B267, 231 (1986).
47. NA(10): Phys. Letts. 193B, 368 (1987).
48. R. Bickerstaffe, M. Birse, and G. Miller, Phys. Rev. Lett. 52, 2532 (1984).
49. CDMS: H. Abramowicz et al., Z. Phys. C25, 29 (1987); E. Berger and F. Coester, Argonne report ANL-HEP-PR-87-13 (to appear in Ann. Rev. of Nuclear and Particle Science).
50. EMC: J. Aubert et al., Phys. Letts. 152B, 433 (1985); Nucl. Phys. B213, 1 (1983); F. E. Close, R. G. Roberts, and G. G. Ross, Z. Phys. C26, 515 (1985).
51. E691, M. D. Sokoloff et al., Flab 86/120E.
52. E537, S. Katsanevas et al.,
53. E672, reported by A. Zieminski in Proceedings of St. Croix Workshop on QCD, October 1987.
54. J. Moss et al., E772 collaboration.
55. C. Gerschel (NA38), p. 443 in Hadrons, Quarks, and Gluons, Editions Frontiers (J. Tran Thanh Van, Ed.) 1987.
56. T. Matsui and H. Satz, Phys. Lett. B178, 416 (1986).

Early Universe and Big Bang Nucleosynthesis

N.C. Rana

Tata Institute of Fundamental Research, Homi Bhabha Road,
Bombay 400005, India

This is a series of six one-hour lectures tuned to the level of a graduate course covering basically the background required for understanding the phenomenon of the big bang nucleosynthesis. It begins with a brief introduction to the geometry, dynamics and thermodynamics of the universe as a whole, followed by one lecture on the discovery, properties and implications of the 3 K microwave background radiation. Then we move on to the thermodynamical properties of the early universe, effects of pair annihilation, the role of the weak interactions in creating a neutrino background and freezing the ratio of the available free neutrons to protons. In the fourth lecture, we describe the process of the big bang nucleosynthesis leading to the formation of deuterium, helium and lithium. The methods of the observational estimations of these primordial abundances are discussed in the fifth lecture, and finally in the sixth, their comparison with the predictions of the standard model and the inadequacy of the standard model, if any. It is in this respect that primordial nucleosynthesis provides a testing ground for one of the possible cosmological consequences of the quark-hadron phase transition in the early universe.

1. Geometry and Dynamics of the Universe

Universe is defined to be a closed system which encompasses everything that is accessible to our studies. It has always been a profound question to ask about the nature of the universe, particularly how big it is, and its age and physical contents here and elsewhere.

1.1 Homogeneity, Isotropy and Expansion of the Universe

For ages, earth was considered to be the centre of the universe. It was only after Copernicus that earth had to lose this privileged position to sun. Very soon the process of decentralisation went further out, and it was conjectured that no point in the universe is preferential, a principle known as the Copernican principle. Newtonian mechanics also supported this viewpoint in the following way. We know that for studying the motion of projectiles, we consider the effect of earth alone and neglect the existence of anything outside it. Similarly, for the motion of planets, the effect of sun alone suffices and we neglect the existence of any star outside it. Thus for the motion of stars, we neglect the effect of other galaxies. All this is possible because of the Newtonian result that a homogeneous distribution of matter in a spherical shell does not produce any gravitational force at any interior point. It really does not matter whether the universe is finite or infinite in extension, so long as the distribution of matter is homogeneous and isotropic, the distant shells of matter can not produce any observable effect at the origin. The very fact that both the heliocentric and geocentric frames of reference can explain all the observable local effects, the universe ought to be homogeneous and isotropic with respect to sun as well as earth, and hence all other points in space. So from the mechanics point of view, no point in the universe is more privileged than the other.

By Birkhoff's theorem, for such frames of reference any scalar point function of coordinates alone must be same everywhere, for example, the matter density ρ_m must satisfy $\rho(\vec{r}, t) = \rho(\vec{r}', t)$, where $\vec{r}$ and $\vec{r}'$ are the position vectors of any two arbitrary points at time t. Homogeneity and isotropy do not necessarily mean that the universe is static, it can in fact be expanding. In fact, any vector point function $\vec{A} = f(t)\vec{r}$, defined in the model of $\vec{r}$ with an arbitrary but constant scale factor $f(t)$ for the whole space at a given instant, ought to satisfy the properties of homogeneity and isotropy. So the velocity of any point in the universe can be given by

$$\vec{v}(\vec{r},t) = f(t)\vec{r}, \qquad (1)$$

without violating the conditions of homogeneity and isotropy. Thus, if $f(t) = 0$, the universe is *static*, if $f(t) > 0$, the universe is *expanding*, and if $f(t) < 0$, the universe is *contracting*. Now, the question is, which solution has been chosen by the nature.

The definitive answer has been provided by the resolution of the so-called Olbers' paradox (Olbers, 1826), or in other words, the reason why the night sky appears to be dark. The argument goes as follows. Suppose the universe is uniformly filled with stars with a number density $n_\star$, each star emitting a power $L_\star$ in say, the optical band. Although both gravitation and apparent intensity obey inverse square law of distance, gravity as a vector quantity finds a chance to cancel for each spherical shell about the given point, but the intensity of light can not, as a result of which the intensity (I) grows with the upper limit of integration for distance (r_o), namely

$$I = \int_0^{r_o} \int_0^{\pi/2} \frac{L_\star \cos\theta}{4\pi r^2} 2\pi n_\star r^2 \sin\theta \, d\theta \, dr = \frac{L_\star n_\star r_o}{4}. \qquad (2)$$

Obviously, as $r_o \to \infty$, $I \to \infty$, leading to infinite brightness of the sky for an infinite homogeneous universe!

So some people suggested that the distribution of stars can not be infinite. However, it can easily be shown that in an expanding (not static or contracting) universe, this problem does not arise. Classically speaking, if the universe is expanding according to the law given by (1) and light is traveling with a finite speed c, the light emitted from any region beyond $r_o = c/f(t)$ can never reach the observer sitting at the origin, and thus defining the *world horizon* of the universe. r_o is finite and positive, only if $f(t) > 0$, that is, only if the universe is expanding. But it could not say how fast the expansion should be.

The other proof along with a quantitative measure of the speed of expansion came from Edwin Hubble (1929), who noticed from the redshift of spectral lines of all the nearby 29 galaxies sampled by him that they are all moving away from us with the speed of recession directly proportional to their distance, the relation being exactly given by (1). Even though his original estimates of distance to galaxies are now modified, the relation still holds good, and the constant of proportionality $f(t)$ is now denoted by H, the *Hubble's constant*, after his name. This remarkable discovery firmly establishes the expanding nature of the universe. The measured value of the Hubble constant at the present epoch $H_o = H(t_o)$ with its uncertainties in measurement ranges between 50 and 100 km/s/Mpc, pc being a unit of measuring distances in astronomy ($= 3.085678 \times 10^{16}$m or approximately 3.261633 light years). Because of its uncertainty, one usually denotes H_o by h_o in units of 100 km/s/Mpc, where $0.4 \lesssim h_o \lesssim 1.0$

Supposing that $h_o = 0.5$, it would mean that as we move out radially by a distance of 1 Mpc ($= 3.26$ million light years), the speed of expansion increases by 50 km/s. So the classically defined world horizon has a radius $r_o = c/H_o = 3000 h_o^{-1}$Mpc. The value of the Hubble constant can be looked at in another way. The galaxies on the spherical shell of radius 1 Mpc are now receding at a speed of $100h$ km/s. It means that these galaxies as well as the radius of the shell was smaller in the past. In fact this shell should have originated on earth at a time

$$t_o = r_o/v_o = H_o^{-1} \simeq 10 h_o^{-1} \text{Gy} \qquad (3)$$

ago. This is also true for all shell as t_o is independent of r_o, as if the entire universe originated from earth only a finite time ago. The same conclusion will be drawn by the observers on any other galaxy that the universe began from that point. The state of the universe at this particular instant is usually referred to as the *Big Bang*.

So, are left with two questions: did the universe begin only finite time ago and, is the universe finite or infinite? The existence of a finite world horizon does not necessarily imply its finiteness.

The first question is answered by various estimates of the ages of different objects in the universe. As we see it, the universe is made up of matter. Since all matter is potentially radioactive, nothing can exist forever. Unless there is continuous creation, the universe ought to have a finite age. Suppose, in the beginning the two isotopes of uranium, say ^{238}U and ^{235}U, formed in equal proportion, the present observed ratio in the terrestrial rocks (139 : 1) with their known mean life times, 6.5 and 1.0 Gy respectively, suggest that the terrestrial uranium was synthesised no earlier than about 5.8 Gy (by solving for t in $e^{\lambda_{235} t} : e^{\lambda_{238} t} = 139 : 1$). Slightly more sophisticated analysis of the relative abundances of their final radioactive products, ^{206}Pb and ^{207}Pb respectively, in Allende meteorites, the age of the earth has been determined to be 4.553 ± 0.004 Gy. The age of the Galaxy by such methods of nuclear cosmochronology turns out to be 13.6 ± 1.2 Gy. The age of certain oldest members of the Galaxy, called globular clusters is found to fall in the range of 13 – 18 Gy. All these estimates are consistent with the estimate of the Hubble age, given the uncertainties in the value of H_o.

However, proving finiteness of the age of the universe does not mean that the expansion cannot go forever. This issue can be studied using results of classical mechanics. For a homogeneous and isotropic universe, we can fix any point to serve as the origin of a frame. At a distance $R(t)$ at time t, the radial speed due to expansion is $\vec{\dot{R}} = H\vec{R}$, so the kinetic energy per unit mass $= \frac{1}{2}\dot{R}^2$, the potential energy per unit mass apart from a constant $= -GM/R = -\frac{4\pi}{3}G\rho R^2$, where G is Newtonian constant of gravitation and M is the mass enclosed inside the radius R, with mass density ρ which is uniform. Now, by the energy theorem, $\dot{R}^2 - 2GM/R=$ constant, say $= -k$, which immediately gives

$$\dot{R}^2 + k = \frac{8\pi G}{3}\rho R^2, \qquad (4)$$

for the evolution of R.

Now, if $k > 0$, the total energy per unit mass is negative, the value of R can not increase indefinitely with time, and hence, the galaxy now sitting at a distance of R should go to a maximum value sometime in future and then would begin to return towards the origin. This statement, if true for one galaxy, would be true for all galaxies, since there is assumed to be only radial motion of the Hubble expansion. Similarly, if $k < 0$, the total energy per unit mass is positive and $\dot{R}$, if positive once, remains so forever allowing no return to its original location. Such a universe must continue to expand for ever. The case for $k = 0$, is a critical case for never-ending expansion, the galaxy having just the sufficient amount of kinetic energy to take it to infinity. Since in all cases, $\dot{R} = v = HR$, we can rewrite (4) in the form

$$H^2 + \frac{k}{R^2} = \frac{8\pi G\rho}{3}, \qquad (5)$$

a relation among H, ρ and k, which gives an opportunity to determine the sign of k from the local measurement of H and ρ. In particular, when $k = 0$, we must have from (5) the so-called *critical density* of the universe for the present epoch

$$\rho_c = \frac{3H_o^2}{8\pi G} = 1.8788 \times 10^{-26} h_o^2 \text{ kg m}^{-3}, \qquad (6)$$

and obviously, $k < 0$ implies $\rho < \rho_c$, and $k > 0$ implies $\rho > \rho_c$. The precise measurements of the values of the Hubble constant and the present density of the universe (ρ_o) can answer to the question of whether the expansion of the universe will continue for ever or not. An observable parameter called the *density parameter* $\Omega_o = \rho_o/\rho_c$ is also defined to replace the role of k. Sometimes another parameter called the *deceleration parameter* q_o is defined through $q_o =$

$-\ddot{R}_o R_o/\dot{R}_o^2$, which plays a role similar to Ω_o. It can easily be checked that for $\Omega_o = 1$, $k = 0$ and $q_o = 0.5$, and that $\Omega_o \gtrless 1$, $q_o \gtrless 0.5$.

We now move to finding an answer to the second question posed earlier.

1.2 Is the Universe Finite or Infinite?

This profound question has hardly left the regime of philosophical arguments even in the age of positivistic reasoning and scientific objectivity. Even today, the simplest answer is that we do not know.

Newton thought that space and universe were infinite and that the star system was finite, because an infinite star system was shown to be unstable. Leibniz, a contemporary of Newton, argued for an infinite space and star system, as introduction of a finite star system implied a preferred centre, which violated the Copernican principle. So Kant, using the rationalistic argument, namely, if the universe has to satisfy that (i) the system of stars has no centre, (ii) the system of stars is stable, and (iii) the space is infinite, he concluded that there is no consistent solution to this problem and suggested not to discuss about it any further. Kant's logic remained unchallenged until Gauss, Lobachevski and Riemann probed deeper into the wisdom of Kant, which, of course, assumed the following unproven hypotheses to be true: (a) finiteness ↔ having a boundary, and (b) infiniteness ↔ having no boundary.

Even the Greeks, as early as fifth century BC, knew that earth is finite but not bounded. But it was certainly very difficult to extend the same idea to decide the geometry of the universe, nevertheless Riemann (1854) proved 'unboundedness of space does not imply infiniteness of space and a space endowed with positive curvature can be finite but unbounded, whereas the one with negative curvature remains infinite and unbounded.' If the distance between any two neighbouring points in space is given by

$$d\sigma^2 = R^2(t) \left[\frac{dr^2}{1 - kr^2} + r^2(d\theta^2 + \sin^2\theta d\phi^2) \right], \tag{7}$$

where (r, θ, ϕ) are the spherical polar coordinates with their standard relations to rectangular Cartesian coordinates, and $R(t)$ serving as the scale factor for measuring true distances in relation to the dimensionless r-coordinate, then $\delta \int_1^2 d\sigma = 0$, defines the 'straight line(s)' or *geodesics* between any two given points, and the possible sign of the 'curvature' of any such line (k/R) implies the finiteness of the space. For $k = 1$, the space is thus obviously finite but unbounded; for $k = -1$, the space is infinite and unbounded; whereas for $k = 0$, the space is critically infinite and unbounded.

Equation (7) satisfies homogeneity, because the coordinate differentials are invariant under any arbitrary translation of the origin to any point. It satisfies isotropy, because the coefficient of the radial part of the metric is independent of θ and ϕ and that of the angular part remains Euclidean. The expansion of space is incorporated in the functional dependence of $R(t)$ on t, so that the coordinates (r, θ, ϕ) of any particular galaxy can remain fixed for all time and $R(t)$ would describe its changing true (metric) distances to all other galaxies. For this reason, these coordinates are called *comoving* (moving with the cosmic fluid) coordinates. The space is expanding and carrying the cosmic fluid with it without changing the coordinates (r, θ, ϕ) of any particle in it, unless there is some peculiar velocity, thermal or chaotic.

The meaning of finiteness for the case of $k = 1$ would be clear if we try to compute the total volume available in the space defined by (7), which turns out to be $2\pi^2 R^3(t)$ (it is infinity for $k = -1$). The meaning of unboundedness for any 'straight line' (say along any radial line) becomes obvious as it traces a closed contour in $k = 1$ universe and an open ended curve in $k = -1$ universe. Hence, the former class of spaces is christened as *closed* spaces and the latter *open* spaces. We can also see that as we go out radially along the metric (7), the metric distance to a given galaxy increases as

$$\sigma(r) = \int_0^r R(t)\frac{dr}{\sqrt{1-kr^2}}, \tag{8}$$

giving

$$\begin{aligned}
\sigma(r) &= R(t)\sin^{-1}(r) \simeq R(t)\{r + \frac{r^3}{3!} - \cdots\} \quad \text{for} \quad k = 1, \\
&= R(t)r, \quad \text{for} \quad k = 0, \\
&= R(t)\sinh^{-1}(r) \simeq R(t)\{r - \frac{r^3}{3!} - \cdots\} \quad \text{for} \quad k = -1.
\end{aligned} \tag{9}$$

The metric distances are larger for the closed universe than the Euclidean or the open ones, while the separation between two neighbouring galaxies at the same radial distance reads identical in all the three cases.

The form of the 4-dimensional metric as the one given by Robertson (1935) and Walker (1936) for the expanding universe, to be called as Friedmannian universe (Friedmann 1922), can now easily be constructed by defining

$$ds^2 = c^2 dt^2 - d\sigma^2 = c^2 dt^2 - R^2(t)\left[\frac{dr^2}{1-kr^2} + r^2(d\theta^2 + \sin^2\theta d\phi^2)\right], \tag{10}$$

which when put in the Einsteinian field equations for a dust-particle-filled universe

$$R^\mu_\nu - \frac{1}{2}\delta^\mu_\nu R = \frac{8\pi G}{c^4}T^\mu_\nu$$

with the energy-momentum tensor

$$T^\mu_\nu = \left(\rho + \frac{p}{c^2}\right)u^\mu u_\nu - \delta^\mu_\nu p = (\rho c^2, -p, -p, -p)$$

leads identically to the equation (4) and to one more equation

$$\frac{d}{dt}\left(\rho R^3\right) + \frac{p}{c^2}\frac{dR^3}{dt} = 0. \tag{11}$$

This gives us an opportunity to understand the Newtonian meaning of the Friedmannian curvature parameter k. The *open* universe therefore stands for the ever-expanding *infinite unbounded* universe, and the *closed* one obviously for finitely (in time) expanding *finite* but *unbounded* universe. So the question of finiteness can only be answered if we can figure out the sign of k, or in other words, by measuring the value of the cosmological density parameter Ω_o.

The first equation namely, equation (4) corresponds to the conservation of energy in any comoving frame, and the second, the equation (11) can be rewritten as

$$\frac{dE}{dt} + p\frac{dV}{dt} = 0,$$

where E is the total energy $E = \rho V$ inside a comoving volume $V \propto R^3$. This is the first law of thermodynamics for adiabatic systems! So the expansion of the universe is not only *conservative* but also *adiabatic* in nature. Therefore, we can say that the first equation represents the dynamical evolution of the universe and that the second equation the thermodynamical evolution of the expanding universe.

1.3 Dark Matter Problem

The most direct way to measure the average mass density of the universe would be to sample a reasonably large volume of space. If there are n_G galaxies per Mpc^3, each having an average mass M_{11} in units of $10^{11} M_\odot$ ($1\ M_\odot = 1.989 \times 10^{30}$ kg), we must have

$$\rho_o = 6.5 \times 10^{-27} n_G M_{11}\ \text{kg m}^{-3} \simeq 2.5 \times 10^{-28}\ \text{kg m}^{-3}, \qquad (12)$$

for $n_G \simeq 0.02$ and $M_{11} \simeq 2$, which accounts for no more than a few percents of the value of the critical density derived in equation (6). Of course, the major uncertainty lies in the determination of the mass of individual galaxies. Assuming that galaxies are mostly populated by sun-like stars, the total luminosity of a galaxy gives an estimate of the mass of its luminous portion. Another way to estimate the mass of a gravitating system is to see how fast its various components are moving around its centre much like the speed of revolution of planets at known distances indicating the mass of the central body, sun. In late thirties, Zwicky (1938) noticed that the dynamical mass obtained from the measurement of the speeds of revolution stars around galaxies (which is plotted to give what is called a *rotation curve* of a galaxy) was in general many times higher than the total mass indicated by its optical luminosity. For example, the luminous part of our Galaxy suggests its mass to be about 10 times smaller than its dynamical mass estimated from the speed of revolution (about 465 km/s) of a distant star called R15 at a distance of about 60 kpc from the centre of the Milky Way. This means that matter might exist mostly in nonluminous form, often dubbed as the cosmic *dark matter*. Such discrepancies or evidences for dark matter are widespread in the universe on various scale. Applying similar dynamical procedures to entities called groups of galaxies, binary galaxies and cluster of galaxies, it is now believed that more than 90% of matter in the universe *possibly* exist in the *dark* form.

The next problem comes as to the nature of the dark matter. However, they can not be in the form of neutral hydrogen existing in the intergalactic space due to reported lack of electromagnetic emission in significant quantities at its characteristic wavelength of 21 cm. They can neither be in the form of hot ionised hydrogen due to lack of the x-ray bremsstrahlung radiation, nor in the form of dust particles. Other baryonic candidates for dark matter could be the dead remnants of stars in the form of dead neutron stars or non-accreting black holes, or failed stars (brown dwarfs) that never found a chance to begin thermonuclear power generation in them. There are of course a huge number of proposed nonbaryonic candidates, mostly elusive and exotic particles, namely, neutrinos, axions, photinos, gravitinos, etc. A speculative solution to the problem of dark matter also comes from the quarter of the nuclear-particle physics, which has made us assembled here. One of the aims of the present meeting is to critically review the situation as to how likely it is that the baryonic form of matter can still be passed as dark matter, posing no threat to the nice results of the primordial nucleosynthesis, which is the main subject of my talk.

1.4 Thermodynamics of Expanding Universe

In the universe, matter exists in different physical states because of a wide range in the distributions of density and temperature. Given sufficient time, any such a system will try to achieve a state of thermal equilibrium. The second law of thermodynamics suggests that the universe as a whole must have been continually and irreversibly degrading from order to chaos, from nonuniformity to uniformity and from structure to structurelessness. This is alarming because the universe will be approaching a state of what is known as 'heat death'. All thermal engines work due to the *ab initio* differences of temperature between two parts. If that keeps on decreasing, all engines and therefore work producing agents will slowly stop functioning. If the present state of the universe exhibit lot of structures and highly nonuniform distribution of temperature, the second law of thermodynamics would suggest that the past was more full of structures and nonuniformities in temperature and that the future will be a doomed one. Would it have to be really true for an expanding universe?

The expanding universe was obviously more dense in the past as the total number of galaxies inside a comoving sphere of radius R was forced to share a smaller volume of metric space, implying that $n_G R^3(t)$ = constant, and hence the rest mass density of the galaxies

(inclusive of any such dark matter)
$$\rho_m \propto R^{-3}. \tag{13}$$

If we keep on increasing the density of matter indefinitely, depending on its temperature, the matter will either become highly degenerate due to degeneracy pressure of electrons or assume a state of nondegenerate thermal bath. In either state, the system will approach a state of uniformity in both density and temperature, implying that the past was more uniform than the present, or in other words, structure and order in the universe should have appeared and developed with time. This is in clear contradiction with the second law of thermodynamics. Should then the thermodynamics of the expanding universe be different from the thermodynamics of the laboratory systems? Furthermore, had the universe been totally homogeneous both in density and temperature in the past, it is perhaps impossible, it seems, to create any structure ever out of this state.

In order to resolve this issue, let us assume that the expanding universe at some stage is found to be made up of matter and radiation both in thermal equilibrium at a uniform temperature T. The density and pressure of radiation are $\rho_r = aT^4/c^2$ and $p_r = aT^4/3$ respectively, where a is Stefan's constant. If nonrelativistic, the pressure of matter is, $p_m = nkT$, n being number density of matter particles each having rest mass m_p and therefore the total energy density of matter $= m_p n c^2 + 3nkT/2$. Since the universe is expanding and there is nothing exterior to it, the expansion must be adiabatic, for which the first law of thermodynamics suggests that the change in total energy dE must equal to work done on the system $-pdV$, V being the comoving volume of any region, across the boundary of which no matter is exchanged. If we take this region to be a sphere of radius R, total volume is $V = \frac{4\pi}{3}R^3$. So, for radiation under adiabatic expansion, we have by the first law of thermodynamics (11),

$$\rho_r \propto R^{-4}, \quad \text{or,} \quad T \propto R^{-1}. \tag{14}$$

Similarly, for matter under adiabatic expansion,

$$\rho_m \propto R^{-3} \quad \text{or,} \quad T_m \propto R^{-2} \propto T^2. \tag{15}$$

This means that the temperature of nonrelativistic matter will be falling faster than that of radiation, giving rise to a difference of temperature between the matter and radiation components. However, this difference will fail to manifest until the rate of thermalisation falls short of the rate of expansion. When this happens, the two components become thermally isolated from each other and each component evolves on its own following the respective laws of adiabatic evolution.

Again, it can be shown that gravitating systems can have negative specific heat, which can make the system thermodynamically unstable. Given a small perturbation in temperature, the perturbation keeps on growing in gravitating systems. In short, it is the queer nature of gravitation that has made the universe expanding and reversed the usual direction of thermodynamic evolution.

Furthermore, if at any stage the universe was in full thermal equilibrium with matter, then from (14) and (15) it can be shown that the universe ought to have an early phase of radiation domination, because the total density $\rho = \rho_m + \rho_r \simeq \rho_r$ for all $R < \tilde{R} =$ the value of $(\rho_r R/\rho_m)$ evaluated at the present epoch, see Fig 1. ρ_r is always rising faster than ρ_m for smaller values of R. For an early enough epoch satisfying $R < \tilde{R}$, the universe must be *radiation-dominated*, followed by its transition to $R > \tilde{R}$ when the universe becomes *matter-dominated*. This general result led Gamow to expound his theory of the *hot big bang* in 1946. Incidentally, the phrase *big bang* was introduced by Hoyle while delivering a talk over the radio BBC in 1950.

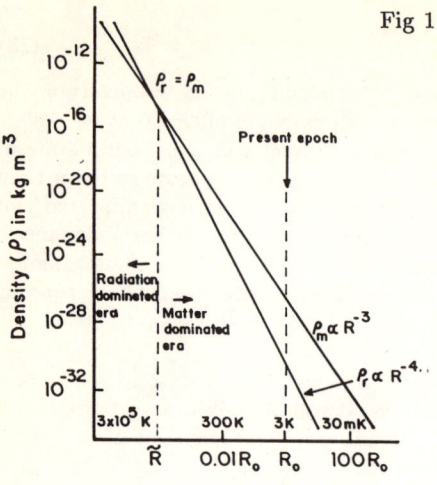
Fig 1

1.5 Dynamical Evolution of the Universe

The equation (4) describes the the dynamical evolution of the universe in the sense how the scale factor of the universe or the comoving radial distances to galaxies evolve with time. The equation can be solved subject to the knowledge of the R-dependence of the total density, $\rho = \rho_m + \rho_r$, which can be taken from (14) and (15). Since we shall be more interested in the early universe, the density term on the right hand side of (4) will become much more dominant over $k\ (=\pm 1)$ and hence can be neglected, giving

$$\left(\frac{1}{R}\frac{dR}{dt}\right)^2 = \frac{8\pi G\rho}{3}. \tag{16}$$

On integration, one finds for the age of the universe (Peebles, 1968)

$$t = \left(\frac{3}{2\pi G\rho_m}\right)^{1/2}\left[\frac{1}{3}\left(1+\frac{\rho_r}{\rho_m}\right)^{3/2} - \frac{\rho_r}{\rho_m}\left(1+\frac{\rho_r}{\rho_m}\right)^{1/2} + \left(\frac{\rho_r}{\rho_m}\right)^{3/2}\right]. \tag{17}$$

However, if the universe is totally dominated by either matter or radiation, the use of the two component density can be avoided, and the integration becomes then much simpler to carry out. For example, for the *radiation-dominated early universe*, we must have $\rho \propto R^{-4} \propto T^4$, and hence the equation (16) must read as $\dot{R}^2 = 8\pi G\rho R^2/3 \propto R^{-2}$. It is a standard practice to express the total density ρ as some g_{eff} times the pure radiation density ρ_r, the latter being given by Stefan's law, $\rho_r = aT^4/c^2$, where $a = 7.56591 \times 10^{-16}$ J m^{-3} K^{-4} is the value of Stefan's constant. The dynamical equation is exactly solvable with the condition for adiabaticity, namely $RT =$ constant, giving finally for the age of the radiation-dominated universe

$$t = \left(\frac{3c^2}{32\pi G g_{\text{eff}} a}\right)^{1/2} T^{-2} = 2.3049 \times 10^{19} g_{\text{eff}}^{-1/2} T^{-2} \text{ sec}, \tag{18}$$

and other auxiliary relations, such as

the scale factor $R \propto t^{1/2}$,   the Hubble constant $H = (2t)^{-1}$, (19)

and the horizon radius, Rindler (1956),

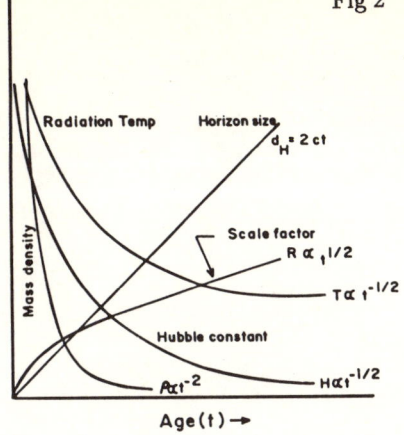

Fig 2

$$d_H = cR(t)\int_0^t \frac{dt'}{R(t')} = 2ct = \frac{c}{H}. \qquad (20)$$

These dependences are graphically demonstrated in Fig 2. We leave it as an exercise for deriving the respective formulae for the matter-dominated era.

2. Microwave Background Radiation

In the previous lecture, we have analysed how much matter is there in the universe. Is there really any contribution to mass density that might come from electromagnetic radiations?

2.1 Discovery of Microwave Background Radiation

Like a collapsing balloon getting hotter and hotter, Gamow contended in the late forties that the big bang had possibly occurred in a hot state, thus the term '*hot* big bang', predicting a uniform background of cold relic radiation of 5 K (Alpher and Herman, 1950).

Even before Gamow talked about it, McKellar (1941) had recorded existence of a cold radiation field in the interstellar space, which constantly kept the interstellar CN molecules excited to higher states of rotations. Thus, when light from a star (ζ Bootes) behind passed through this CN bearing cloud on its way to earth, the molecular absorption lines other than R(0), such as R(1), P(1) and P(2), were also formed. The measurement of the relative strengths of these absorption lines helped him derive the temperature of the radiation field that caused the excitation to be about 2.3 K.

This fact remained unnoticed till its accidental rediscovery by two radio astronomers of the Bell Laboratory, Penzias and Wilson in 1965. They were puzzled by an additional radio background at their working wavelength 7.35 cm, apparently coming uniformly from all directions of space. They ruled out the possibility of solar contribution by seeing no diurnal variation in the reception of radio waves through their horn antennae pointed vertically upward. Similarly, the galactic contribution was also ruled out in absence of any yearly variation. They contacted people at Princeton for a possible explanation. It turned out that Dicke (1962) and others at Princeton were planning for discovering the relic radiation from the big bang. So the Astrophysical Journal published Dicke's search for the relic radiation *followed* by the report on the accidental discovery of the same by Penzias and Wilson. The latter two were honoured with the Nobel prizes in physics for the year 1978.

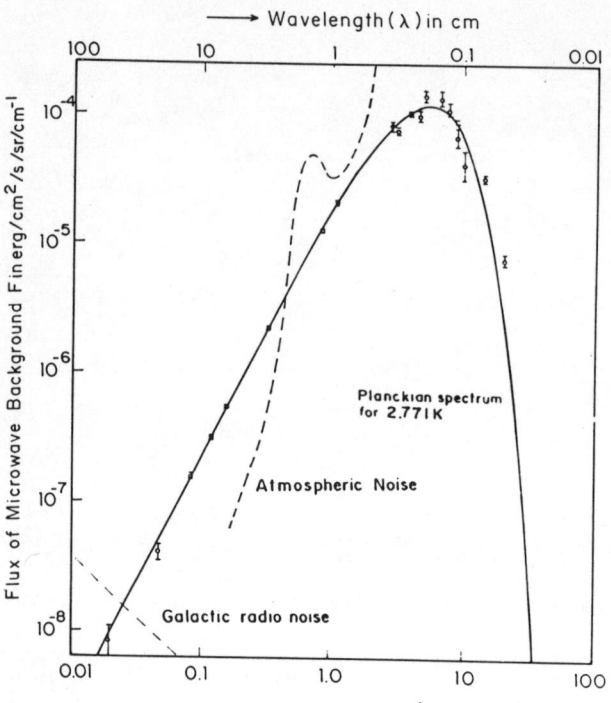

Fig 3

Penzias and Wilson measured the intensity of the background radiation only at one wavelength. In order to establish its thermal character, one has to measure the intensities at several wavelengths along with its polarisation properties. The strength of this signal was rather extremely weak compared to the atmospheric emissions at these wavelengths, the Galactic emission interfered quite strongly on the longer wavelength side (see the dotted lines in Fig 3). Soon a large number of independent measurements both on the ground and on board balloons, were scanned at several wavelengths and it fitted nicely with the expected intensity pattern on the long wavelength side of a black body emitting at temperature 2.7 K. The Wien's law of radiation thermodynamics suggests that the peak of intensity must lie at the wavelength $\lambda_m = 0.29$ cm $/T$ (K) $\simeq 1.1$ mm, a region where atmospheric interference was too much. Finally the peak as well as parts of the other side of the spectrum was explored for the first time in 1979 by Woody and Richards from Berkeley and established its overall Planckian shape. A compilation of the existing data taken mostly from Crane et al (1989) is shown in Fig 3, which suggests a nice spectral fit to a single black body temperature of 2.771 ± 0.012 K. Some people refer to this background radiation as Microwave Background Radiation (in short MBR), others call it Cosmic Microwave Background (CMB) or Cosmic Background Radiation (CBR) or simply 2.7 K or 3 K Background Radiation.

2.2 Observed Properties of MBR

That this radiation is thermal in nature was further supported by the measurement of its polarisation. Lubin and Smoot (1980) found no sign of polarisation down to the level of 0.3 mK, fully consistent with the expectation.

The radiation was referred to as a background radiation, which required that its origin must not be associated with any foreground source(s). Proving this was no simple. The excitation of the interstellar CN molecules proved that this radiation extends at least up to interstellar spaces. Its profound interaction with the Galactic cosmic rays and with ultrahigh energy electrons in extragalactic radio sources proves its existence on extragalactic scales. In fact, the verification of the predicted Zeldovich and Sunyaev (1969) effect, namely slight cooling of the radiation during passage through hot intracluster medium of the rich Abell clusters, seems to be most direct proof of its existence beyond these clusters of galaxies.

Most important property of MBR that led to developing viable scenarios of galaxy formation was its absolutely high degree of isotropy. The matter distribution on large scales seems to be more than thousand times more clumpy than the distribution of MBR over different angular scales.

Suppose, we try to fit the temperature distribution of MBR with respect to absolute directions in space by

$$T_\gamma(\theta,\phi) = \sum_{l,m} a(l,m) Y_{lm}(\theta,\phi) + T_{\text{noise}} = T_o + T_1 \cos\theta + \sum_{m=-l}^{l} T_{2m} Y_{2m}(\theta,\phi) + \cdots + T_{\text{noise}},$$

where T_o is the uniform part of the MBR = 2.77 K, T_1 = amplitude of the dipole component, T_{2m} = three amplitudes of the quadrupole components, and T_{noise} is the noise temperature which includes all fine scale anisotropies.

The dipole component of the MBR was measured latest and most accurately by Strukov et al. (1987) having the amplitude $T_1 = 3.16 \pm 0.12$ mK, the axis of the dipole pointing in the sky towards $RA = 11^h.3 \pm 0^h.24$ and declination $\delta = -7°.5 \pm 2°.5$. The standard interpretation of the dipole variation has been that the observer that is, earth is moving with respect to the cosmic frame of reference, the Robertson-Walker comoving frame in which the MBR together with the whole universe remains isotropic and homogeneous by definition. If the earth is moving with a velocity $\vec{v}$ with respect to the cosmic frame, then all radiations will be Doppler shifted by the standard formula for the special relativistic Doppler shift given by

$$\nu \to \nu' \text{ where } \nu = \nu' \left(1 - \frac{v}{c}\cos\theta\right)(1 - v^2/c^2)^{-1/2} \simeq \nu'(1 - \frac{v}{c}\cos\theta).$$

Since for a given direction all frequencies of the incoming MBR will be shifted by the same factor, it will effectively keep the Planckian shape intact but the temperature of the MBR will seem to change from

$$T_o \to T' \simeq T_o(1 - \frac{v}{c}\cos\phi),$$

which mimicks like a dipole term with the coefficient

$$T_1 = T_o(v/c). \tag{21}$$

If $T_1 = 3.16 \pm 0.12$ mK and $T_0 = 2.77$ K, the earth's motion with respect to cosmic frame becomes 350 ± 15 km/s. Correcting for the earth's orbital motion around the sun at the time of measurement sun's motion with respect to cosmic frame can be obtained as 316 ± 12 km/s. Similarly correcting for the motion of sun around the centre of the milky way, our galaxy is supposed to be moving with a velocity 510 ± 50 km/s with respect to the cosmic frame in a direction RA $= 10^h.7 \pm 0^h.5$ and declination $= -23° \pm 17°$. People try to interpret the peculiar motion of the galaxy as a free fall towards the Virgo cluster of galaxy, the later being only $45° \pm 10°$ away from the dipole axis of the MBR.

Anisotropy measurements of the quadrupole and octupole components and over angular scales as small as $6°$ at wavelength $\lambda = 8$ mm have been carried out by Klypin et al (1987) and are given by only the upper limits of order $\delta T/T \lesssim 2 \times 10^{-5}, \lesssim 7 \times 10^{-5}, \lesssim 5.6 \times 10^{-6}$ respectively. Therefore virtually no anisotropy has been reported above a few parts in 10^{-5}. The latest measurement at $\lambda = 1.5$ cm by the Owen Valley group (1989) gives $\delta T/T < 1.7 \times 10^{-5}$ on $2'$, $< 9.4 \times 10^{-5}$ on $12''$ and $< 3 \times 10^{-4}$ on $25'$. These results have put stringent constraints on the models of galaxy formation.

2.3 Interpretation of MBR and implications

The standard interpretation of the origin of MBR is that it simply came with the hot Big Bang. The knowledge of its present temperature fixes the entire temperature evolution of the early universe given by equation (14). Since the radiation mimicks a blackbody of temperature 2.77 K, the early universe must have been in an equilibrium state at some stage of its evolution.

We know that the spectral distribution of these photons per unit volume is given by

$$\tilde{n}_\gamma(E_\gamma) = \frac{1}{\pi^2 c^3 \hbar^3} \frac{E_\gamma^2}{e^{E_\gamma/kT_\gamma} - 1}, \qquad (22)$$

where E_γ is the energy of photon and T_γ the equilibrium temperature of photons. On integration, the total number of photons per unit volume becomes

$$n_\gamma = \int_0^\infty \tilde{n}_\gamma(E_\gamma)\, dE_\gamma = \frac{2}{\pi^2}\left(\frac{kT_\gamma}{c\hbar}\right)^3 \zeta(3) \simeq 2.028719 \times 10^7 T_\gamma^3 \text{ m}^{-3}, \qquad (23)$$

giving about 3.99×10^8 $(T_\gamma/2.7 \text{ K})^3$ photons per m^3, which is extremely large compared to the average number density of baryons in a universe having the critical density, $11.23\, h_o^2$ m^{-3}, h_o being the present value of the Hubble constant in unit of 100 km/s/Mpc. So the universe is literally full of photons only. There is hardly one matter particle for every hundred million photons. The ratio of these two numbers, called the *baryon to photon number density ratio* η is obviously related to the baryonic component of the density parameter Ω_b by

$$\Omega_b = 3.5569 \times 10^7 \eta h_o^{-2} \left(\frac{T_\gamma}{2.7\text{K}}\right)^3. \qquad (24)$$

However, if we now compare the mass densities of these two components, photons will lose its leading position, for every proton outweighs an average photon of the background by the energy factor 938.5 MeV$/10^{-4}$ eV $\simeq 10^{13}$ so that at present

$$\frac{\rho_m}{\rho_r} = 4.20 \times 10^4 \frac{\Omega_b h_o^2}{T_{2.7}^4}, \qquad (25)$$

assuming that all matter is baryonic in form, that is, $\Omega_o = \Omega_b$. So when the scale factor R of the universal expansion was $\sim 10^4$ times smaller than its present value, the universe passed through a transition from the radiation dominated to matter dominated era (see Fig 1). For example, if we assume $\Omega_o = \Omega_b \approx 1$ and $h_o \approx 0.5$, the temperature of background radiation was then about 30,000°K. The transition temperature $\tilde{T}$ would be lower for smaller value of Ω.

At such high temperature matter was obviously completely ionised, in which case electrons alone will be in a position to scatter the radiation very strongly, and the process of ionisation and recombination of hydrogen will operate such that the radiation couples with matter through intensive radiative transfer. Now from (17), the age of the universe around that time could be given by $t \simeq 10^{12} T_4^{3/2}$ sec, T_4 being the temperature of radiation in unit of 10^4 K, and the number density of baryons for a critically dense universe $n_b \simeq 5.3 \times 10^{11} T_4^3$ m^{-3}. So the optical depth due to electron scattering over the Hubble distance at that time was $\kappa_{\text{el}} \sim n_b \sigma_T (ct) \sim 1.3 \times 10^4 T_4^{3/2}$, which is $\gg 1$ if $T_4 > 0.01$ or $T > 100K$. However, the exact variation of matter temperature was given by Peebles (1969):

$$\frac{dT_m}{dt} = \frac{8\sigma_T a T^4}{3 m_e} x_e (T - T_m) - \frac{2T_m}{R}\frac{dR}{dt}, \qquad (26)$$

x_e = ionisation fraction of matter assumed to exist in the form of hydrogen only. If $x_e = 0$, the equilibrium matter temperature $T_m \propto T^2$ as expected, (see equation (15)).

Table 1. Decoupling of Matter Temperature from Radiation Temperature

Radiation temperature T(K)	Fractional ionisation (x_e)	Matter temperature T_m K	Radiation temperature T(K)	Fractional ionisation (x_e)	Matter temperature T_m K
5000	0.996	5000	2000	0.000123	1920
4500	0.920	4500	1500	0.000053	1280
4000	0.400	4000	1000	0.000032	680
3500	0.072	3500	500	0.000022	197
3000	0.0098	3000	200	0.000017	33
2500	0.00092	2500			

However, for the exact behaviour of T_m as a function T, one has to solve the equation (26). Thus, on solving equation (26), one can find the fractional ionisation x_e and the matter temperature T_m as a function of the radiation temperature T, and the result is shown Table 1. When the value of x_e fell to 0.5, the temperature of the universe was about 4100 K, which is taken to be the *temperature of decoupling of radiation from matter*, by definition.

However the entries in the 3rd column of Table 1 show that the thermal coupling continued until the radiation temperature fell below 2200 K. The reason for this extended thermal contact is due primarily to the processes such as

$$H^* \rightarrow H + \gamma_1 + \gamma_2, \quad H^* + \text{soft } \gamma \rightarrow p + e,$$
$$H + H \rightarrow H_2 + \gamma, \quad H + p \rightarrow H_2^* + \gamma, \quad \text{etc.}$$

which could operate at lower temperature with more soft photons than those of the Lyman continua.

So we expect that the matter in the space between the galaxies, called the intergalactic medium should be at the present epoch in the neutral phase, that is, hydrogen and helium not ionised. However the present day intergalactic medium is found not to contain any trace of gas in the neutral phase but in the completely ionised state with a characteristic kinetic temperature exceeding 10^4 K. A typical evolution of the matter temperature in recent times is shown in Fig 4. The radiation temperature drops continuously as R^{-1} or $\propto (1+z)$ where z is called the *cosmological redshift*, defined through the relation

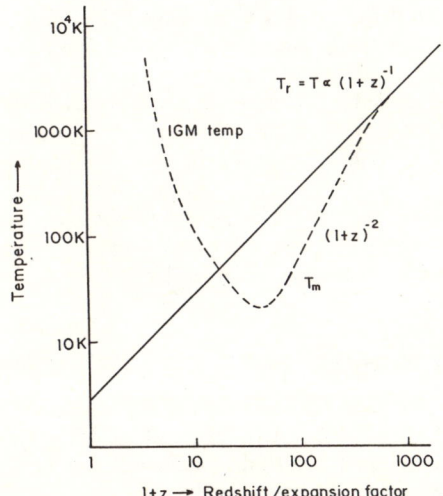

Fig 4

$$1 + z = R(t_o)/R(t), \tag{27}$$

where t_o is the present epoch and t any arbitrary epoch.

Since T_m is initially falling as T^{-2} and at present T_m is $\gg T$, T_m must pass through a minimum, between $z = 10$ and 100. The latest rise in T_m is possibly due to the *reionisation* of matter during the process of galaxy formation. In any case, assuming that the intergalactic medium has remained almost fully ionised since the redshift $z \approx z_o$, the optical depth due to Thomson scattering of the low energy MBR photons (appropriately blueshifted) by these lately released free electrons is given by

$$\kappa_{el} \simeq 0.07 \Omega_b h_o \int_0^{z_o} \frac{(1+z)dz}{\sqrt{1+\Omega_o z}} \simeq 0.05 \Omega_b h_o (1+z_o)^{3/2} \qquad \text{for } \Omega_o = 1.$$

If $\Omega_b h_o \simeq 0.05 - 0.5$, $\kappa_{el} = 1$ corresponds to $z_o \simeq 50$.

This means that if the galaxy formation has taken place earlier than $z_o = 50$, the MBR is bound to be recoupled with hot intergalactic medium through Thomson's scattering by free electrons, a phase often termed as *reionisation* of the universe. This will certainly isotropise the background radiation provided the intergalactic medium was homogeneous enough at the time of reionisation, if any.

2.4 Implications for the Development of Structures

First we consider the small scale anisotropy of MBR. Any such anisotropy reflects the degree of density inhomogeneity at the time of decoupling of MBR. Since the latest measurement on the small scale MBR is found to show $\delta T/T < 2 \times 10^{-5}$ on $2'$, it implies certain density contrast $\delta \rho_m / \rho_m$ at the time of recombination. The age of the universe at the time of recombination was about 300,000 years.

If the density contrast arises due to *adiabatic* fluctuations, it ought to satisfy $\rho_m \alpha T^{-3}$ (the condition of adiabaticity, see equation(15)), or

$$\frac{\delta \rho_m}{\rho_m} \simeq \frac{3 \delta T}{T} \lesssim 6 \times 10^{-5}. \tag{28}$$

Since in the post-recombination era $\delta \rho_m / \rho_m$ grows as $(1+z)^{-1}$, it fails to grow up to ~ 1 at the present epoch. Therefore the class of *adiabatic models* fails to produce the inhomogeneity in large scale structure of the universe observed at present.

Another type of density contrast called the *isothermal* or *isocurvature* fluctuation is nowadays preferred to adiabatic ones. In this case the temperature remains uniform and the density contrast represents real clumpiness of matter in space. Usually the dark matter in the form of neutrinos or other particles is favoured for generating isothermal fluctuations at very early epochs. Readhead *et al* (1989) have scrutinised the prospect of explaining the small scale anisotropy of MBR in different classes of models generating fluctuations. It seems according to them that the isocurvature models can be made consistent with the observed limits of anisotropy of MBR as small and intermediate angular scales.

The large scale anisotropies also have implications for the probable size of the particle horizon at the time of recombination. The particle horizon at an epoch characterised by the cosmological redshift z_s subtends an angle θ_s at the present epoch to be given by

$$\sin\left(\frac{\theta_s}{2}\right) = \frac{\Omega_o \sqrt{1+\Omega_o z_s}}{\Omega_o z_s + (\Omega_o - 2)\left(\sqrt{1+\Omega_o z_s} - 1\right)} \tag{29}$$

or, $\theta_s \simeq 2\sqrt{\Omega_o/z_s}$, which corresponds to $\theta_s \sim 3°$ at $z_s \sim 1000$. Any structure of MBR on angular scales $\gtrsim 3°$, the adjacent density contrast has to be causally disconnected from each other. In such cases also, the density fluctuations have to be isothermal in nature.

3. Thermal History of the Early Universe

We now shift our interest towards the thermal and dynamical evolution of the universe prior to the recombination of MBR. The scale factor R of the universal expansion was smaller and smaller in the past. The temperature of radiation field was higher following the rule $T_r \propto R^{-1}$, the density of matter component was higher according to $\rho_m \propto R^{-3}$ and the density of radiation component ρ_r was higher according to $\rho_r \propto R^{-4}$. At some stage the universe reached a radiation dominated phase. Given the present day temperature of the MBR, this phase of radiation domination has possibly occurred sometime before the decoupling of radiation. So the universe had a single temperature with $T_m = T_r = T$ prior to this epoch. When $T > 10^4 K$, all matter (which consisted of mainly hydrogen and helium) was ionised forming a soup of primordial plasma.

When the temperature of the radiation field increases up to a level of 10^9 K, or in terms of the average energy of photons becomes ~ 100 KeV, the nuclear processes begin to operate. It is around this time that the universe underwent the process of primordial nucleosynthesis. The age of the universe was then about few minutes. If we go further back in time, say when the temperature of the universe was about 10^{10} K or the average energy of the photons $E_\gamma \sim 2.7kT \sim 1$ MeV, the photons interacting with matter could produce $e^\pm$ pairs (rest mass of electron ~ 0.510999 MeV). Since the number density of photons at any temperature is given by the Planck law (23), under thermal equilibrium almost an equivalent number of $e^\pm$ pairs would be produced, irrespective of the original number density of electrons or protons; the only conservation it respects being the charge conservation, given by

$$n_{e^-} - n_{e^+} = N_A \rho_b \sum_i X_i \frac{Z_i}{A_i}, \qquad (30)$$

where N_A = Avogadro's number, $i = i^{\text{th}}$ nuclear species, X_i = its abundance by mass fraction, Z_i = its charge number, and the sum extends over all the nuclear species. We have already seen that in the present day universe there are no more than 1 or 2 electrons for every 10 billion photons. This ratio had got elevated to almost unity when the pair production processes became operative at temperature $T \sim 10^{10}$ K. The rule of thumb for such a process to set in is very simple: whenever the temperature of radiation is higher than a certain value $T = m_A c^2/k$, a new species of particle, say of type A and rest mass m_A can be produced in pairs by the process

$$\gamma + \gamma \rightleftharpoons A + \bar{A}, \qquad (31)$$

where $\bar{A}$ is the antiparticle of A. Under thermal equilibrium, the number densities on both the sides ought to become comparable to each other, provided the chemical potentials are negligible compared to the average energy of the particles. For example, the muons could have been copiously produced at a temperature $\gtrsim 10^{12}$ K (~ 100 MeV), since the rest mass of $\mu^\pm \sim 100$ MeV. In fact, all possible types of particle species can show up in the early universe, because as the temperature $T \to \infty$ and $t \to 0$, and the reaction channels (31) can have sufficient cross sections for back and forth reactions. Even the extremely weakly interacting neutrinos try their luck and succeed, as we shall see later.

3.1 Energy and Momentum Distribution of Particles in Thermal Equilibrium

When the time scale of interaction of any species of particles with the radiation field is much shorter than that of the dynamical evolution of the system as a whole, the system achieves an overall thermal equilibrium with the energy and momentum distributions of the individual species of particles given by their statistically equilibrium distributions. The distributions come only in two varieties, one for the boson type of particles (with integral spins) and the other for the fermionic type of particles (with half-integral spins).

For any bosonic type of particles in thermal equilibrium at temperature T, the momentum distribution is given by

$$\tilde{n}_b(p) = \frac{p^2}{2\pi^2 \hbar^3} g_b \left(\frac{1}{e^{(E-\mu_b)/kT} - 1}\right), \tag{32}$$

where μ_b = chemical potential of the particular type of boson b, g_b = spin multiplicity factor for the same bosons (for example, for photons $g_\gamma = 2$ and $\mu_\gamma = 0$) and $E = \sqrt{p^2c^2 + m_b^2 c^4}$ = the energy of the bosons of momentum p and rest mass m_b. Similarly for the fermion type of particles the momentum distribution in given by

$$\tilde{n}_f(p) = \frac{p^2}{2\pi^2 \hbar^3} g_f \left(\frac{1}{e^{(E-\mu_f)/kT} + 1}\right), \tag{33}$$

g_f, μ_f and m_f being the spin multiplicity factor, the chemical potential and the rest mass of the given species of fermions denoted by the suffix f.

The momentum distributions for their antiparticles, be they bosons or fermions, are identical to theirs own except for a flip in the sign for the chemical potential, that is,

$$\mu_{\bar{f}} = -\mu_f \quad \text{and} \quad \mu_{\bar{b}} = -\mu_b.$$

This follows simply from the very nature of the thermalising reactions $f + \bar{f} \rightleftharpoons \gamma + \gamma$ and $b + \bar{b} \rightleftharpoons \gamma + \gamma$, combined with the fact that photons do not have any chemical potential as their numbers are not conserved, allowing processes like $e^- + p \rightleftharpoons e^- + p + \gamma$. One may remember that if there is any reaction $A + B \rightleftharpoons C + D$ then under thermal equilibrium their chemical potentials satisfy

$$\mu_A + \mu_B = \mu_C + \mu_D. \tag{34}$$

Thus, μ is additively conserved in all reactions.

In the standard hot big bang model, we take $\mu_e \neq 0$, the value of which for a given temperature is derived by solving the equation (30) after inserting the expressions for $n^\pm$, and the chemical potentials for all other species are set to zero. One should also remember that the value of μ_i once determined at a given temperature does not remain same at other temperatures. Unlike the energy and momentum, chemical potential increases as the temperature decreases; the dependence becomes exponential during the transition from the relativistic to the nonrelativistic regime of the particle energies (for example, see the last column of Table 2 for the evolution of μ_e in the early universe).

The number and energy densities and the partial pressures of the particle species in thermal equilibrium are given by

$$n_i = \int_0^\infty \tilde{n}_i(p)\, dp, \quad c^2 \rho_i = \int_0^\infty \tilde{n}_i(p) E_i(p)\, dp, \quad \text{and} \quad P_i = \int_0^\infty \frac{p^2 c^2 \tilde{n}_i(p)\, dp}{3 E_i(p)}, \quad i = b, f. \tag{35}$$

These integrals are generally not expressible in closed form except for the following special cases.

When the temperature of a thermally equilibrium system is so high that the average energy of the particles greatly exceeds the rest mass energy of the particles, that is, $kT \gg E \simeq pc \gg m_o c^2$, all bosonic distributions become identical to that of photons except for the generally nonzero chemical potential factor (μ_b) and for the difference in the spin multiplicity factor (g_b), if any. Under ultrarelativistic conditions with negligible chemical potentials, the integrated number and energy densities and the partial pressures of bosons and fermions at equilibrium are related to each other by

$$n_f = \frac{3}{4}\left(\frac{g_f}{g_b}\right) n_b, \quad \rho_f = \frac{7}{8}\left(\frac{g_f}{g_b}\right)\rho_b \quad \text{and} \quad P_{b,f} = \frac{1}{3}\rho_{b,f} c^2. \tag{36}$$

However, if the chemical potentials (particularly, of the fermions) are non-zero, in the ultarelativistic limit, the following results are expressible in closed form:

$$n_f - n_{\bar{f}} = \frac{g_f}{2\pi^2}\left(\frac{kT_f}{c\hbar}\right)^3\left[\frac{\pi^2}{3}\left(\frac{\mu_f}{kT_f}\right) + \frac{1}{3}\left(\frac{\mu_f}{kT_f}\right)^3\right],$$

$$\rho_f + \rho_{\bar{f}} = \frac{g_f}{2\pi^2}\frac{(kT_f)^4}{(c\hbar)^3}\left[\frac{7\pi^4}{60} + \frac{\pi^2}{2}\left(\frac{\mu_f}{kT_f}\right)^2 + \frac{1}{4}\left(\frac{\mu_f}{kT_f}\right)^4\right], \quad (37)$$

and $\quad P_f + P_{\bar{f}} = \frac{1}{3}(\rho_f + \rho_{\bar{f}})c^2.$

Otherwise, one has either to evaluate (35) numerically or use approximate series expansions in terms of modified Bessel functions (Wagoner *et al* 1967), which finally reduces in the non-relativistic limit, to the form

$$n_i = \left(\frac{m_i kT}{2\pi\hbar^2}\right)^{3/2} g_i \exp\left(\frac{\mu_i - m_i c^2}{kT}\right). \quad (38)$$

Therefore, the total mass density of a system at thermal equilibrium is given by

$$\rho = \sum_b \rho_b + \sum_f \rho_f \equiv \rho_\gamma g_{\text{eff}}, \quad (39)$$

which defines g_{eff} as the effective increase in the spin multiplicity factor of the system compared to that of the photons. In general, the temperature of the individual bosons (T_b) and fermions (T_f) might not be identical to that of photons (T), in which case

$$2g_{\text{eff}} = \sum_b g_b \left(\frac{T_b}{T}\right)^4 + \frac{7}{8}\sum_f g_f \left(\frac{T_f}{T}\right)^4. \quad (40)$$

It is customary to express the temperature in units of 10^9 K denoted by T_9. The mass density of photons for a given T, is by Stefan's law

$$\rho_\gamma = a\frac{T^4}{c^2} = 8.4182 \times 10^3\, T_9^4 \text{ kg/m}^3,$$

where a is the usual Stefan's radiation constant. It is now easy to find out the mass density of $e^\pm$ in thermal equilibrium with photons. The mass density of $e^\pm$ pairs in thermal equilibrium with photons at the ultrarelativistic temperatures $T \gg m_e c^2/k = 5.92986 \times 10^9$ K, and with negligible chemical potential $\pm\mu_e$ for electrons and positions becomes

$$\rho_{e^\pm} = 2 \times \rho_e = 2 \times \frac{7}{8}\rho_\gamma = 1.4732 \times 10^4 T_9^4 \text{ kg/m}^3.$$

We show in Table 2 how the mass densities of the $e^\pm$ pairs $(\rho_{e^\pm})$ and the neutrinos $(\rho_{\nu_e \bar{\nu}_e})$ fell with time (or temperature) in the early universe.

Table 2. Density Evolutions during Pair Annihilation

T_9	t	$\rho_{e^\pm}/\rho_\gamma$	$\rho_{e^\pm}/\rho_{\nu_e \bar{\nu}_e}$	$\rho_\gamma/\rho_{\nu_e \bar{\nu}_e}$	ϕ_e/η *
100.0	$0^s.010$	1.750	2.00	1.143	1.080
34.5	$0^s.084$	1.747	2.00	1.146	1.222
6.88	$2^s.17$	1.650	2.02	1.222	1.721
3.29	$10^s.37$	1.334	1.99	1.489	2.185
2.22	$25^s.32$	0.967	1.84	1.897	2.930
1.36	$82^s.47$	0.405	1.19	2.946	6.708
1.04	$152^s.7$	0.180	0.65	3.644	15.65
0.88	222^s	0.089	0.36	4.003	33.78
0.73	$335^s.4$	0.032	0.135	4.245	107.0
0.41	1065^s	0.0002	0.0009	4.404	254.4
0.15	7940^s	1.69×10^{-9}	7.3×10^{-9}	4.406	4.7×10^{10}

*$\phi_e = \mu_e/kT$ and η = present value of baryon to photon number density ratio.

3.2 Thermal Distribution of Neutrinos

Neutrinos are weakly interacting particles that have practically zero rest mass and are electrically neutral and fermionic in nature with spin $=\pm\frac{1}{2}\hbar$. They are believed to be of the Majorana type with neutrinos having spin only $+\frac{1}{2}\hbar$, and antineutrinos having spin only $-\frac{1}{2}\hbar$. This suggests that their spin multiplicity factors are $g_\nu = g_{\bar\nu} = 1$. In case, they are of the Dirac type, like electrons, they would have some finite rest mass, and both particles and antiparticles can be left-right symmetric, thus allowing $g_\nu = g_{\bar\nu} = 2$. Recent experiments on double beta decays favour the Majorana identities of the neutrinos. Neutrinos are believed to come in three families, namely, in electron type, muon type and tauon type. Recent measurements of the width of the Z^o bosons have provided a direct means to calculate the total number of families of light neutrinos (N_ν), the width increasing by about 6% per family of light neutrinos. As can be seen from Table 3, it is by now almost certainly established that $N_\nu = 3$. The knowledge of this number is important as it contributes directly to the total energy density of the universe, and hence to the age and the rate of expansion of the universe.

Table 3. Number of Light Neutrino Species (N_ν)

Experiments	N_ν	number of events used
L3 (Adeva et al)	3.29±0.17	17,000
ALEPH (De Camp et al)	3.27±0.30	3112
OPAL (Akrawy et al)	3.12±0.42	4359
DELPHI (Aarnio et al)	2.40±0.40(±0.50)	1066
SLAC (Abrams et al)	2.80±0.60	5114
Weighted mean value of $N_\nu = 3.22 \pm 0.16$		

Since the number density of protons and neutrons is practically negligible compared to that of the $e^\pm$ pairs before annihilation, the thermal equilibrium of the light neutrino species, in absence of muons and tauons, can be maintained through the following weak interactions

$$e^\pm + \nu_i(\bar\nu_i) \longrightarrow e^\pm + \nu_i(\bar\nu_i), \quad \text{and} \quad e^- + e^+ \rightleftharpoons \nu_i + \bar\nu_i, \quad i = e,\mu,\tau, \qquad (41)$$

and therefore, the cross section for interaction between neutrinos and rest of the primordial soup will be primarily determined by the cross section for their interaction with $e^\pm$, given by

$$\sigma_\nu = O\left(\frac{2\pi G_w kT}{h^2 c^2}\right)^2 \simeq 10^{-49} T_9^2 \text{ m}^2,$$

where G_w is the Fermi coupling constant for weak interactions. We take the number density of the ultrarelativistic $e^\pm$ pairs in thermal equilibrium with the photons

$$n_{e^\pm} \simeq \frac{1}{3}\left(\frac{kT}{\hbar c}\right)^3 \simeq 3 \times 10^{34} T_9^3 \text{ m}^{-3},$$

the age of the universe before the $e^\pm$ pair annihilation

$$t = \left(\frac{8\pi G\rho}{3}\right)^{-1/2} \simeq 100\, T_9^{-2} \text{ sec}, \qquad (42)$$

for $g_{\text{eff}} = 43/8$, as would be derived later, and derive the optical depth for neutrinos over the scale size of the then universe

$$\kappa_\nu = \lambda(e^\pm \nu)t \simeq \sigma_\nu n_{e^\pm} ct \simeq 10^{-4} T_9^3, \tag{43}$$

which increases as the cube of temperature and of course, $\lambda(e^\pm \nu)$ as T_9^5. Obviously, κ_ν can exceed unity provided the temperature of the universe exceeds $T_9 \gtrsim 10^{4/3} \simeq 21$, that is,

$$T \gtrsim T_{d\nu_e} = 2.1 \times 10^{10} \text{ K} \simeq 1.85 \text{ MeV}. \tag{44}$$

So at temperature exceeding 10^{10} K, photons and the $e^\pm$ pairs in the primordial soup will produce copious number of neutrinos and antineutrinos and establish the full thermal contact with the rest of the universe. The number densities of neutrinos and antineutrinos will reach the same order as those of $e^\pm$ pairs or photons. Coming from the high temperature side the story will be just the opposite; ν and $\bar{\nu}$ will loose the thermal contact with the primordial soup when the temperature of the universe drops below the so-called *neutrino decoupling temperature* $T_{d\nu_e} = 2.1 \times 10^{10}$ K. More exact calculations by Herrera and Hacyan (1989) also show that the decoupling of ν_e-$\bar{\nu}_e$ should have taken place at the same temperature. In Fig 5, the dashed line corresponds to the weak reaction rates (normalised to the beta decay rates of free neutrons) corresponding to the processes (41) as a function of the continuously dropping temperature of the universe. Initially the rate falls as T_9^5 as expected from (44), and then it becomes faster due to fast annihilation of $e^\pm$ pairs and hence due to fast decrease in the population of $e^\pm$ (see Table 2). On its way the curve hits the line for the comparatively slowly falling rate of the expansion of the universe, which goes as T_9^2. $T_{d\nu_e}$ corresponds to the abscissa for the point of intersection of these two curves. Following a similar procedure for the muon type neutrinos, it can be shown that the decoupling of ν_μ-$\bar{\nu}_\mu$ should take place at slightly higher temperature $T_{d\nu_\mu} \sim 4 \times 10^{10}$ K, simply because of a lower total rate for the less numerous ν_μ-channels compared to those for the ν_e's.

Once the neutrinos get decoupled, their distributions are still given by (33) with T_ν dropping continuously from its value at decoupling $T_{d\nu}$ in the same proportion as the scale factor R of the universal expansion increases. This is inevitable, simply because the de Broglie wavelength of the freely streaming neutrinos has to increase in proportion with R, and hence,

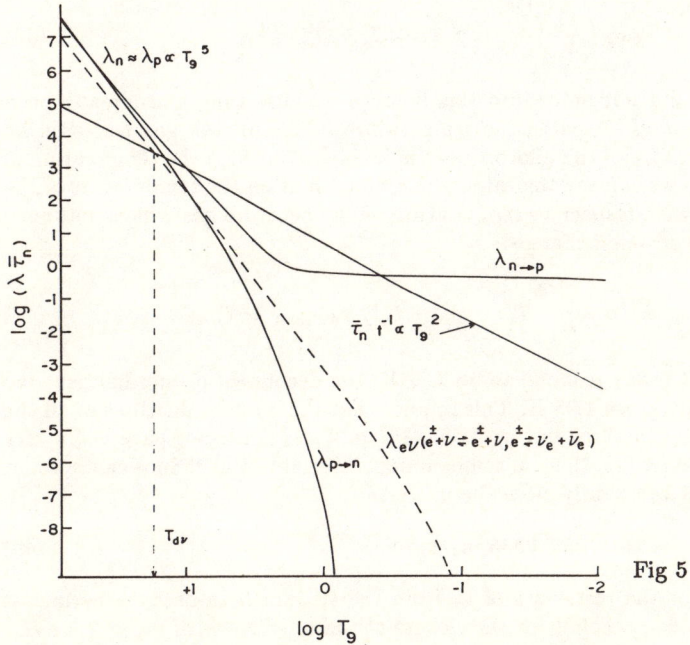

Fig 5

their momenta p_ν must fall as R^{-1}. It keeps the form of the momentum distribution unchanged provided we scale the temperature T_ν satisfying the relation $RT_\nu =$ constant. Since at one stage T_ν was equal to T_γ, the present day temperature of the neutrino gas must not be very different from 3 K. In reality, a slight difference arises due to the $e^\pm$ pair annihilation, which heats up the photon gas but not the neutrinos as the neutrinos were by then decoupled from the rest of the universe.

3.3 Effect of Pair Annihilation on the Entropy Content of Photons

Since $e^\pm$ pairs are annihilated into photons at $T \lesssim m_e c^2/k = 5.93 \times 10^9$ K and since the number and energy densities of $e^\pm$ pairs were almost equivalent to those of photons before annihilation, the entropy of photons after the annihilation must increase considerably. Before the pair annihilation the universe contained photons and $e^\pm$ in perfect thermal equilibrium with one another and all the light families of the neutrinos $\nu_e, \bar{\nu}_e, \nu_\mu, \bar{\nu}_\mu, \nu_\tau, \bar{\nu}_\tau$ (each having $g_\nu = 1$) continued with the same temperature as that of photon even though the latter were thermally decoupled from $e^\pm$ and photons. Therefore the value of g_{eff} for photons, electrons, and neutrinos before the $e^\pm$ annihilation is given by

$$g_{\text{eff}} = \frac{1}{2}\left[g_\gamma + \frac{7}{8}\left(g_{e^+} + g_{e^-} + g_{\nu_e} + g_{\bar{\nu}_e} + g_{\nu_\mu} + g_{\bar{\nu}_\mu} + g_{\nu_\tau} + g_{\bar{\nu}_\tau}\right)\right] = 1 + \frac{35}{8} = \frac{43}{8}. \quad (45)$$

Expansion of the universe remained adiabatic throughout. Therefore entropy inside a comoving volume remained always constant. If any species of particles and antiparticles annihilate into photons via $A + \bar{A} \to 2\gamma$, their reserve of entropy will be transferred to photons and heating will take place. So,

$$S_A + S_{\bar{A}} + S_\gamma \longmapsto S'_\gamma = S_A + S_{\bar{A}} + S_\gamma.$$

Let the initial temperature of photon gas before the $e^\pm$ annihilation be T_i and final temperature after annihilation of photon gas be T_f. Since the entropy density of any species $= (P + \rho T)/T \propto T^3$, we can have $(g_{e^+} + g_e + g_\gamma)T_i^3 = g_\gamma T_f^3$, or, $(11/4)T_i^3 = T_f^3$, or,

$$T_f = \left(\frac{11}{4}\right)^{1/3} T_i \simeq 1.401\, T_i. \quad (46)$$

The entropy of the neutrinos did not enter into this balance because they had already been decoupled. So after annihilation of $e^\pm$ pairs, the temperature of the photon gas increased by about 40%. The initial temperature of the photon gas can be identified with the temperature of the decoupled neutrino gas. So we can say that after the $e^\pm$ annihilation the temperature of the photon gas increased by 40% with respect to temperature of the neutrino gas which continued without its total entropy content being changed:

$$T_\gamma|_{T \ll 6 \times 10^9 \text{ K}} = \left(\frac{11}{4}\right)^{1/3} T_\nu, \quad \text{and} \quad T_\gamma|_{T \gg 6 \times 10^9 \text{K}} = T_\nu. \quad (47)$$

Since the temperature of MBR today is found to be 2.77 K, the decoupled primordial neutrino gas must have its present temperature 1.98 K. This implies that the number densities of all the neutrinos and their antiparticles in three flavors 3.50×10^8 m^{-3}. If neutrinos have got a very little rest mass, say, few electron volts, this enormous number density of neutrinos can account for so high a mass density that can nearly close the universe:

$$\Omega_\nu h_o^2 = 0.0335\, m_\nu(\text{in eV}), \quad (48)$$

the present day upper limit for the rest mass of ν_e from the tritium beta-decay experiments being 13 eV. For $h_o = 0.5$, Ω_ν can account for the closure of the universe with $m_\nu = 7.5$ eV!

The gradual rise in the ratio $\rho_\gamma/\rho_{\nu_e \bar{\nu}_e}$ is clearly demonstrated in the Table 2. The value of g_{eff} at the end of $e^\pm$ pair annihilation becomes

$$g_{\text{eff}} = \frac{1}{2}\left[g_\gamma + \frac{7}{8}\left(\frac{T_\nu}{T_\gamma}\right)^4 \left(\sum_i g_{\nu_i}\right)\right] = 1 + \frac{21}{8}\left(\frac{4}{11}\right)^{4/3} = 1.6813, \quad (49)$$

which, when put in equation (18), gives the age-temperature relation for the early universe after the pair annihilation

$$t = 177^s.8 \, T_9^{-2}. \quad (50)$$

3.4 Freezing of the Neutron and Proton Ratio

Like neutrinos, neutrons and protons also maintain the thermal contact through the weak interactions via $e^\pm$ pairs, making use of the following channels of weak interactions:

$$n \rightleftharpoons p + e^- + \bar{\nu}_e, \quad n + e^+ \rightleftharpoons p + \bar{\nu}_e, \quad \text{and} \quad n + \nu_e \rightleftharpoons p + e^-. \quad (51)$$

The reactions in the forward direction destroy neutrons and create protons, while in the backward direction the reverse happens. The total rate of destruction of neutrons, $\lambda(n \rightarrow p) \equiv \lambda(n \rightarrow p + e^- + \bar{\nu}_e) + \lambda(n + e^+ \rightarrow p + \bar{\nu}_e) + \lambda(n + \nu_e \rightarrow p + e^-)$ depends only on the densities of the reactants, that is, e^+ and ν_e, the cross sections for these weak reactions and the availability of the phase space for the new products which are all fermions by nature, and therefore, the knowledge of the temperatures and chemical potentials of $e^\pm$ and $\nu_e \bar{\nu}_e$ would suffice apart from of course the cross sections. Same is true for the destruction of protons leading to formation of neutrons.

If the temperature of the photon and electron gas is T, temperature of the neutrinos T_ν, mass of the electrons m_e, mass difference between protons and neutrons parametrised by

$$q = \frac{m_n - m_p}{m_e} = 2.530$$

and the chemical potential of the neutrinos (μ_ν) and electrons (μ_e), if any, the explicit form of the reaction rates are given by (Wagoner et al 1967)

$$\lambda(n) \equiv \lambda(n \rightarrow p) = \lambda(n + e^+ \rightarrow p + \bar{\nu}_e) + \lambda(n + \nu_e \rightarrow p + e^-) + \lambda(n \rightarrow p + e^- + \bar{\nu}_e)$$

$$= A\left[\int_1^\infty \frac{(\epsilon+q)^2 \sqrt{\epsilon^2-1} \, \epsilon \, d\epsilon}{[1+e^{-(\epsilon+q)z_\nu - \phi_\nu}][1+e^{\epsilon z + \phi_e}]} + \int_1^\infty \frac{(\epsilon-q)^2 \sqrt{\epsilon^2-1} \, \epsilon \, d\epsilon \, F(\epsilon)}{[1+e^{-\epsilon z + \phi_e}][1+e^{(\epsilon-q)z_\nu - \phi_\nu}]}\right], \quad (52)$$

where $z = m_e c^2/kT = 5.93/T_9$, $z_\nu = m_e c^2/kT_\nu = 5.93/T_{9\nu}$, $\phi_\nu = \mu_\nu/kT_\nu$, $\phi_e = \mu_e/kT$, $F(\epsilon) =$ Coulomb factor $= 2\pi\chi/\{1 - \exp(-2\pi\chi)\}$, $\chi = \epsilon/\{137.03\sqrt{\epsilon^2-1}\}$ and $A = m_e^5 c^4(g_V^2 + 3g_A^2)/2\pi^3\hbar^7$, $g_V = 1.4146 \times 10^{-50}$ J m^{-3} and $g_A = -1.262 g_V$ are the vector and axial vector coupling constants of charged current electroweak interactions, giving the theoretical value of $A = 6.515 \times 10^{-4}$ s^{-1}.

The reverse reaction rate, namely, protons transmuting into neutrons, the latter being heavier than the former (that is, endothermic), $\lambda(p \rightarrow n)$ is usually smaller than the forward reaction rate, given by

$$\lambda(p) \equiv \lambda(p \rightarrow n) = \lambda(p + e^- \rightarrow n + \nu_e) + \lambda(p + \bar{\nu}_e \rightarrow n + e^+) + \lambda(p + e^- + \bar{\nu}_e \rightarrow n)$$

$$= A\left[\int_1^\infty \frac{(\epsilon-q)^2 \sqrt{\epsilon^2-1} \, \epsilon \, d\epsilon \, F(\epsilon)}{[1+e^{\epsilon z - \phi_e}][1+e^{(q-\epsilon)z_\nu + \phi_\nu}]} + \int_1^\infty \frac{(\epsilon+q)^2 \sqrt{\epsilon^2-1} \, \epsilon \, d\epsilon}{[1+e^{-\epsilon z - \phi_e}][1+e^{(\epsilon+q)z_\nu + \phi_\nu}]}\right]. \quad (53)$$

However, the constant A is also directly related to the mean lifetime of free neutrons ($\bar{\tau}_n$) against the usual β-decays given by

$$\frac{1}{A} = f^R \bar{\tau}_n, \quad f^R = \int_1^q (\epsilon - q)^2 \sqrt{\epsilon^2 - 1} \, F(\epsilon) \, \epsilon \, d\epsilon + \text{radiative corrections, etc.} = 1.71465. \tag{54}$$

Since we have not put the radiative and other corrections in (52) and (53), we shall use $f^R = 1.68573$ and the measured value of $\bar{\tau}_n$ in order to evaluate A, rather than the theoretical expression involving the basic coupling constants.

Listed in Table 4 are the values of the laboratory measurements of $\bar{\tau}_n$ over the past 30 years which seem to converge to its theoretically estimated value determined from the measurements of the coupling constants. We take $\bar{\tau}_n = 891.6$ sec for the calculations in the present work.

Table 4. Laboratory Measurements of Mean Life Time of Neutrons ($\bar{\tau}_n$)

Year	Author(s)	$\bar{\tau}_n$	Year	Author(s)	$\bar{\tau}_n$
1959	Sosnovskii et al	1013±26	1980	Byrne et al	936±17
1972	Christensen et al	919±14	1984	Bopp et al	889±11
1975	Krohn and Ringo	907±18	1984	Byrne et al	914±6
1978	Bondarenko et al	877±16	1986	Kosvintsev et al	903±13
1978	Strataura et al	902±20	1988	Last et al	876±22
1979	Erozolimskii et al	905±14	1989	Mampe et al	887±3

Expected theoretical value of $\bar{\tau}_n = 894^s.3 \pm 6$

In the high temperature limit, that is, for $T \gg 10^{10}$ K, the two reaction rates are nearly equal

$$\lambda(n) \gtrsim \lambda(p) \simeq 4 \times 10^{-6} \, T_9^5 \, \text{sec}^{-1},$$

as can be seen in Fig 5. Notice that at all temperatures, $\lambda(n) > \lambda(p)$ and $\lambda(n) > \lambda(e^\pm \nu_e)$, (see the equation (43)). At lower temperatures $\lambda(p)$ drops to practically zero as $T_9 \to 1$, due to proton's inability to obtain the required mass of a neutron by colliding with low energy neutrinos and electrons, but $\lambda(n)$ does quickly settle at $(\bar{\tau}_n)^{-1}$. The straight line shown in Fig 5 corresponds to the rate of decline of the universal expansion and intersects the lines for the weak interaction rates at $T_9 > 10$ as expected. $\lambda(p)$ falls off faster than T_9^{-5} at lower temperatures because of the initially gradual, and later on the fast, annihilation of the $e^\pm$ pairs.

So long as the condition that both $\lambda(n)$ and $\lambda(p) > t^{-1}$, t being the age of the universe, neutrons and protons can continue to maintain the thermal equilibrium with the primordial soup, in which case

$$0 = \frac{d}{dt} X_n = \lambda(p) X_p - \lambda(n) X_n, \tag{55}$$

where X_n and X_p are respectively the abundances of neutrons and protons by mass fractions. So the equilibrium ratio of the abundances of neutrons and protons, usually represented by $n/p \equiv X_n/X_p$ becomes

$$\frac{n}{p} \equiv \frac{X_n}{X_p} = \frac{\lambda(p)}{\lambda(n)} \simeq e^{-(m_n - m_p)c^2/kT} = e^{-15.008/T_9}, \tag{56}$$

which is in fact independent of both X_n and X_p and hence on η, the baryon to photon number density ratio.

When the temperature drops below 10^{10} K, neutrons and protons gradually falls short of equilibrium. Had there been no channel for free β-decay of neutrons, the n/p ratio would

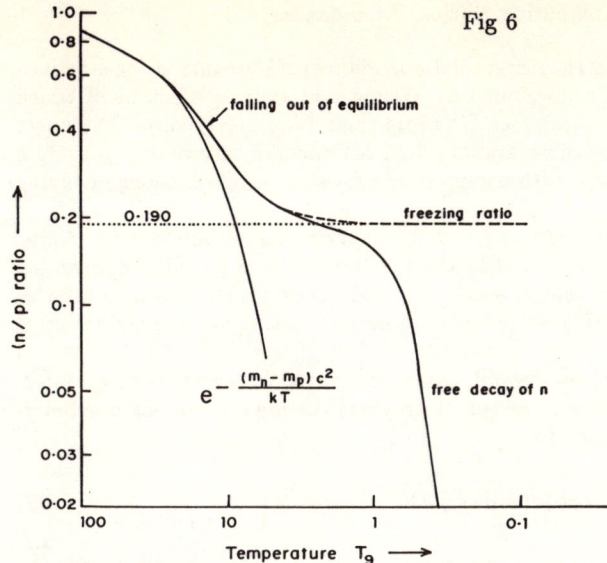

Fig 6

have gradually approached the so called *freezing ratio* at $T_9 < 1$ and this canonical freezing ratio of n/p is found to be

$$\left(\frac{n}{p}\right)_o \simeq 0.190, \tag{57}$$

which should be virtually independent of η.

Subsequently, because of the free decay of neutrons, the n/p ratio should keep on falling from its freezing value 0.190 following the simple radioactive law

$$n = n_o \, e^{-t/\bar{\tau}_n}$$

or,
$$\frac{n}{p} = \left(\frac{n}{p}\right)_o e^{-t/\bar{\tau}_n} \left[1 + \left(\frac{n}{p}\right)_o \left(1 - e^{-t/\bar{\tau}_n}\right)\right]^{-1}. \tag{58}$$

The free decay of neutrons continues till the neutrons become bound inside the nuclei where they can no longer decay freely. The evolution of the n/p ratio with the falling temperature of the universe is shown in Fig 6. Initially, the ratio drops exponentially with T_9^{-1} as given by (56), then it approaches the freezing ratio of 0.190, and finally falls off almost exponentially with time given by (58). This continues till T_9 drops to $T_9^\star$ at which temperature the free neutrons become mostly bound inside the newly synthesised nuclides.

4. The Big Bang Nucleosynthesis

The first step in the process of nucleosynthesis is the formation of deuterium, a combination of a proton and neutron, rather than anything else simply because two protons and two neutrons do not have any bound state by themselves. Being electrically neutral, a neutron can easily approach indefinitely close to a proton and form the nucleus of deuterium without requiring any threshold for energy. Before we move on to the synthesis of deuterium in the early universe, we need a brief introduction to the method of computing the growth of nuclear abundances, given the rates of nuclear reactions.

4.1 A Brief Outline of the Method of Computing Nuclear Abundances

The numerical code for computing the primordial abundances of elements in the standard model of the hot big bang was originally developed by Wagoner in 1967, the details of which can be found in Wagoner et al (1967), Fowler *et al* (1967), and Wagoner (1969). The exact expressions for the nuclear reaction rates of several hundred reactions in their most up-to-date form are available in Caughlan *et al* (1988) with a supplement for some vital reactions involving lithium in Malaney and Fowler (1989).

The nuclear species are first numbered as $i = 1, 2, \cdots$, their mass numbers (A_i) charge numbers (Z_i) and mass defects (Δ_i) are listed, then the reactions are numbered and classified according to their types, such as photonuclear, weak, two-body type with neutron, proton or alpha as partners, three-body type, etc. The abundances of nuclear species are handled in terms of their mass fractions, defined in (30).

Now, a particular nuclear species, say the ith, can be destroyed (or created) by an interaction with either a photon or a lepton, forming (or destroying) the nuclear species number j. The reaction rate of such a process is given by

$$\lambda_w(j) = \int_0^\infty \tilde{n}_w(E,T,\phi)\sigma(E)v(E)\,dE, \qquad w = \gamma, e, \nu \tag{59}$$

where E and v are respectively the CM energy and speed of the photon or lepton, and $\sigma(E)$ the cross section for the interaction. This formula can be used to derive the expressions (52) and (53), for example.

When the ith species is destroyed (or created) as a result of the interaction between the species j and k, the rate of reaction is calculated as

$$[jk] = \rho_b N_A <\sigma v>_{jk}, \quad \text{where} \quad <\sigma v> = \int_0^\infty f(v,T)\sigma(v)v\,dv, \tag{60}$$

$f(v,T)=$ the Maxwell-Boltzmann distribution function for velocities of nuclei. Similarly, for three-body interactions, $j + k + l \rightleftharpoons i+$ anything, the rate is given by

$$[jkl] = \rho_b^2 N_A^2 <\sigma v>_{jkl}. \tag{61}$$

Thus the growth of abundance of the ith species can be formulated as

$$\frac{1}{A_i}\frac{dX_i}{dt} = \pm \sum_j \frac{X_j}{A_j}\lambda_w(j) \pm \sum_{j \geq k}\frac{X_j}{A_j}\frac{X_k}{A_k}[jk] \pm \sum_{j \geq k \geq l}\frac{X_j X_k X_l}{A_j A_k A_l}[jkl], \tag{62}$$

$\pm$ sign to be chosen according to the destructive/constructive nature of the process. The above equation obviously represents a highly nonlinear matrix equation of the form

$$\dot{x}_i = a_{ij}x_j + a_{ijk}x_j x_k + a_{ijkl}x_j x_k x_l.$$

The time steps Δt are usually chosen short enough to enable one to reduce the above nonlinear matrix differential equation into a tractable linear one and at the same time long enough to ensure chemical equilibrium among the species within the chosen time step. Under these conditions, one can write $\dot{x}_i = 0$ for the chemical equilibrium,

$$0 = a_{ij}x_j + a_{ijk}x_j x_k^{(o)} + a_{ijkl}x_j x_k^{(o)} x_l^{(o)} = \left[a_{ij} + a_{ijk}x_k^{(o)} + a_{ijkl}x_k^{(o)}x_l^{(o)}\right]x_j \equiv b_{ij}x_j,$$

where $x_k^{(o)}$ correspond to the abundances obtained during the previous time step. On solving this linearised matrix equation, new set of x_i's are obtained. A few iterations are performed before moving to the next time step.

4.2 Formation of Deuterium

The nuclear reaction

$$p + n \rightleftharpoons d + \gamma \, (2.225 \text{ MeV})$$

is an exothermic reaction producing a photon of energy 2.225 MeV. This means that by nuclear standards the binding energy of ^{2}H is quite low. Since the abundance of photons is about 10^9 times greater than those of neutrons and protons, a high energy photon of this energy is readily available to induce the back reaction, that is, the dissociation of deuterium back into protons and neutrons takes place as soon as it is formed. In fact, the rate of formation of deuteron under the physical conditions of the early universe is given by

$$[pn]_\gamma = 4.40 \times 10^4 \, \rho_b \left[1 - 0.860\sqrt{T_9} + 0.429 \, T_9\right] \text{ sec}^{-1}$$

where ρ_b is the mass density of baryons and is given by

$$\rho_b = 3.34 \times 10^7 \, \eta \, T_9^3 \text{ kg/m}^3,$$

thus giving

$$[pn]_\gamma = 1.47 \times 10^9 \, \eta \, T_9^3 \left[1 - 0.860\sqrt{T_9} + 0.429 \, T_9\right] \text{ sec}^{-1}. \tag{63}$$

The rate of the inverse process, namely the photodissociation of ^{2}H is given by

$$\lambda_\gamma(d) = 2.07 \times 10^{14} \, T_9^{3/2} \left[1 - 0.860\sqrt{T_9} + 0.429 \, T_9\right] \exp\left[-\frac{25.82}{T_9}\right] \text{ sec}^{-1}. \tag{64}$$

These may be compared with the rate of expansion of the universe $H = (2t)^{-1} = 2.8 \times 10^{-3} \, T_9^2 \text{ sec}^{-1}$, where t is the age of the universe.

At $T_9 \gg 1$, it is easily seen that

$$\lambda_\gamma(d) \gg [pn]_\gamma \gg t^{-1}. \tag{65}$$

So the process of the formation of ^{2}H can remain in statistical equilibrium with the expanding universe, the equilibrium abundance of deuterium X_d by mass fraction is reached within each time step in accordance with the equation (essentially the Saha equation):

$$0 = \dot{X}_d = [pn]_\gamma X_p X_n - \lambda_\gamma(d) X_d. \tag{66}$$

Strictly speaking, on the RHS of (66) we must add all terms representing the possible channels for destruction and production of deuterium, but so long as only n, p and d are available among the nuclear species, the equation (66) remains valid, and we get the equilibrium abundance of deuterium at any stage

$$\frac{X_d}{X_p X_n} = \frac{[pn]_\gamma}{\lambda_\gamma(d)},$$

which can approach O(1) until the temperature drops below a certain value, say, $T_9^\star$, given by

$$\frac{3}{2} \ln T_9^\star + \frac{25.82}{T_9^\star} + \ln \eta \geq 11.85 \tag{67}$$

The equation (67) shows that $T_9^\star$ is an increasing function of η, that is, higher the baryon content of the universe (η), higher is the temperature $T_9^\star$ at which the formation of ^{2}H in significant amount takes place as can be seen in the following Table

η	$T_9^\star$
10^{-11}	0.72 (0.47)
10^{-10}	0.77 (0.73)
10^{-9}	0.82 (0.88)
10^{-8}	0.88 (1.02)

The quantities in the parenthesis are the actual values computed by detailed numerical codes which contains all the relevant nuclear reactions for the formation and destruction of ^{2}H. The comparison fails at lower values of η because the condition of (67) fails due to incompleteness of the list of reactions on the RHS of the equation (66). Inclusion of more channels would increase the gradient of $T_9^\star$ with respect to η.

However, the implication of the exact value of $T_9^\star$ for the subsequent progress of nucleosynthesis is that the free decay of neutrons will continue until T_9 drops to $T_9^\star$. The final n/p ratio for the nucleosynthesis at the time of formation of ^{2}H is therefore given by equation (58) with $n/p = 0.190$ and $t = t(T_9^\star)$ to be calculated from the post-pair-annihilation age-temperature relation (50).

The growth of deuterium in the early universe for $\eta = 2.5 \times 10^{-10}$ is shown in Fig 7. The abundance of deuterium leads over that of all other elements down to about $T_9 \simeq 1.4$. So the other reaction channels involving formation of ^{3}H, ^{3}He and ^{4}He take over before the deuterium reaches its peak.

In stellar conditions, ^{2}H is almost always destroyed near the surface of the star provided the temperature is about a few million degree Kelvin. So ^{2}H is almost universally destroyed during the Hayashi contraction of the prestellar nebulae before the star lands on main sequence. The stellar nucleosynthesis operates in the central region of the star and it begins with the formation of ^{2}H from p because free neutrons are absolutely non-existent in stars. Since two protons, both of which bear positive electrical charges, have to be united to form ^{2}H, they have to overcome the Coulomb repulsive barrier by quantum mechanical tunneling, even though the process is intrinsically exothermic. So the stellar rate of formation of ^{2}H is so slow even at temperature $T \gtrsim 10^7$ K that it takes billions of years to destroy a few percent of all protons. The ^{2}H thus synthesised, is readily destroyed forming ^{3}He and ^{3}H, as it also happened in the early universe. So the essential difference between stellar and big bang nucleosynthesis is that, in the former a proton has to be converted into its heavier counterpart neutron for forming ^{2}H, whereas in the latter, free neutrons are readily available to form ^{2}H. As a result, nucleosynthesis in big bang is practically over in a few minutes time compared to several million years to billion years in the case of a star.

4.3 Synthesis of Helium

In the beginning, only p, n and d are available. As soon as T_9 drops below 3.0, ^{3}H and ^{3}He begin to form and their abundances grow at an extremely fast rate because of the existence of a long chain of reactions leading to the synthesis of ^{3}H and ^{3}He, given as follows:

$$p + d \rightleftharpoons {}^3\text{He} + \gamma \ (\ 5.494 \ MeV\ )$$

$$n + d \rightleftharpoons {}^3\text{H} + \gamma \ (\ 6.257 \ MeV\ )$$

$$d + d \rightleftharpoons n + {}^3\text{He} \ (\ Q = 3.269 \ MeV\ )$$

$$\rightleftharpoons p + {}^3\text{H} \ (\ Q = 4.033 \ MeV\ )$$

$$n + {}^3\text{He} \rightleftharpoons p + {}^3\text{H} \ (\ Q = 0.7638 \ MeV\ )$$

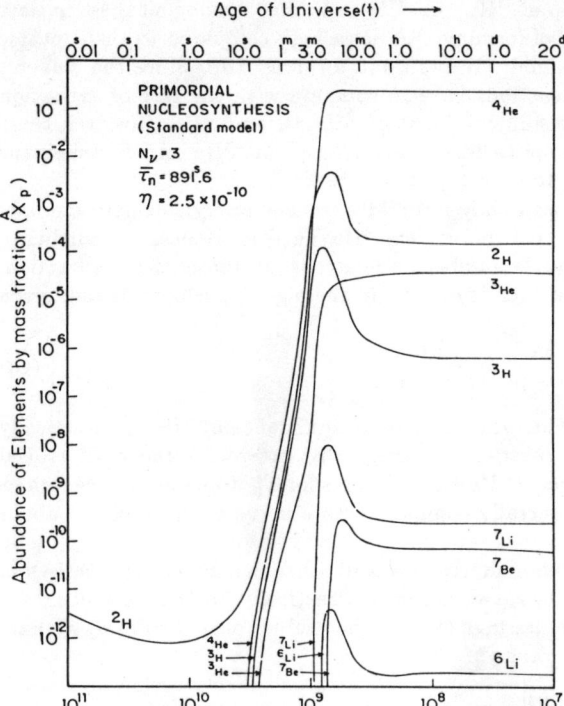

Fig 7

These are immediately followed by a large network of the ^{4}He forming reactions, namely,

$$p + {}^3\text{H} \rightleftharpoons {}^4\text{He} + \gamma \ (19.81 \text{ MeV})$$
$$n + {}^3\text{He} \rightleftharpoons {}^4\text{He} + \gamma \ (Q = 20.58 \text{ MeV})$$
$$d + d \rightleftharpoons {}^4\text{He} + \gamma \ (23.85 \text{ MeV})$$
$$d + {}^3\text{He} \rightleftharpoons {}^4\text{He} + p \ (Q = 18.35 \text{ MeV})$$
$$d + {}^3\text{H} \rightleftharpoons {}^4\text{He} + n \ (17.59 \text{ MeV})$$
$${}^3\text{He} + {}^3\text{He} \rightleftharpoons {}^4\text{He} + 2p \ (Q = 12.86 \text{ MeV})$$
$${}^3\text{H} + {}^3\text{H} \rightleftharpoons {}^4\text{He} + 2n \ (Q = 11.33 \text{ MeV})$$
$${}^3\text{He} + {}^3\text{H} \rightleftharpoons {}^4\text{He} + p + n \ (Q = 12.10 \text{ MeV})$$
$${}^3\text{He} + {}^3\text{H} \rightleftharpoons {}^4\text{He} + d \ (Q = 14.32 \text{ MeV})$$

Practically all of ^{3}H and ^{3}He are converted into ^{4}He, so that the growth of ^{4}He follows very closely to that of ^{3}H and ^{3}He. Very soon the abundance of ^{4}He exceeds all others except of course the remnant hydrogen. From the pattern of the growth of various elements in Fig 7, it seems that all ^{2}H go into ^{4}He and all ^{3}H into ^{3}He , leaving only a trace of the former in each case.

There is a well-known gap in the periodic table for the mass number 5, because of which the addition of neutrons and protons to ^{4}He becomes inhibitive. Even the fusion of a ^{2}H with ^{4}He is energetically unfavourable. So, it becomes practically impossible to cross the barrier of the mass number 5 in order to continue the ladder of nucleosynthesis beyond ^{4}He. Essentially all the available free neutrons are locked up in ^{2}H, ^{3}H, ^{3}He and ^{4}He. Among all these nuclear species, ^{4}He is most bound and the rates of forming ^{4}He is so high compared to their reverse reactions at $T_9 < T_9^\star$ that *almost* all ^{2}H, and ^{3}H end up in forming ^{4}He. Since the universe is expanding at a very fast rate, all the reactions have to compete with the rate of expansion of the universe, there must remain some residual amount of ^{2}H, ^{3}H and ^{3}He. However, these residual amounts are always found to be much less than 1% of all ^{4}He and H unless the baryonic matter density of the universe is too low ($\eta < 1 \times 10^{-11}$).

Now, if all the available free neutrons finally form ^{4}He, we can easily calculate the final abundance of ^{4}He. Two neutrons are required to form one ^{4}He nucleus. Hence, the number of ^{4}He nuclei should be the half of the available number on neutrons. In terms of mass fraction, the primordial abundance of ^{4}He after the nucleosynthesis in the big bang is over, is thus given by

$$Y_p = \frac{2n}{n+p} = \frac{2(n/p)}{(1+n/p)}, \qquad (68)$$

where n/p is given by equation (57). Thus the primordial abundance of ^{4}He is practically independent of all the details of nuclear reactions forming ^{4}He except for the most crucial reaction leading to the just stable formation of ^{2}H which decides for $T_9^\star$ to be used in equations (50) and (58). The synthesis of ^{4}He is generally complete before universe the becomes about 15 minutes old.

In Table 5, we summarise the final abundances ^{4}He and other elements as synthesised in the standard hot big bang universe for different values of η. Practically no ^{4}He is synthesised below $\eta = 10^{-12}$ and its abundance becomes higher for higher values of η. In Fig 8, we have plotted the final abundances for a smaller range of η.

Table 5. Primordial Abundances of Light Elements by Mass Fraction

η	^{2}X	^{3}X	$^{2+3}$X	^{4}He	^{6}Li	^{7}X	^{9}Be	$\geq^{12}$X
1.0×10^{-14}	1.9(-4)	4.6(-9)	1.9(-4)	3.8(-12)	$<10^{-25}$	6.0(-23)	$<10^{-25}$	1.0(-21)
1.0×10^{-13}	1.9(-3)	2.8(-6)	1.9(-3)	4.0(-7)	7.0(-20)	7.0(-20)	$<10^{-25}$	1.0(-21)
1.0×10^{-12}	1.4(-2)	1.5(-4)	1.4(-2)	2.6(-3)	1.9(-14)	1.0(-11)	1.3(-17)	1.6(-21)
1.0×10^{-11}	1.0(-2)	3.6(-4)	1.0(-2)	0.082	1.6(-12)	5.2(-9)	8.1(-15)	7.8(-18)
1.0×10^{-10}	7.9(-4)	7.1(-5)	8.6(-4)	0.221	8.2(-13)	1.4(-9)	1.2(-16)	1.2(-15)
2.5×10^{-10}	1.6(-4)	3.9(-5)	2.0(-4)	0.238	1.9(-13)	3.2(-10)	4.0(-18)	2.6(-15)
1.0×10^{-9}	1.4(-5)	2.0(-5)	3.4(-5)	0.252	2.0(-14)	6.2(-9)	5.8(-20)	7.2(-15)
1.0×10^{-8}	1.0(-11)	8.7(-6)	8.7(-6)	0.272	1.9(-20)	6.0(-8)	$<10^{-25}$	8.3(-14)
1.0×10^{-7}	2.4(-21)	4.9(-6)	4.9(-6)	0.292	$<10^{-25}$	3.6(-7)	$<10^{-25}$	8.4(-12)
1.0×10^{-6}	2.5(-22)	7.0(-7)	7.0(-7)	0.311	$<10^{-25}$	6.2(-7)	$<10^{-25}$	1.7(-9)
1.0×10^{-5}	2.5(-22)	2.5(-9)	2.5(-9)	0.331	$<10^{-25}$	8.5(-8)	$<10^{-25}$	7.5(-8)

4.4 Abundance of ^{2}H, ^{3}H and ^{3}He

As explained earlier, the abundance of these three nuclear species results from the residue of the nucleosynthesis of ^{4}He. Since the rates of all the helium forming reactions are proportional to ρ_b or equivalently to η, for higher value of η the final residual abundances of ^{2}H, ^{3}H and ^{3}He become smaller and smaller (see Table 5 and Fig 8).

Among these three elements, ^{3}H is unstable against β-decay with the half life for ^{3}H $\rightarrow$ ^{3}He $+ e^- + \bar{\nu}_e$, about 12 years. So after about 100 years since the big bang, all ^{3}H will decay into ^{3}He and therefore the actual primordial amount of ^{3}He should be regarded as the sum of the abundances of ^{3}H and ^{3}He. The range of the final abundances of ^{2}H varies extremely widely with respect to the wide range of η. It assumes a peak value of about 0.015 by mass fraction for about $\eta = 7 \times 10^{-12}$, but practically no ^{2}H is left above $\eta = 3 \times 10^{-9}$.

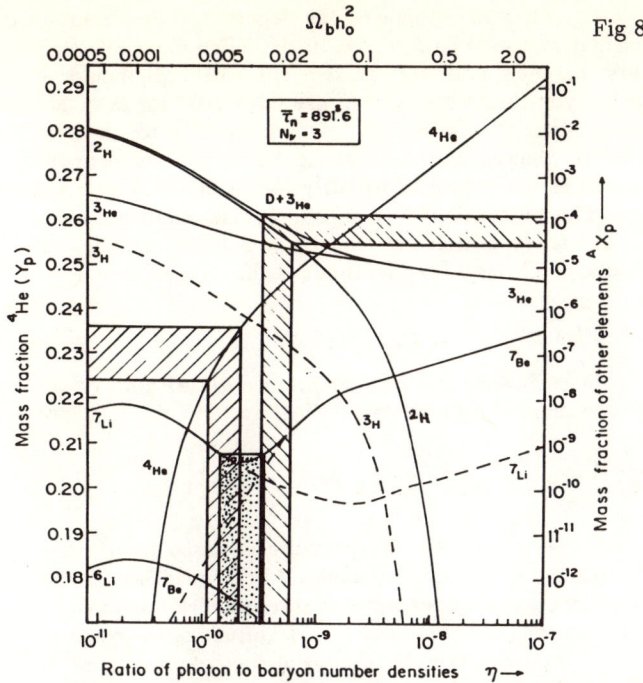
Fig 8

Since stars also destroy ^{2}H and astrophysical synthesis of ^{2}H in interstellar conditions due to cosmic ray interactions do not quite succeed in producing enough amount of ^{2}H, it is always highly desirable to have all the required ^{2}H synthesised in the big bang. In fact, only if $\eta < 5 \times 10^{-10}$, enough amount ^{2}H is produced in the standard hot big bang model of the universe, and the same is true for ^{3}He, as we shall see later.

4.5 Synthesis of Elements Heavier than ^{4}He

As stated earlier because of the gap at mass number of 5 in the periodic table, the synthesis of the subsequent elements of the periodic table, namely, lithium and beryllium is highly suppressed in the big bang. Some of the relevant reactions that help forming post-^{4}He elements are as follows:

$$^3\text{He} + {}^4\text{He} \rightleftharpoons {}^7\text{Be} + \gamma \ (\ 1.587 \text{ MeV }),$$
$$^3\text{H} + {}^4\text{He} \rightleftharpoons {}^7\text{Be} + \gamma \ (\ 2.467 \text{ MeV }).$$

Even though they are exothermic, their Coulomb barriers are too high. So the reaction rates are too slow and a very little of ^{6}Li, ^{7}Li and ^{7}Be is synthesised, which being highly reactive, mostly end up in disintegrating into two α-particles. The final residual abundances of ^{7}Li and ^{7}Be are in the range of $10^{-10} - 10^{-7}$ by mass fraction. The final abundances of ^{6}Li and ^{7}Li maintain practically a constant ratio for a wide range of η. The growth of lithium peaks around $\eta = 2 \times 10^{-11}$ and it falls off both at larger and smaller values of η.

The growth of ^{7}Be behaves quite opposite to that of ^{7}Li, its abundance being higher at higher values of η reaching $\sim O\left(10^{-7}\right)$ for $\eta \sim O(10^{-6})$. However, ^{7}Be is radioactive against β–decays, ^{7}Be $\rightarrow$ ^{7}Li $+ e^+ + \nu_e$ with a half life $\simeq 53$ days. So within about a year since the big bang, practically all ^{7}Be will decay into ^{7}Li. Thus it is meaningful to consider the total primordial abundance of ^{7}Li and ^{7}Be, denoted by ^{7}X by mass fraction. Since the individual functional dependence of ^{7}Be and ^{7}Li on η are rather opposite, the dependence of ^{7}X on η assumes a peculiar shape (see Fig 8). There is a peak at $\eta = 2 \times 10^{-12}$, followed by a trough at $\eta \simeq 2.5 \times 10^{-10}$, and finally the curve assumes another peak at $\eta = 5 \times 10^{-7}$. It should be remembered that the closure density for baryons implies a value of $\eta = 6.5 \times 10^{-9}$ for $h_o = 0.5$.

Surprisingly enough there occurs another significant gap in the periodic table at mass number 8. Because of this second nuclear bottleneck, the synthesis of ^{9}Be or ^{12}C is virtually impossible in the big bang. Their abundances are restricted to fairly less than 10^{-7} by mass fraction for all values of $\eta < 10^{-5}$. The production of ^{12}C and heavier elements must therefore rely absolutely on stellar type of nucleosynthesis, as the observed stellar abundances are about 2% by mass fraction. Basically, ^{12}C has to be synthesised through a three-body reaction, namely,

$$^4\text{He} + {}^4\text{He} + {}^4\text{He} \rightleftharpoons {}^{12}\text{C} + \gamma\ (\ 7.274\ \text{MeV}\ ).$$

Being a three-body reaction, three α-particles have to assemble simultaneously within a range of 10^{-15} m or so. Its rate will thus be exceedingly low, given by

$$[\ 3\ ^4\text{He}\ ] = 1.80 \times 10^{-8}\ \rho_b^2\ T_9^{-3}\ \left[\exp\left(-\frac{4.32}{T_9}\right) + 30.3\ \exp\left(-\frac{27.4}{T_9}\right)\right]\ \text{s}^{-1}.$$

The rate actually changes by 93 orders of magnitudes the temperature is raised from 10^6 to 10^9 K. It was Sir Fred Hoyle who predicted that such a resonant 3-body reaction for the synthesis ^{12}C must have to be adopted by the nature in order to synthesise adequate amount of ^{12}C which is vital for our existence. In fact, he predicted the right magnitude of the reaction rate purely from the above condition of necessity. However, for the present exercise we must be satisfied with the fact that the standard model of the hot big bang fails to synthesise carbon and heavier elements by at least 12 orders of magnitude compared to what we observe today in the stars.

5. Observational Determinations of Primordial Abundances

We have already noted that the big bang does not produce any heavy element whereas the abundances of all elements heavier than ^{4}He in the sun for example amounts to about 0.02 by mass fraction. This quantity is usually denoted by Z, the mass fraction for ^{4}He by Y and that for H by X, implying by definition, $X + Y + Z = 1$. The heavy elements that are found in sun and other stars must have been synthesised in the stars of previous generation or in other exotic astrophysical sources such as in the hot accretion discs around the black holes.

5.1. General Guidelines for Estimating Primordial Abundances

We have to subtract all the contributions made by various astrophysical sources to the primordial abundances of any given element from their present day observed values. Any star that synthesises heavier elements must also synthesise ^{4}He. Normally, the astrophysical contribution of ^{4}He by mass fraction (ΔY) is *assumed* to be proportional to the amount of heavy elements synthesised by the object, say ΔZ by mass fraction, that is to say,

$$\frac{\Delta Y}{\Delta Z} \equiv \alpha, \tag{69}$$

where α is the ratio of the two yields.

Stars of different masses are found to result in a wide spectrum of values of α. Nevertheless, if we assume the mass function of stars to be quite universal, the average value of α seems to lie between 4 – 6. By noting the metallicity Z of an object, we can estimate the total astrophysical contribution of ^{4}He to be given by $\Delta Y \simeq \alpha Z$, and subtract this amount of ΔY from the total observed abundances Y_{obs} in order to derive the primordial abundance of ^{4}He, Y_p, that is,

$$Y_p = Y_{\text{obs}} - \Delta Y \simeq Y_{\text{obs}} - \alpha Z. \tag{70}$$

So the general rule is to select the oldest possible objects so that all the abundances would be closest to their primordial values and the model-based corrections also become least important. Obviously, all metal-poor objects must be very old. In fact, we know of many stars in our Galaxy and of emission clouds elsewhere, which have their iron abundances as low as 10^{-2} to 10^{-6} times lower than that of the sun.

It is known that massive stars have a convective core and a radiative envelope and that nucleosynthesis takes place in the core region. Because of the convection in the core, the composition of the core is well mixed. Since the envelope is not convective, no mixing is possible between the core and the envelope and therefore all massive stars retain their original surface composition throughout their quiet evolution on the main sequence. The intermediate mass stars have, however, a radiative core and a convective envelope. So in this case too, the original surface composition remains unaltered throughout the main sequence evolution. It is only for the low mass stars that the convective envelope extends fairly deep into the core region and we normally do not consider low mass stars for abundance determinations. The net result is that the surface composition of intermediate and high mass stars give directly the composition of their parent interstellar cloud from which they were born. Once again, the surface composition of the older stars show directly their original composition. For example, the present solar surface composition stands for the fossilised composition of the presolar nebulae, that is, the composition of the local interstellar medium about 4.6 Gy ago. So the solar system abundances have to be corrected for the possible contributions made during the previous 8 Gy or so, in order to derive the primordial abundances from the solar system abundances.

However, this rule is not valid for deuterium and lithium, as both of them are easily burnt at low temperature $O(10^6)$ K, due to massive convection during the pre-mainsequence evolution. So for these elements, special corrections are needed for possible depletions.

The details of the most of the methods are summarised in Boesgaard and Steigman (1985).

5.2. Determination of Primordial Helium

The best determination of Y_p comes from the spectral analysis of the old extragalactic gaseous nebulae that emit the recombination lines of HI, HeI and HeII (= He$^+$). Strömgren zones of HeI and HeII are usually very distinct. The extragalactic HII regions are preferred to Galactic ones as many of them represent the oldest systems retaining closest to the primordial abundances of ^{4}He and heavy elements. The Galactic HII regions, being younger systems, require uncertain corrections for their past history of chemical evolution. The total abundance of ^{4}He is by definition the sum of the abundances of HeI, HeII and HeIII. But in general, He lines are most difficult to excite simply because the electronic transitions corresponding to 380 – 780 nm wavelength range of optical spectroscopy exclude the ground state of ^{4}He and therefore, the transitions have to take place between the two levels. But the excited levels can be populated only at temperatures fairly above 10^4 K, which is why the only spectral classes, namely, OB stars can display the He lines. This is also the reason why one considers the hot gaseous nebulae for the determination of ^{4}He abundance. It may be mentioned that the surface temperature of our sun is too low to excite any He line, thus defying any direct determination of the solar abundance of ^{4}He.

Sometimes the abundances of the individual heavy elements such as C, N and O are available. Assuming a relation of the type

$$Y_p = Y_{\text{obs}} - \left(\frac{dY}{dX_A}\right) X_A,$$

where A is any of the elements C, N and O. The value of Y_p can also be determined separately with reference to C, N, O and their total Z.

Table 6 gives a summary of the important determinations of Y_p from different sources including the ones from the extragalactic emission nebulae. Determinations of Y_p from the Galactic HII regions namely the planetary nebulae involve corrections of greater uncertainty due to the complex nature of the Galactic chemical evolution. Determinations also exist from the Galactic globular clusters, the oldest members of the galaxy and the oldest field stars belonging to the Halo of the Galaxy. Few determinations exist for local OB type of stars, the abundances in the solar system particularly in meteorites, Jupiter, Saturn and the sun correspond to the values of Y_{obs} at the time of the formation of the solar system.

Table 6. Determinations of Y_p

year	Author(s)	Y_p	Remarks
Determinations from Emission Nebulae in the Past			
1976	Peimbert and Torres–Peimbert	0.228	
1980	French	0.216	non-linearity correction needed
1980	Talent	0.216	–do–
1980	Rayo et al	0.216	–do–
1983	Kunth and Sargent	$0.245 \pm 0.003(1\sigma)$	from 13 HII regions
1984	Rana	≤ 0.230	from the same 13 objects
1984	Vidal-Madjar and Gry	$0.221 \pm 0.016(1\sigma)$	from same data
1985	Kinman and Davidson	$0.212 \pm 0.015(1\sigma)$	reanalysing all data
1986	Pagel	$0.234 \pm 0.008(1\sigma)$	same objects and data as K & S minus one object
1986	Pagel et al	$0.236 \pm 0.005(1\sigma)$	same 12 objects but independent data set
1987	Gallagher et al	$Y_p(C) = 0.235 \pm 0.004(1\sigma)$	
1988	Pagel	$Y_p(O) = 0.230 \pm 0.005(1\sigma)$	
		$Y_p(N) = 0.232 \pm 0.004(1\sigma)$	
1989	Torres–Peimbert and Peimbert	$0.230 \pm 0.006(1\sigma)$	3 extragalactic emission nebulae
1989	Pagel	$0.229 \pm 0.004(1\sigma)$	all data
Determinations from Planetary Nebulae (HII region)			
1980	Peimbert and Serrano	0.227	from 6 disk PNs
		0.220	from 16 disk PNs
1983	Peimbert	0.218	from 3 best observed halo PNs
Determinations from Globular Clusters: Directly metal-poor objects			
1979	Carney	0.19 ± 0.04	from halo field stars
1983	Buzzoni et al	0.23 ± 0.02	from $L - T_e$ fittings
Determinations from Abundances of Stellar Atmospheres			
1976	Nissen	0.19	B stars of h and χ Per, Cep OBIII
1983	Nissen	0.28	OB stars

Determinations from Abundances in the Solar System

year	Author(s)	value	Remarks
1967	Lambert	0.20 ± 0.04	solar cosmic rays
1971	Hirayama	0.21 ± 0.03	solar photosphere
1972	Hundausen	≤ 0.17	quiet solar wind
1976	Iben and Mahaffi	0.20 ± 0.01	solar oscillations
1978	Heasly and Milky	0.28 ± 0.05	from solar photosphere
1979	Chistensen-Dalsgaard et al	0.17	–do–
1980	Isaak	¡ 0.17	–do–
1982	Anders and Ebihara	0.24	from meteorites
1982	Pagel	0.20	solar flares
		0.15	quiet solar wind
1982	Bahcall et al	0.25 ± 0.01	solar model fitting
1983	Gough	0.23 ± 0.02	solar oscillations
1984	Conrath et al	0.18 ± 0.04	from Jupiter
		0.12 ± 0.06	from Saturn
1984	Kosovichev and Severnyi	$0.16 - 0.19$	solar oscillations
1986	Lebreton and Maeder	0.279	fitting solar models

Giving due weight to all the diversity in these estimates, the present day consensus on the value of Y_p seems to be

$$Y_p = 0.230 \pm 0.006\,(1\,\sigma). \tag{71}$$

5.3. Abundances of ^{2}H

These are measured mainly from the isotopic shift of the atomic HI line in absorption. In order to estimate the primordial values, suitable corrections are needed for the possible destruction of ^{2}H due to chemical evolution. The convective envelope of stars reaching temperature above 2×10^6 K destroy ^{2}H. So we do not see any trace of ^{2}H in the envelope of the sun. However, massive stars having a radiative envelope, can retain its original abundance of ^{2}H.

Table 7 gives a summary of important determinations of the abundances of ^{2}H in stars, in the interstellar medium (through the absorption features of Lyman–α lines for ^{2}H along the

Table 7. Determinations of Abundances of Deuterium

Stellar Deuterium

year	Author(s)	D/H	Remarks
1980	Ferlet et al	$< 1 \times 10^{-6}$	on Canopus ($m \simeq 9 M_\odot$)

Interstellar Deuterium

year	Author(s)	D/H	Remarks
1976	York and Rogerson	1.6×10^{-5}	towards α-Vir, α-Cru, γVel
1977	Dupree et al	2.4×10^{-5}	towards α Cen A
		3.9×10^{-5}	towards α Aur
1978	Sarma and Mohanty	$< 6 \times 10^{-5}$	towards centre of galaxy
1979	Laurent et al	$0.7 - 1.4 \times 10^{-5}$	towards $\lambda_i, \epsilon, \delta$ Ori
1980	Ferlet et al	$1.3 \pm 0.25 \times -5$	
1983	York	$0.6 - 1.0 \times 10^{-5}$	towards λ Sco
1984	Vidal-Madjar and Gry	$2.0 \pm 1.0 \times 10^{-5}$	towards β CMa

Interstellar $< D/H >_{mean} = 0.8 - 2 \times 10^{-5}$

Solar System Values

year	Author(s)	D/H	Remarks
1984	Hubbard and McFarlane	2×10^{-5}	from D/H directly
1986	deBurgh et al	$4.5 - 18 \times 10^{-5}$	for Uranus (CH_3D/CH_4)
1989	Smith et al	$1.0 - 2.9 \times 10^{-5}$	for Jupiter (HD/H_2)
		$2.6 - 8.4 \times 10^{-5}$	for Saturn (CH_3D/CH_4)

line of stars to distant OB stars in the Galaxy), in the solar system (from the atmospheric composition of Jupiter, Saturn and Uranus of the deuterated molecules) and in the Galaxy by radio observations at 92 cm wavelength (the hyperfine lines for atomic hydrogen corresponds to 21 cm, where as the same line for neutral ^{2}H atom appears at 92 cm, the wavelength at which the 530 m long Ooty radio telescope is tuned to).

In summary, the ^{2}H/H measurements for young stellar atmosphere gives an upper limit of 10^{-6} by number fraction. The pre–solar values of ^{2}H/H, that is, the interstellar value about 4.5 Gy ago seems to be in the range of $1 - 4 \times 10^{-5}$ and the present day interstellar abundance of ^{2}H/H seems to be $0.8 - 2 \times 10^{-5}$. One has to now calculate the approximate depletion factor between the big bang value and its pre-solar values in order to derive the primordial value from the presolar value. The comparison between the present day interstellar and the presolar values suggests that the factor of depletion between the presolar and the primordial values could be between 2 and 3 (Pagel 1982). This leads to

$$(^2\text{H/H})_p = 1.6 - 6.0 \times 10^{-5}, \quad \text{by number fraction.} \tag{72}$$

However, Boesgaard and Steigman (1985) suggest a wider range for the depletion factor, say, $2 - 10$ with their proposed 'conservative' limit

$$(^2\text{H/H})_p = 1.6 - 4 \times 10^{-5}, \tag{73}$$

and a 'liberal' one given by

$$^2\text{H/H} = 8 - 20 \times 10^{-5}. \tag{74}$$

In any case, a firm lower limit of $(^2\text{H/H})_p$ can always be stated as

$$(^2\text{H/H})_p > 1.6 \times 10^{-5}, \quad \text{by number fraction} \tag{75}$$

or,

$$X_p^D > 2.5 \times 10^{-5}, \quad \text{by mass fraction.} \tag{76}$$

5.4 Abundances of ^{3}He

Only few determinations of the abundances of ^{3}He exist. Most reliable data are the ones derived from the stoichiometric analysis of carbonaceous chondrites and a few meteorites, as shown below in Table 8.

The mean pre-solar abundance ratio of (^{3}He/^{4}He) is generally accepted as

$$(^3\text{He}/^4\text{He})_\odot = (1.54 \pm 0.27) \times 10^{-4}. \tag{77}$$

Searches for 8.7 GHz hyperfine line for ^{3}He$^+$ in the interstellar medium have resulted a variety of values for ^{3}He$^+$/H mostly with upper bounds at 6×10^{-5} and two positive measurements both exceeding the pre-solar values. This is not surprising because ^{2}H is universally destroyed and converted into ^{3}He. The primordial value of ^{3}He/^{2}H might be $\simeq 1.0 \times 10^{-5}$, by number fraction.

Noting that ^{2}H is destroyed to form ^{3}He, thus sum total of ^{2}H + ^{3}He should not change much with time. Most reliable estimates for combined abundance in the pre-solar nebulae is taken to be

$$\left(\frac{^2\text{H} + ^3\text{He}}{\text{H}}\right)_\odot = (3.6 \pm 0.60) \times 10^{-5}, \quad \text{by number fraction,} \tag{78}$$

or

$$^{2+3}X_\odot = (8.3 \pm 1.5) \times 10^{-5}, \quad \text{by mass fraction.} \tag{79}$$

Table 8. Determinations of Abundances of ^{3}He

Solar system measurements of ^{3}He/^{4}He

year	Author	$(^3$He/^{4}He$)_\odot$	Remarks
1970	Jeffrey and Anders	$1.43 \pm 0.4 \times 10^{-4}$	Carbonaceous chondrites
1971	Black	$1.5 \pm 1.0 \times 10^{-4}$	–do–
1977	Frick and Moniot	$1.558 \pm 0.055 \times -4$	meteorites
1978	Eberhardt	$1.46 \pm 0.073 \times 10^{-4}$	–do–

Interstellar Measurements of ^{3}He$^+$/H

year	Author	^{3}He$^+$/H	Remarks
1979	Rood et al	$< 5 \times 10^{-5}$	for 6 galactic HII regions
1984	Rood et al	$< 6 \times 10^{-5}$	for W 49
		$< 2 \times 10^{-5}$	for M 175
		$\simeq 8 \times 10^{-5}$	for W 51
		$\simeq 4 \times 10^{-4}$	for W 3

However, ^{3}He is susceptible to stellar processing, and therefore, a correction to the presolar values is needed in order to derive an upper limit to the combined primordial abundances. If g_3 be the fraction of ^{3}He that survives stellar processing, then

$$\left(\frac{^2H + {}^3He}{H}\right)_p < \left(\frac{^2H + {}^3He}{H}\right)_\odot + \left(\frac{1}{g_3} - 1\right)\left(\frac{^3He}{H}\right)_\odot. \quad (80)$$

The value of g_3 is of course uncertain. It is claimed to be greater than 0.25 by Yang et al (1984) and > 0.5 by Truran and Brunish (1985). In any case the result is extremely sensitive to the choices of the values of g_3 which is at present ad hoc. However, assuming $g_3 > 0.5$, one gets

$$\left(\frac{^2H + {}^3He}{H}\right)_p < 6 \times 10^{-5}, \quad \text{by number fraction.} \quad (81)$$

or, $\quad ^{2+3}X_p < 1.3 \times 10^{-4}, \quad$ by mass fraction. $\quad (82)$

In a sense, this upper bound also implies, though only implicitly, an upper bound for the primordial abundance of ^{2}H.

5.5 Abundances of ^{6}Li and ^{7}Li

The solar abundance of ^{7}Li is about 10^{-11} by number fraction. Because of its extreme low abundance, it is quite difficult to measure. Nevertheless its abundance has been measured quite accurately in hundreds of stars, based on the measurements of the equivalent width of LiI resonance doublet at the rest frame wavelengths 670.7761 and 670.7912 nm. For ^{6}Li, this doublet is shifted by only 0.0016 nm redward of the doublet lines of ^{7}Li. Lithium is subject to easy destruction and dilution due to the reaction Li (p,α) ^{4}He, which operates at 2.6×10^6 K. The dilution takes place due to ongoing convective mixing. The solar convective envelope has almost totally destroyed its Li content. One finds higher abundances of Li/H in meteorites with ^{7}Li/H $= 2.6 \times 10^{-9}$ by number fraction, and $O(10^{-9})$ in F and G type of stars in Pleiades implying its present interstellar abundance to be more than 100 times the solar abundance.

Therefore one has to pick up the oldest stars of the Galaxy in order to determine the primordial abundance. Table 9 summarises the recent determinations of Li abundance in several dozens of halo stars by independent groups of observers.

The present day consensus is strongly in the favour of taking the abundance of ^{7}Li in the halo stars as the primordial value, given by

Table 9. Abundance of ^{7}Li in old disc stars (Pop II)

year	Author	12+log(^{7}Li/H)	Remarks
1982	Spite and Spite	2.05 ± 0.15	13 halo stars & 9 disk stars
1984	Spite *et al*	2.05 ± 0.15	13 metal poor halo stars
1987	Hobbs *et al*	2.05 ± 0.15	23 halo stars (remeasured)
1987	Spite *et al*	2.05	2 halo dwarfs
1988	Rebolo *et al*	2.08 ± 0.10	37 metal poor dwarf stars

$$\left(\frac{^7Li}{H}\right)_p = (1.12 \pm 0.38) \times 10^{-10} \quad \text{by number fraction,} \tag{83}$$

or
$$^7X_p = (6.0 \pm 2.0) \times 10^{-10} \quad \text{by mass fraction.} \tag{84}$$

^{7}Li abundance in the Galactic open clusters have recently evoked a lot of controversy which we shall not go into except to note that Hobbs and Pilachowski (1989) suggest $(^7Li/H)_p \simeq 10^{-9}$.

One of the most direct determinations of the interstellar abundance of ^{7}Li/H has recently come from the search for Li doublet lines in absorption along the line of sight of the most bright extragalactic point source of light, namely the Supernova 1987A. Measurements by Sahu *et al* (1989) have placed an upper bound at

$$^7Li / H \leq 0.8 \times 10^{-10} \; (1\,\sigma). \tag{85}$$

We take the primordial abundance to be given by that of the halo stars. Only a couple of measurements on the ratio of ^{6}Li /^{7}Li have recently been reported by Pilachowski *at al* (1989), giving

$$^6Li / ^7Li \lesssim 0.1. \tag{86}$$

5.6 Abundance Measurements of ^{9}Be

The abundance of ^{9}Be is generally found to be one order of magnitude lower than that of ^{7}Li. The measurements are summarised by Rebolo *et al* (1988), clearly demonstrating the continuous growth of ^{9}Be with time from its primordial to the present value. In meteorites, they find, ^{9}Be/H = $(3.6 \pm 1.3) \times 10^{-11}$; in Hyades, sun and low mass disc stars in the range of 1-2 $\times 10^{-11}$, whereas all old stars show the values $O(10^{-12})$ by number fraction indicating that its primordial value is possibly even lower.

6. Observational Estimates of Abundances vs the Predictions of the Hot Big Bang Model

We now discuss how the predictions of the big bang nucleosynthesis and the observational estimates of the primordial abundances compare with each other, and if they do not, what the possible modifications would be like.

6.1 General Success of the Standard model

The comparison between the prediction of the standard Big Bang model and the observational data on the abundances of light elements demonstrate the correctness of the standard models to the first order of approximation. In any case, the normal stars fail to synthesise more than 10% of ^{4}He on a time scale of 15 Gy. Also we know that ^{2}H is very difficult to synthesise in astrophysical conditions. If we want to make about 25% of ^{4}He and $O(10^{-4})$ of ^{2}H by mass fractions, the fitting of these two data in Fig 7 fixes the ratio of the photon to the baryon number densities η to be $O(10^{-10})$, or equivalently, $\Omega_b h_o^2 = O(0.01)$, which is well supported by the current estimates of the average density of the 'actually seen' baryonic matter in the universe. It also suggests that the number of light neutrino species cannot be greater than 4, a result deduced by Schramm *et al* in 1979, years ahead of the laboratory measurements in 1989. These are some of the profound, significant results of the primordial nucleosynthesis calculations based on the assumptions of the *standard* hot big bang model.

6.2 A Sign of Inconsistency ?

Notwithstanding the overall success of the standard model, both the theoretical and the observational primordial abundances of the light elements over the past one decade have improved considerably. On closer scrutiny, reports of inconsistencies have surfaced several times. In Fig 8 we have already shown a typical plot for the abundance distribution as function of η, assuming $\bar{\tau}_n$ = 891.6 sec and N_ν = 3. With the stringent limits on the observational estimates of the primordial abundances as discussed in the previous lecture, we can represent the shaded regions for the admissible solutions for different elements to the range of η. It seems that the bounds resulting from ^{2}H and D + ^{3}He , are consistent with the range of η given by $3.5 - 6.3 \times 10^{-10}$. Similarly, the bounds on Y_p, namely Y_p = 0.230 ± 0.006 (1σ) correspond to $\eta = 1.1 - 2.2 \times 10^{-10}$. These two ranges of η are mutually exclusive. Such inconsistencies were reported in the past by Stecker (1981), Rana (1982), Pagel (1982), Fry and Hogan (1982), Scherrer (1983), Vidal-Madjar and Gry (1983, 1984) and more recently by Riley and Irvine (1989) and Basu and Rana (1990). When the measurements of the primordial abundance of Li became available in 1982, the bound on η became fairly overlapping on one side with the ^{4}He solution on the other side with the ^{2}H +^{3}He solution, given by an overall range of $\eta = 1.4 - 3.5 \times 10^{-10}$ from the ^{7}Li solution. If one takes these narrowly disagreeing results very seriously, one has to perhaps accept that the *standard* model of the hot big bang might not be *absolutely* correct, but at the same time, not very far from the truth either, so that a slight modification might cure all these minor discrepancies. As discussed in the next section, recent works seem to have fulfilled this hope, with a promise even for a baryon-dominated universe with $\Omega = 1$.

6.3 Possible Remedies

The predictions of the standard model are very rigid because there is practically no parameter left to fiddle with except for η. So, people have tried to introduce other free parameters. For example, by introducing one such free parameter called the *neutrino degeneracy*, Rana (1982, 1984) showed that the consistency with respect to the observationally estimated primordial abundances of all light elements can be restored. The degeneracy of neutrinos implies a non-zero value of ϕ_ν. This intended solution required $\phi_{\nu_e} \simeq 0.25 \gg \eta \simeq \phi_e$. It can be shown from equation (37) that the value of ϕ_{ν_e} is directly related to the asymmetry of the distribution of $\tilde{n}_{\nu_e}$ and $\tilde{n}_{\bar{\nu}_e}$, which again directly affects the equilibrium and freezing ratios of n/p through equations (52) and (53), and hence reducing the primordial abundance of ^{4}He without much affecting the abundances of other elements (see Fig 3 in Rana 1984).

Another possible solution together with the above were suggested by Scherrer (1983) by introducing inhomogeneity in the early universe. This second possibility gained a firmer soil when Witten (1984) discussed about the origin of strong density inhomogeneities due to quark-hadron phase transition immediately preceding the era of the big bang nucleosynthesis. The study was carried further by Alcock, Fuller, Scherrer and Hogan over the past 4–5 years in a number of publications. Charles Alcock will be discussing in great length about these models in his series of lectures (see in this volume). Perhaps it may not be out of place if I summarise our recent works in this respect (Rana *et al* 1990, and Basu and Rana 1990).

If $\mathcal{R}$ is defined to be the density inhomogeneity at $T_9 = 100$, and the volume fraction of the high density phase be f_v, there can be three kinds of final states, (a) the high density phase, being originally the quark-gluon phase, fails to be completely hadronised, (b) the high density phase as well as the low density phase hadronises, but diffusion of neutrons is practically over before the nucleosynthesis, and (c) late diffusion plays a significant role. We studied the nucleosynthesis for the case (a), and found that if quark nuggets are allowed to exist, their passive role can allow us to have $\Omega_b = 1$, without of course, affecting the results of the standard model, for any value of $\mathcal{R}$ but

$$1 + f_v(\mathcal{R} - 1) = (\Omega_b h_o^2)^{-1}_{standard} \simeq 0.004 - 0.024. \tag{87}$$

In the other work, we have considered the case (b), and found that for $f_v = 0.5$ and $\mathcal{R} = 2$, the agreement among the solutions for ^{4}He, ^{2}H + ^{3}He and ^{7}X improves for an assumed average value of $\eta = 2.5 \times 10^{-10}$, or equivalently,

$$\Omega_b h_o^2 = 0.010. \tag{88}$$

For the above set of parameters, we in fact get $Y_p = 0.232$, $^2X = 2.2 \times 10^{-4}$, $^3X = 4.1 \times 10^{-5}$, $^{2+3}X = 2.6 \times 10^{-4}$, and $^7X = 4.5 \times 10^{-10}$. If such a solution is to be viable, the depletion factor for deuterium between its primordial and the present interstellar values has to be greater than 7, if not 15 - 20. In conclusion, we can say that the primordial nucleosynthesis in a slightly inhomogeneous universe with a density contrast at the time of nucleosynthesis no greater than about 2, can successfully explain all the observationally determined primordial abundances of the light nuclei.

Acknowledgements: It is a pleasure to thank Dr(s). Bikash Sinha, Sibaji Raha and S. Pal for their kind invitation, and Dr. Banashree Mitra for the help in preparing the manuscript.

References

Aarnio, P. et. al., *Phys. Lett. B*, **231**, 539 (1989) [DELPHI].
Abrams, G. S. et. al., *Phys. Rev. Lett.*, **63**, 2173 (1989) [MARK II SLAC].
Adeva, B. et. al., *Phys. Lett. B*, **231**, 509 (1989) [L3].; **237**, 136 (1990) [L3 revised].
Akrawy, M. Z. et. al., *Phys. Lett. B*, **231**, 530 (1989) [OPAL].
Alcock, C., Fuller, G. M., and Mathews, G. J., *Astrophys. J.*, **320**, 439 (1987).
Alpher, R. A., and Herman, R. C., *Rev. Mod. Phys.*, **22**, 153 (1950).
Anders, E., and Ebihara, M., *Geochim. Cosmochim. Acta*, **46**, 2363 (1982).
Applegate, J. H., Hogan, C. J., and Scherrer, R. J., *Astrophys. J.*, **329**, 592 (1988).
Bahcall, J. N., Huebner, W. F., Lubow, S. H., Parker, P. D., and Ulrich, R. K., *Rev. Mod. Phys.*, **54**, 767 (1982).
Basu, S., and Rana, N. C., submitted to *Astrophys. J.* (1990).
Black, D. C., *Nature Phys. Sci.*, **234**, 148 (1971).
Boesgaard, A. M., and Steigman, G., *Ann. Rev. Astron. Astrophys.*, **23**, 319 (1985)
Bondarenko, L. N., Kurguzov, V. V., Prokofev, Yu. A., Rogov, E. V., and Spivak, P. E., *JETP Lett.*, **28**, 303 (1978).
Bopp, P., Dubbers, D., Klemt, E., Last, J., Schultze, H., Weibler, W., Friedmann, S. J., and Scharpf, D., *J. Phys. C*, **3**, 21 (1984).
Buzzoni, A., Fusi Pecci, F., Buohanno, R., and Corsi, C. E., in *ESO Workshop on Primordial Helium*, eds. P. A. Shaver, D. Kunth, K. Kjär, Garching, p 422 (1983).
Byrne, J., *Nature*, **310**, 212 (1984).
Byrne, J., Morse, J., Smith, K. F., Shaikh, F., Green, K, and Greene, F. L., *Phys. Lett. B*, **92**, 274 (1980).
Carney, B., *Astrophys. J.*, **233**, 877 (1979).
Caughlan, G. R., and Fowler, W. A., *Atomic Data and Nuclear Data Tables*, **40**, 283 (1988).
Christensen, C. J., Nielsen, A., Bahnsen, A., Brown, W. K., and Rustad, B. M., *Phys. Rev.*, **D5**, 1628 (1972).
Christensen-Dalsgaard, J., Gough, D. O., and Morgan, J. G., *Astron. Astrophys.* **73**, 121 (1979); *op. cit.*, **79**, 260 (1979).

Churchwell, E., Smith, L. F., Mathis, J., Mezger, P. G., and Hutchmeier, W., *Astron. Astrophys.*, **70**, 719 (1978).
Conrath, B. J., Gautier, D., Hanel, R. A., and Hornstein, J. S., *Astrophys. J.*, **282**, 807 (1984).
Crane, P., Hegyi, D. J., Kutner, M. L., and Mandoleski, N., *Astrophys. J.*, **346**, 136 (1989).
deBurgh, C., Lutz, B. L., Owen, T., Brault, J., and Chauville, J., *Astrophys. J.*, **311**, 501 (1986).
Decamp, D. et. al., *Phys. Lett. B*, **231**, 519 (1989) [ALEPH].
Dicke, R. H., *Nature*, **194**, 329 (1962).
Dufour, R. J., *Astrophys. J.*, **195**, 315 (1975).
Dupree, A. K., Baliunus, S. L., and Shipman, H. L., *Astrophys. J.*, **218** 361 (1977).
Eberhardt, P., *Proc. Lunar Planet. Sci. Conf., 9th*, 1027 (1978).
Erozolimskii, B. G., Frank, A. I., Mostovoi, Yu. A., Arzumanov, S. S., and Voitzik, L. R., *Sov. J. Nucl. Phys.*, **30**, 356 (1979).
Ferlet, R., Vidal-Madjar, A., Laurent, C., and York, D. G., *Astrophys. J.*, **242**, 576 (1980).
Fowler, W. A., Caughlan, G. R., and Zimmerman, B. A., *Ann. Rev. Astron. Astrophys.*, **5**, 525 (1967).
French, H., B., *Astrophys. J.*, **240**, 41 (1980).
Frick, U., and Moniot, R. K., *Proc. Lunar Planet. Sci. Conf., 8th*, 229 (1977).
Friedmann, A., *Z. Phys.*, **10**, 377 (1922).
Fry, J. N., and Hogan, C. J., *Phys. Rev. Lett.*, **49**, 1873 (1982).
Gallagher, J. S., Steigman, G., and Schramm, D. N., in 13th Texas Symp. on Rel. Astrophys., ed. M. P. Ulmer, World Scientific, Singapore, (1987).
Gamow, G., *Phys. Rev.*, **70**, 572 (1946).
Gautier, D., and Owen, T., *Nature*, **302**, 215 (1983).
Gough, D. O., in *ESO Workshop on Primordial Helium*, eds. P. A. Shaver, D. Kunth, K. Kjär, Garching (1983).
Heasley, J., and Milky, R., *Astrophys. J.*, **221**, 677 (1978).
Herrera, M. A., and Haeyan, S., *Astrophys. J.*, **336**, 539 (1989).
Hirayama, T., *Solar Phys.*, **19**, 384 (1971).
Hobbs, L. M., and Duncan, D. K., *Astrophys. J.*, **317**, 796 (1987).
Hobbs, L. M., and Pilachowski, C., *Astrophys. J.*, **334**, 734 (1988).
Hubbard, W. B., and McFarlane, J. J., *Icarus*, **44**, 676 (1980).
Hubble, E. P., *Proc. Nat. Acad. Sci.*, **15**, 168 (1929).
Hundausen, A. J., Coronal Expansion and the Solar Wind, Springer, New York (1972).
Iben Jr., I., and Mahaffi, J., *Astrophys. J.*, **209**, L39 (1976).
Isaak, G. R., *Nature*, **283**, 644 (1980).
Jeffrey, P. M., and Anders, E., *Geochim. Cosmochim. Acta*, **34**, 1175 (1970).
Klypin, A. A., Sahzin, M. V., Strukov, I. A., and Skulachev, D. P., *Pisma Astr. Zh.*, **13**, 259 (1987).
Kosovichev, A. G., and Severnyi, A. B., *Sov. Astron. Lett.*, **10**, 284 (1984).
Kosvintsev, Yu. Yu., Morozov, V. I., and Terekhov, G. I., *JETP Lett.*, **44**, 571 (1986).
Krohn, V. E., and Ringo, G. R., *Phys. Lett. B*, **55**, 175 (1975).
Kunth, D., and Sargent, W. L. W., *Astrophys. J.*, **273**, 81 (1983).
Lambert, D. L., *Observatory*, **960**, 199 (1967).
Last, J., Arnold, M., Döhner, J., Dubbers, D., and Feedman, S. J., *Phys. Rev. Lett.*, **60**, 995 (1988).
Laurent, C., Vidal-Madjar, A., and York, D. G., *Astrophys. J.*, **229**, 923 (1979).
Lebreton, Y., and Maeder, A., *Astron. Astrophys.*, **161**, 119 (1986).
Lequeux, J., *Astron. Astrophys.*, **71**, 1 (1979).
Lequeux, J., Peimbert, M., Rayo, J. F., Serrano, A., and Torres-Peimbert, S., *Astron. Astrophys.*, **80**, 155 (1979).
Lubin, P. M., and Smoot, G. F., *Phys. Rev. Lett.*, **42**, 129 (1979).

Malaney, R. A., and Fowler, W. A., *Astrophys. J.*, **333**, 14 (1988); *op. cit.*, **345**, L5 (1989).
Mampe, W., Ageron, P., Bates, C., Pandlebury, J. M., and Steyerl, A., *Phys. Rev. Lett.*, **63**, 593 (1989).
McKellar, A., *Publ. Dom. Astrophys. Obs.*, **7**, 251 (1941).
Nissen, P. E., *Astron. Astrophys.*, **36**, 57 (1976).
Nissen, P. E., in *ESO Workshop on Primordial Helium*, eds. P. A. Shaver, D. Kunth, K. Kjär, Garching, p 163 (1983).
Olbers, H. W. M. O., *Bode's Jahrbuch*, 111 (1826).
Pagel, B. E. J., in *Structure and Evolution of Normal Galaxies*, ed. M. Fall and D. Lynden-Bell, Cambridge Univ. Press, Cambridge (1981).
Pagel, B. E. J., *Phil. Tran. R. Soc. Lon. Ser. A*, **307**, 19 (1982).
Pagel, B. E. J., in *Advances in Nuclear Astrophysics*, ed. E. Vangioni-Flam *et al*, Gif-Yvette: Editions Frontieres, p. 53 (1986).
Pagel, B. E. J., RGO Preprint, (1989).
Pagel, B. . J., Terlevich, , Melnick, , *Publ. Astron. Soc. Pac.*, **98**, 1005 (1986).
Peebles, P. J. E., *Astrophys. J.*, **153**, 1 (1968).
Peimbert, M., in *ESO Workshop on Primordial Helium*, eds. P. A. Shaver, D. Kunth, K. Kjär, Garching, p 267 (1983).
Peimbert, M., and Torres-Peimbert, M., *Astrophys. J.*, **203**, 581 (1976).
Penzias, A. A., and Wilson, R. W., *Astrophys. J.*, **142**, 419 (1965).
Pilachowski, C. A., Hobbs, L. M., and De Young, D. S., *Astrophys. J.*, **345**, L39 (1989).
Rana, N. C., *Phys. Rev. Lett.*, **48**, 209 (1982).
Rana, N. C., *Il Nuovo Cim.*, **84 B**, 53 (1984).
Rana, N. C., Datta, B., Raha, S., and Sinha, B., *Phys. Lett. B*, **240**, 175 (1990).
Rayo, J. F., Peimbert, M., and Torres-Peimbert, S., *Astrophys. J.*, **255**, 1 (1982).
Readhead, A. C. S., Lawrence, C. R., Myers, S. T., Sargent, W. L. W., Hardebeck, H. E., and Moffet, A. T., *Astrophys. J.*, **346**, 566 (1989).
Rebolo, R., Molaro, P., Abia, C., and Beckman, J. E., *Astron. Astrophys.*, **193**, 193 (1988b).
Rebolo, R., Molaro, P., and Beckman, J. E., *Astron. Astrophys.*, **192**, 192 (1988).
Riemann, G. F. B., *Göttingen lectures*, (1854).
Riley, S. P, and Irvine, J. M., *Preprint M/C - TH* 8928 (1989).
Rindler, W., *Mon. Not. Roy. astron. Soc.*, **116**, 663 (1956).
Robertson, H. P., *Astrophys. J.*, **82**, 284 (1935).
Rood, R. T., Bania, T. M., and Wilson, T. L., *Astrophys. J.*, **280**, 629 (1984).
Rood, R. T., Wilson, T. L., and Steigman, G., *Astrophys. J. Lett.*, **227**, L97 (1979).
Sahu, K. C., Sahu, M., and Pottasch, S. R., *Astron. Astrophys.*, **207**, L1 (1988).
Sarma, N. V. G., and Mohanty, D. K., *Mon. Not. R. astron. Soc.*, **184**, 181 (1978).
Scherrer, J., *Mon. Not. R. astron. Soc.*, **205**, 683 (1983).
Smith, W. H., Schempp, W. V., and Baines, K. H., *Astrophys. J.*, **336**, 967 (1989).
Sosnovskii, A. N., Spivak, P. E., Prokofev, Yu. A., Kutikov, I. E., and Dobrynin, Yu. P., *Zh. Eksp. Teor. Fiz.*, **35**, 1059 (1959).
Spite, M., Maillard, J. P., and Spite, F., *Astron Astrophys.*, **141**, 56 (1984).
Spite, F., and Spite, M., *Astron. Astrophys.*, **115**, 357 (1982).
Spite, M., Spite, F., Peterson, R. C., and Chaffee, F. H., *Astron. Astrophys.*, **172**, L9 (1987).
Stecker, F. W., *Phys. Rev. Lett.*, **46**, 17 (1981).
Stratawa, Chr., Dobrozemsky, R., and Weinzierl, P., *Phys. Rev. D*, **18**, 3970 (1978).
Talent, D. L., *Ph. D. Thesis*, Rice University, (1980).
Thum, C., *Vistas Astron.*, **24**, 355 (1981).
Torres-Peimbert, S., Peimbert, M., and Fierro, J., *Astrophys. J.*, **345** 186 (1989).
Truran, J. W., and Brunish, W., referred to in Boesgaard and Steigman (1985).
Vidal-Madjar, A., and Gry, C., *Astron. Astrophys.*, **124**, 99 (1983); *op. cit.*, **138**, 285 (1984).
Wagoner, R. V., *Astrophys. J. Suppl. Ser.*, **18**, 247 (1969).

Wagoner, R. V., Fowler, F. A., and Hoyle, F., *Astrophys. J.*, **148**, 3 (1967).
Walker, A. G., *Proc. Lond. Math. Soc.* (2), **42**, 90 (1936).
Wilkinson, D. H., *Nucl. Phys. A*, **377**, 474 (1982).
Witten, E., *Phys. Rev. D.*, **30**, 272 (1984).
Woody, D. P. and Richards, P. L., *Phys. Rev. Lett.*, **42**, 925 (1979).
Yang, J., Schramm, D. N., Steigman, G, and Rood, R. T., *Astrophys. J.*, **227**, 697 (1979).
Yang, J., Turner, M. S., Steigman, G., Schramm, D. N., and Olive, K. A., *Astrophys. J.*, **281**, 493 (1984).
York, D. G., *Astrophys. J.*, **264**, 172 (1983).
York, D. G., and Rogerson, J. B., *Astrophys. J.*, **203**, 378 (1976).
Zeĺdovich, Ya. B., and Sunyaev, R., *Astrophys. Space Sci.*, 4, 301 (1969).
Zwicky, F., *Publ. Astron. Soc. Pacific*, **50**, 218 (1938).

The Astrophysics and Cosmology of Quark–Gluon Plasma

C. Alcock

Lawrence Livermore National Laboratory, Livermore, CA 94551, USA

I. Introduction

The physics of the quark–gluon plasma is not well understood. None of the important physical attributes of this putative state of matter can be computed with confidence from an elementary theory. The quark–gluon plasma has, most probably, not been created yet in any laboratory; certainly, there is no observable that has indicated the presence of quark–gluon plasma in any experiment that has already been performed. Indeed, as the other reviews in this volume make clear, it is difficult to devise observable consequences for the quark–gluon plasmas that will be produced in future experiments at CERN and Brookhaven.

The extraordinary prospect of producing and observing a radically new state of matter motivates large numbers of theoreticians and experimenters to strive towards overcoming these obstacles. There is the very real prospect that good theoretical models of the quark–gluon plasma will become available, and, more important, experimental demonstration of the existence of quark–gluon plasma will occur. In the meantime there is plenty of work to do!

Since there is such great uncertainty regarding the properties of the quark–gluon plasma, one might ask what astrophysicists and cosmologists are doing in this field. Astrophysicists customarily apply reasonable extrapolations of known physics to the modelling of observed celestial phenomena. Cosmologists apply similar extrapolations to models of the early universe. When there is little known physics, what is there to extrapolate?

Astrophysicists and especially cosmologists are highly motivated to study quark–gluon plasma, because there is overwhelming reason to believe that the universe was filled with this material at early times ($\leq 10\ \mu S$) when the universe was very hot ($\geq 100\ MeV$). Furthermore, the passage from quark–qluon plasma to hadron plasma must have occurred. Clearly it behooves us to investigate the possible consequences of this transition for cosmology.

II. Review of Elementary Thermodynamics

Given a situation where the motivation to study the phase is strong, but the knowledge of the underlying physics is weak, astrophysicists must choose their approach with care. Models for the physics must be devised; since these models are not derived from an underlying theory, their adoption is to some extent arbitrary. Under these circumstances, the following rules should govern the choice of thermodynamic models in this field:

(1) The models should be simple and clear.
(2) The underlying degrees of freedom should be explicit in the mathematics.
(3) The models should be consistent with all that is known.
(4) It must be possible to describe what is not known in terms of a few, physically meaningful parameters.

Working in the spirit of these rules, we now turn to the study of thermodynamic equilibria. It is important to understand why so much of our time will be spent studying these equilibria in spite of the facts that the universe is not in equilibrium now, and further that equilibria contain no information about history; the properties of an equilibrium state were independent of the path that led to the equilibrium.

The first reason for studying equilibria is that they are relatively easy to compute! The powerful formalisms of equilibrium thermodynamics and statistical mechanics are at our disposal, and usually one can create viable models of systems in equilibrium. The second reason is that many astrophysical systems are nearly in equilibrium. When a system is near equilibrium, its properties may often be described in terms of small departures from the nearest equilibrium state. A third reason that is central to many discussions in cosmology is that the universe was closer to thermodynamic equilibrium in the past than it is at present. While it is clear that the universe today is very far from equilibrium, in the distant past it was very close to being in an equilibrium state. We will discuss the reasons for this in section V.

The bulk properties of matter in thermal equilibrium can be described using one of a number of thermodynamic potentials. For example, the properties of a gas may be derived from the energy E via the expressions:

$$dE = TdS - pdV + \sum_i \mu_i dN_i \;, \tag{1}$$

which may be integrated to obtain

$$E = E\left(S, V, \{N_i\}\right) . \tag{2}$$

In these expressions T is temperature, p is pressure, μ_i is the chemical potential for species i, S is entropy, V is volume and N_i is the number of species i. It is further possible to show that, since T, p, and $\{\mu_i\}$ are *intensive* (i.e., if the system is divided into sub-volumes, these quantities take on the same values in each sub-volume) and the quantities S, V, $\{N_i\}$ are *extensive* (i.e., proportional to the volume of each sub-element), the integral of equation (1) is simply:

$$E = TS - pV + \sum_i \mu_i N_i . \tag{3}$$

All gross properties of a system may be derived from $E(S, V, \{N_i\})$. There are many circumstances, however, where this is not convenient; for instance, it is often more straight forward to specify the temperature than the entropy. It is possible to make a series of Legendre transformations on the thermodynamic potentials to obtain more convenient expressions.

The first of these is the Helmholtz Free Energy

$$F = E - TS = -pV + \sum_i \mu_i N_i , \tag{4}$$

which in differential form is

$$dF = -SdT - pdV + \sum_i \mu_i dN_i . \tag{5}$$

Equation (5) shows that $F = F(T, V, \{N_i\})$, a function conveniently of temperature instead of entropy.

Alternatively, one could transform V to p to obtain the enthalpy

$$H = E + pV = TS + \sum_i \mu_i N_i , \tag{6}$$

with

$$dH = TdS + Vdp + \sum_i \mu_i dN_i . \tag{7}$$

The enthalpy is especially useful in hydrodynamics since the entropy of an element of fluid is often (more or less) constant, and the pressure is a more useful variable than the volume.

The Gibbs Free Energy is obtained by transforming V to p and S to T, to obtain

$$G = E - TS + pV = \sum_i \mu_i N_i , \tag{8}$$

with

$$dG = -SdT + Vdp + \sum_i \mu_i dN_i \ .\qquad(9)$$

The Gibbs Free Energy $G = G(T, p, \{N_i\})$ is especially useful in laboratory chemistry, where in most experiments T and p are held fixed, and reactions between the constituents i are considered. This potential is also convenient for the study of heat engines, which are controlled by adjusting T, and allowing the pressure to perform work.

The last potential we discuss is often called the Landau potential (even though it was known to Gibbs).

$$\Omega = E - TS - \sum_i \mu_i N_i = -pV \ ,\qquad(10)$$

so that

$$d\Omega = -SdT - pdV - \sum_i N_i d\mu_i \ .\qquad(11)$$

Furthermore, since $\Omega = \Omega(T, V, \{\mu_i\})$ and $\Omega = -pV$, we have

$$p = -\Omega/V = p(T, \{\mu_i\}) \ .\qquad(12)$$

Equation (12) is known as the Gibbs–Duhem relation. Expressions (10), (11) and (12) will prove the most useful in the study of phase transitions.

Having gone through this somewhat pedantic, brief review, it is worth stressing a few facts. Foremost among these is that the various potentials are entirely equivalent; one chooses between them on the basis of convenience. Second, each potential must be calculated in terms of the appropriate independent variables (for instance, computing $E(T, V, \{\mu_i\})$ would give only a partial description of the properties of a system; for the complete description $E(S, V, \{N_i\})$ would be needed). Finally, the expressions given here apply only to the bulk properties of gases; surface effects are not included in these equations, and resistance to shear (exhibited by solids) must be treated using more general expressions.

In what follows the Gibbs–Duhem expression (12) will most often be useful. What is important to remember about this relation is that the pressure is determined by the temperature and the chemical potentials, not the temperature and the various particle number densities. The reason for this is that the number densities can be changed by reactions among the species, and the chemical potentials tell us how to determine the equilibria among these reactions.

We must now ensure that the meaning of chemical potential is clear. To do this imagine an adiabatic enclosure divided into two parts by a permeable membrane. A gas is allowed to come into equilibrium in the enclosure, at which point there are N_1 atoms in chamber 1, and N_2 atoms in chamber 2. The condition of equilibrium is that the thermodynamic potential, in this case the energy $E = E_1 + E_2$, must be minimized with respect to all allowed changes. In particular:

$$\delta E = \delta E_1 + \delta E_2 = 0 \ . \tag{13}$$

Under exchange of particles, noting that $\mu = (\partial E / \partial N)_{SV}$ from equation (1), we have

$$\delta E = \mu_1 \delta N_1 + \mu_2 \delta N_2 = 0 \ . \tag{14}$$

Conservation of particles tells us that $\delta N_1 + \delta N_2 = 0$, and hence

$$\mu_1 = \mu_2 \ . \tag{15}$$

Equation (15) is the most general expression describing the distribution of particles among partitions of a system: the condition of equilibrium is that the chemical potentials be equal everywhere. An illuminating example of how this works is equilibrium among reacting species (as in chemistry; our own examples will come later), e.g.:

$$A + B \to C + D \ , \tag{16}$$

with the energy given by $E = E(S, V, N_A, N_B, N_C, N_D)$. At equilibrium, in a closed system as before:

$$\delta E = \mu_A \delta N_A + \mu_B \delta N_B + \mu_C \delta N_C + \mu_D \delta N_D = 0 \ , \tag{17}$$

but equation (16) implies:

$$\delta N_A = \delta N_B = -\delta N_C = -\delta N_D \ , \tag{18}$$

whence

$$\mu_A + \mu_B = \mu_C + \mu_D \ . \tag{19}$$

Equation (19) will appear familiar to chemists, except that they would assert that the binding energy is missing. This is because in the energy budget for the thermodynamic potentials we will always use the relativistically correct expressions which include the rest mass energy. In conventional (low temperature) thermodynamics the rest mass energy is left out of the expressions, and then equation (19) must be explicitly modified to include a term $(M_C + M_D - M_A - M_B)$, the binding energy, on the right hand side.

Chemical potentials are introduced to keep track of conserved numbers; usually these are numbers of particles. Objects whose numbers are not conserved do not have chemical potentials; there is no chemical potential for photons because they can be created and destroyed $(e.g.\ e^- + p \rightarrow e^- + p + \gamma)$.

The relativistically correct expressions for the chemical potentials are adopted for more than aesthetic reasons. When $T \geq m$, where m is the lightest non-zero rest mass in the system (this will generally be $m = 0.511\ MeV$ for the electron), particle-antiparticle pairs begin to appear in the system. These pairs can be produced by a number of reactions, such as

$$\gamma + \gamma \rightarrow e^+ + e^- \ . \tag{20}$$

There is no violation of conserved number here, so long as the antiparticles are counted negatively. The rule of thumb is that the number density of thermal pairs of any species is comparable to the number density of photons when $T > m$, the rest mass of that species.

With copious pair production at $T > m$, the pairs contribute very significantly to the thermodynamics, increasing the pressure, entropy density, etc. It is important to keep track of all thermally generated species. This can become a tiresome bookkeeping exercise in the strong interaction because of the large number of hadronic states. The task is simplified greatly by associating the chemical potential with baryon number, which is the conserved object, and then

$$\mu(p) = -\mu(\bar{p}) = \mu(n) = \ldots = \mu_B \ . \tag{21}$$

where μ_B is the baryon chemical potential.

Another thermodynamic degree of freedom is the phase. Examples of different phases include solids, liquids and vapors. When a system has more than one phase to choose from, the equilibrium phase is that which minimizes the thermodynamic potential. Two phases can reach equilibrium when the thermodynamic potentials are the same in the two phases.

The Landau potential, or more simply the pressure, is overwhelmingly the best potential to work with in discussing phase equilibria. When two phases are in equilibrium, they clearly must be at the same temperature. Furthermore, equilibrium under exchange of particles means that the chemical potentials must be the same in the two phases. Mechanical equilibrium requires the pressures be identical. Thus, equilibrium between phases 1 and 2 is expressed by

$$p_1(T, \{\mu_i\}) = p_2(T, \{\mu_i\}) , \qquad (22)$$

Heat is evolved during passage through the phase transition. If this is done slowly (reversibly, without generating entropy) the heat generated per unit volume is given by the discontinuity in entropy densities at the phase transition. This is called the latent heat

$$L = T_c \Delta S = T_c \left\{ \frac{\partial p_1}{\partial T} - \frac{\partial p_2}{\partial T} \right\} , \qquad (23)$$

where T_c is the temperature at which equation (22) is satisfied; we will call T_c the critical temperature.

There is the possibility that $L=0$ in equation (23). When all of the first-order derivatives are continuous at a phase transition, the phase transition is of "higher order". We will discuss only "first order phase transitions", where $L \neq 0$. First order phase transition have the interesting property of phase separation, where macroscopic regions of different phase are in equilibrium with each other. Phase separation does not occur in higher order transitions.

III. Model of Strong Interaction Thermodynamics

(a) The Phase Diagram

We will now turn briefly to the phase diagram for the strong interaction. A phase diagram is a graphical means for showing which phase exists under which physical conditions. Phase diagrams are widely used in materials science and in geophysics, where they are constructed from experimental measurements. In contrast, the phase diagram for QCD is entirely theoretical. Given the primitive state of our understanding of the thermodynamics of QCD, any phase diagram must be regarded as tentative.

A "reasonable guess" at the phase diagram for QCD is shown in Figure 1. The axes are baryon chemical potential and temperature. The stable phase at any given (T, μ) is the phase which maximizes the pressure (this minimizes the Landau potential, see equation [12]). A solid line separates the hadronic region (at comparatively low temperature or chemical potential) from the quark-gluon plasma. Along this line the two phases are in equilibrium with each other.

It is worthwhile to observe how little detailed knowledge is required to obtain this sketched phase diagram [1]. All that is required to assume are: (i) that hadrons have finite sizes, and are composed of quarks; and, (ii) that at low temperature (<< 100 MeV) and low density (<< nuclear density) the stable phase of QCD is ordinary hadronic matter. Assumption (i) is founded

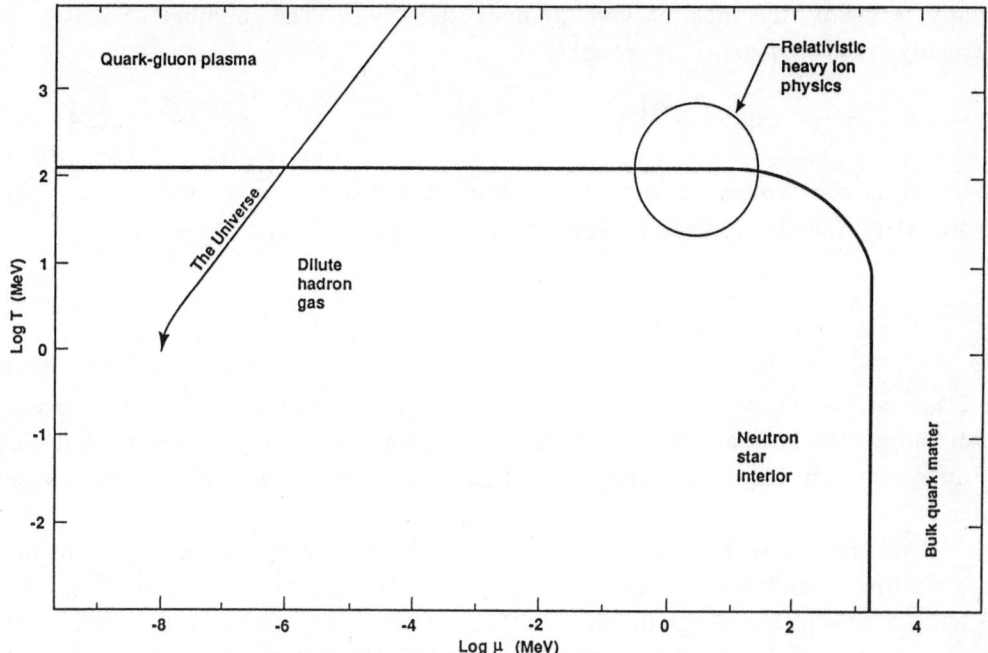

Fig. 1 Phase diagram for QCD in the standard model, in the temperature-chemical potential plane. The trajectory followed by the early universe is shown.

upon abundant experimental and theoretical evidence. Assumption (ii) appears to be validated by everyday experience, but we will see below that it is not as secure as one would hope.

Since hadrons have finite size (~1 Fm), it is clear there is a "maximum density" of ordinary hadronic matter. When the baryon number density exceeds 1 Fm^{-3}, it no longer makes sense to talk about hadrons. Clearly the hadronic identity will, at least to some extent, be lost. Considerations of asymptotic freedom suggest that at high enough density a Fermi gas of free quarks results [2]. A reasonable guess is that the transition from hadron gas to quark gas is via a first order phase transition. This is the phase transition indicated by the vertical line at $\mu \geq 1$ GeV in Figure 1.

It is of great interest to determine whether or not the conditions inside neutron stars overlap the phase boundary. Should this happen, neutron stars would have cores of quark matter [3,4].

A similar argument motivates the transition from dilute hadron gas to quark–gluon plasma in the upper left portion of Figure 1. The first step in this argument is to estimate the number of hadronic particle–antiparticle pairs. This is because at low chemical potential, the pair number density greatly exceeds the net baryon number density. The number density of thermally produced pairs is roughly

$$n \sim p^3 \exp\left\{-\frac{E+\mu}{T}\right\} , \qquad (24)$$

where E is the typical energy of a hadron and p the typical momentum. At $T \sim 100$ MeV, for the typical baryon momentum $p \approx \sqrt{mT}$, and $E \approx m$, thus

$$n \sim (mT)^{3/2} \exp\left\{-\frac{m}{T}\right\} . \qquad (25)$$

We have set $\mu = 0$, a good approximation. With $m = 1$ GeV we find that $n \approx 1 Fm^{-3}$ at $T \sim 100$ MeV. At higher temperatures the density of objects packs them more closely than their size permits; a reasonable hypothesis is that the identity of individual hadrons is lost and the hot quark–gluon plasma is formed.

The phase structure in the region $\mu \sim T$ is an interpolation between the high T, low μ and low T, high μ regions. Also shown on Figure 1 is the region of the phase diagram that will be probed in the central regions of relativistic heavy ion collisions. Note that relativistic heavy ion physics will probe the hot quark–hadron phase transition, which will be useful in cosmology, but will shed less light on the high density phase structure.

Finally, the trajectory of the universe across the phase boundary is shown. Note how very small the chemical potential is, $\mu \sim 1 ev$ at $T \sim 100$ MeV. This is reflected today in the very low baryon to photon ratio in the universe.

Using the phase diagram as a framework, we can now proceed to make model calculations of the Landau potential. It will turn out that for astrophysics the high T, low μ and the low T, high μ regions are of greatest interest, so we will confine our attention to these regions.

(b) The Hot Quark-Hadron Phase Transition

In the spirit of simplicity, the model of the hot quark–gluon plasma that we adopt is the ideal gas. We will model the thermodynamics of the vacuum in terms of the "MIT Bag" model [5], by introducing a phenomenological Bag constant B. In this model confinement can be thought of as a rule which allows quarks to inhabit only regions of a "false vacuum", which has energy density B. Ordinary hadrons are small "bags" of false vacuum containing triplets or pairs of quarks. In the quark–gluon plasma, all of space is endowed with this energy density B. Estimates of the magnitude of B vary, but a typical number is $B = (145\ MeV)^4 = 56\ MeV\ Fm^{-3}$. Some more elaborate models, notably the "chiral bag model", invoke a much larger magnitude, $B \sim (300\ MeV)^4$.

The thermodynamic potential is

$$\Omega_{QGP} = \Omega_Q + \Omega_G + BV \ . \tag{26}$$

We compute Ω_Q using the Grand partition function:

$$Z_G = \Pi \left(1 + \exp\left[\frac{\mu - \varepsilon}{T}\right]\right) , \tag{27}$$

where the product is taken over the single particle states. The thermodynamic potential is obtained using $\Omega = -T \ \ell n \ Z_G$. Converting the resulting sum to an integral, one obtains

$$\Omega_Q = -T \ N_c \ N_f \ \frac{2}{(2\pi)^3} \ V \int_0^\infty 4\pi \ p^2 dp$$

$$x \left[\ln\left\{1 + \exp\left(\frac{\mu - \varepsilon}{T}\right)\right\} + \ln\left\{1 + \exp\left(\frac{-\mu - \varepsilon}{T}\right)\right\} \right] . \tag{28}$$

In this expression N_c is the number of color degrees of freedom ($N_c = 3$), N_f is the number of flavor degrees of freedom ($N_f = 2$ if we include only u and d quarks). In this equation μ is the chemical potential of the quarks, which in magnitude is 1/3 the baryon chemical potential. The two logarithmic terms refer to the quarks and the antiquarks, respectively.

For massless quarks (a good approximation for u and d) we have $\varepsilon = p$, and equation (28) reduces to:

$$\Omega_Q = -N_c N_f V \left\{ \frac{7\pi^2}{180} T^4 + \frac{1}{6}\mu^2 T^2 + \frac{1}{12\pi^2}\mu^4 \right\} . \tag{29}$$

A similar computation for the gluons, but using boson statistics and no chemical potential, yields:

$$\Omega_G = -\frac{\pi^2}{45} N_g \ T^4 \ V \ . \tag{30}$$

Another important note is that $\mu \leq 10^{-8} T$, which means that the contribution of μ to Ω_B is only one part in 10^{16}! However, we must include this term, because it keeps track of the baryon number. The baryon number density is very low compared to all other number densities; for instance there are $\sim 10^6$ $p\bar{p}$ pairs for each unit of baryon number at $T \sim 100$ MeV! It is also important to perform the sum over the various baryonic states, since the sum is $\sim 10\times$ greater than the contribution of the protons alone.

These expressions are valid throughout the quark–gluon plasma region of Figure 1. It is not possible to obtain similarly general expressions for the hadron gas. A reasonable approach to the dilute, hot hadron plasma is once again to assume that the gas is ideal. It is not reasonable, as with the quarks, to assume the rest masses are zero. The thermodynamic potential contains terms due to mesons and terms due to baryons. The term due to each meson is:

$$\Omega_M = -\frac{g_I g_J}{2\pi^2} V m^2 T^2 \sum_{n=1}^{\infty} n^{-2} K_2\left(\frac{nm}{T}\right) , \qquad (31)$$

where g_I and g_J are the isospin and spin degeneracies, respectively, and K_2 is a modified Bessel function. Equation (31) should be summed over all independent meson states [6,7]. However, for $T < 200$ MeV, the pions dominate because of their low mass, and a good approximate expression (useful for $100 MeV \leq T \leq 200 MeV$) is:

$$\Omega_M \approx -\frac{\pi^2}{30} V T^4 . \qquad (32)$$

The thermodynamic potentials for the baryons are calculated as for the quarks, but explicitly including the masses, to yield:

$$\Omega_B = -\frac{g_I g_J}{\pi^2} V m^2 T^2 \sum_{n=1}^{\infty} \frac{(-1)^{n+1}}{n^2} \cosh\left(\frac{n\mu}{T}\right) K_2\left(\frac{nm}{T}\right) . \qquad (33)$$

where μ is now the baryon chemical potential. This expression must be summed over all independent baryon states [6,7]. An approximate expression is:

$$\Omega_B \approx -\frac{V}{\pi^2} \sum_{BARYONS} g_I g_J \, m^{3/2} T^{5/2} \, e^{-m/T} \left(1 + \frac{\mu^2}{T^2}\right) . \qquad (34)$$

Note that the magnitude of Ω_B is small compared to that of Ω_M. The pion gas dominates the entropy and the pressure of the hot hadron gas. Of course, at relatively low temperatures there are no thermal pairs of pions. The expressions (equations [32] and [34]) are only useful for $T \geq 50$ MeV.

Now that we have expressions for the pressure in each phase, we can find the critical temperature T_c. Since $\mu \ll T$, we can set $\mu = 0$, with no loss of accuracy. Figure 2 shows $p(T, \mu = 0)$ curves for both the quark–gluon plasma and the hadron plasma, with $B = (145 \ MeV)^4$. (The pressure due to the photons and leptons is not included here, since this adds identically to both phases.) The equilibrium phase is the phase with higher pressure. The two curves cross at $T_c = 106$ MeV. There is a discontinuity in the slopes at T_c, hence the latent heat (cf equation [23]) is non–zero; this is a first order phase transition.

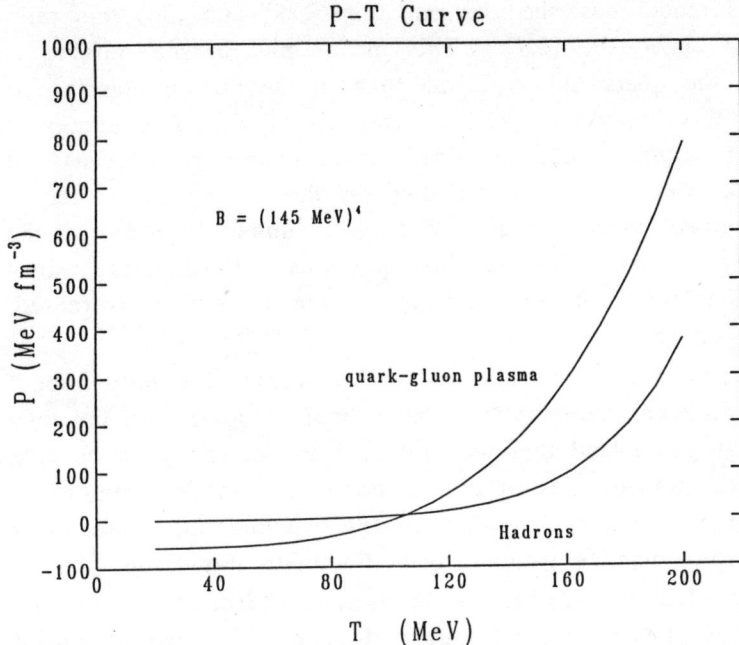

Fig. 2 Pressure versus temperature for the quark-gluon plasma and for the hadron gas. The curves intersect at the critical temperature.

Even though the baryon number is so small that terms containing μ are insignificant in computing pressure and latent heat, it is the behavior of the baryon number than will lead to long term consequences when we discuss the cosmic quark-hadron phase transition in Section VI. The baryon number density in either phase is given by:

$$N_B = \frac{\partial P}{\partial \mu} = \left(\frac{\partial^2 P}{\partial \mu^2}\right)_{\mu=o} \mu \ , \tag{35}$$

where the facts that $\mu \ll T$ and that $(\partial P/\partial \mu)=0$ when $\mu=0$ were exploited in deriving the second derivative expression. When equilibrium between the two phases obtains, the chemical potential μ is the same in both phases. The baryon number density N_B will not, in general, be the same in both phases. We define [7] the equilibrium ratio $R=N_B$ (quark)/N_B (hadron), which is given by:

$$R = \frac{\partial^2 P_Q / \partial \mu^2}{\partial^2 P_H / \partial \mu^2} \ , \tag{36}$$

In the ideal gas model described by equations (29) and (34) this ratio generally turns out to be $R \sim 100$ [6,7]. This means that baryon number is 100× more soluble in the quark–gluon plasma than in the hadron plasma. It is easy to understand this important result. A high penalty in free energy is paid when one unit of baryon number is added to the hadron gas, because all baryons have mass $\geq 1\ GeV$. This is reflected in the Boltzmann factor in equation (34). In contrast, when the unit of baryon number is added to the quark–gluon plasma it is carried by massless particles. Furthermore, since three quarks will carry the unit, the entropy of the system is increased, further reducing free energy.

It is important to recall at this point that these are only model calculations. Since different prescriptions have been adopted for the two phases, we have almost guaranteed that we will find the transition to be first order. The order of the transition is of crucial importance in determining the cosmological consequences, and it is worth asking whether there are other approaches to the thermodynamics which might illuminate this issue.

The great hope for elucidating a theoretical description for the themodynamics of QCD lies in lattice gauge theory. Substantial recent progress has been made in this area of research, but at enormous cost in personal effort (of the investigators) and computer time. There is not yet a clear agreement on the order of the phase transition.

Much more is known about the pure gauge system, in which dynamical quarks are absent, than for full QCD. (This is equivalent to letting the quark masses tend to infinity.) Two recent examinations of the pure gauge, which were carried out on special purpose computers which were built for this problem, have reached different conclusions regarding the order of the transition. The Columbia group [8] studied the behavior of the Gibbs potential near the critical point, and found evidence for a discontinuity. For instance, their data on a $24^3 \times 4$ lattice (24 grid points in the spatial directions, 4 grid points in the time directions) with $\sim 10^5$ sweeps indicate a clear discontinuity at the critical point. They did find that the discontinuity was not as sharp as previously thought, with some "rounding" of the Gibbs potential near the critical point. This rounding is more pronounced in simulations on a $16^3 \times 4$ lattice, and leads them to believe this to be a finite lattice size effect.

The Italian APE collaboration [9] studied the order of the transition using a careful investigation of the correlation length in their system. This approach was adopted because they found evidence for metastability in Monte-Carlo simulations of the 2-D Ising model, which is known to have a second order phase transition. Their comment on the Ising model has been criticized by Fukugita [10], however, who claims their "metastability" was only migration between domain walls.

In their simulations, the APE group found evidence that the correlation lengths in their system diverged as the critical point was approached. This divergence would be expected if the transition were second order, not first order.

The apparent disagreement between these two groups regarding the order of the phase transition suggests that this is a subtle problem, yet to be resolved. The situation is much more complex with dynamical quarks, because enormous technical difficulties associated with their introduction. Preliminary evidence indicated that, as the quark masses are lowered, the phase transition turns from first order, to second order at m~T, to first order at m=o. At m=o the transition is associated with chiral symmetry breaking. Nothing approaching a realistic calculation has been performed. The current situation has been reviewed by Fukugita [10].

In the light of the uncertain conclusions in lattice gauge theory, we will proceed to use the elementary models described here. The reader is cautioned that all of the conclusions regarding primordial nucleosynthesis that are described in section VIII will have to be revised if the phase transition turns out to be second order.

(c) Cold Quark Matter and Strange Matter

The thermodynamic potential for cold quark matter is given by equations like (26), (29) and (30), but evaluated at $T=0$. There is one important issue that remains to be addressed—the role of the strange quarks. The mass of the strange quark is dynamically significant ($m_s \sim 100\ MeV - 300\ MeV$), so the integral of equation (28) must be performed with $\varepsilon^2 = p^2 + m_s^2$. At zero temperature, the expression for the thermodynamic potential for the strange quarks is [11,12]:

$$\Omega_s = -\frac{V}{4\pi^2}\left\{\mu_s(\mu_s^2 - m_s^2)^{1/2}\left(\mu_s^2 - \frac{5}{2}m_s^2\right) + \frac{3}{2}m_s^4 \ln\left[\frac{\mu_s + (\mu_s^2 - m_s^2)^{1/2}}{m_s}\right]\right\}. \quad (37)$$

This somewhat complicated equation yields a simple expression for the number density of strange quarks:

$$n_s = \frac{1}{\pi^2}(\mu_s^2 - m_s^2)^{3/2}. \quad (38)$$

Equations (37) and (38) are only valid if $\mu_s \leq m_s$; otherwise $\Omega_s = n_s = 0$.

There are a variety of possible consequences of cold quark matter for astrophysics, mostly concerned with its possible existence inside neutron stars. By far the most radical possibility was raised by Witten [14], who suggested that cold three flavor quark matter might be absolutely stable. This notion has the most far reaching consequences for astrophysics, and at the same time reminds us of how little is known about the phase diagram of the strong interaction.

Consider the pressure of two flavor (up and down) quark matter. In the model we are using

$$p = \frac{1}{2\pi^2}\mu^4 - B \ . \tag{39}$$

This has a zero pressure solution at some μ_{II}. The baryon chemical potential for zero pressure two flavor quark matter is then $\mu_B = 3\mu_{II}$. Referring back to equation (8), we see that the Gibbs potential per baryon for this state is $3\mu_{II}$.

The Gibbs potential per baryon for cold, catalyzed nuclear matter at zero pressure is M/A, the mass per baryon for Fe^{56}. That ordinary hadronic matter is stable with respect to two flavor quark matter means

$$3\mu_{II} > \frac{M}{A} \ . \tag{40}$$

Now suppose that $m_s \ll \mu_{II}$. The pressure of three flavor quark matter would be

$$p = \frac{3\mu^4}{4\pi^2} - B \ , \tag{41}$$

and at zero pressure $\mu = \mu_{III} = \left(\frac{2}{3}\right)^{1/4}\mu_{II}$. Witten noted the possibility that

$$3\mu_{III} < \frac{M}{A}\left(Fe^{56}\right) < 3\mu_{II} \ . \tag{42}$$

Equation (42) might be satisfied even for dynamically significant mass m_s; a careful exploration of the parameter space by Farhi and Jaffe [12] showed that equation (42) was quite plausible.

The meaning of equation (42) is profound. It means that the true ground state of the strong interaction is three flavor quark matter, and not the hadronic world with which we are familiar. This state is known as "strange matter". Strange matter is, by hypothesis, absolutely stable, because it has lower energy per baryon than any other configuration.

If strange matter is absolutely stable, then ordinary nuclei are metastable with respect to small lumps of strange matter. The transition rate for a large nucleus to transform into a "strangelet" is vanishingly small, because a very high order weak interaction is needed to transform the excess

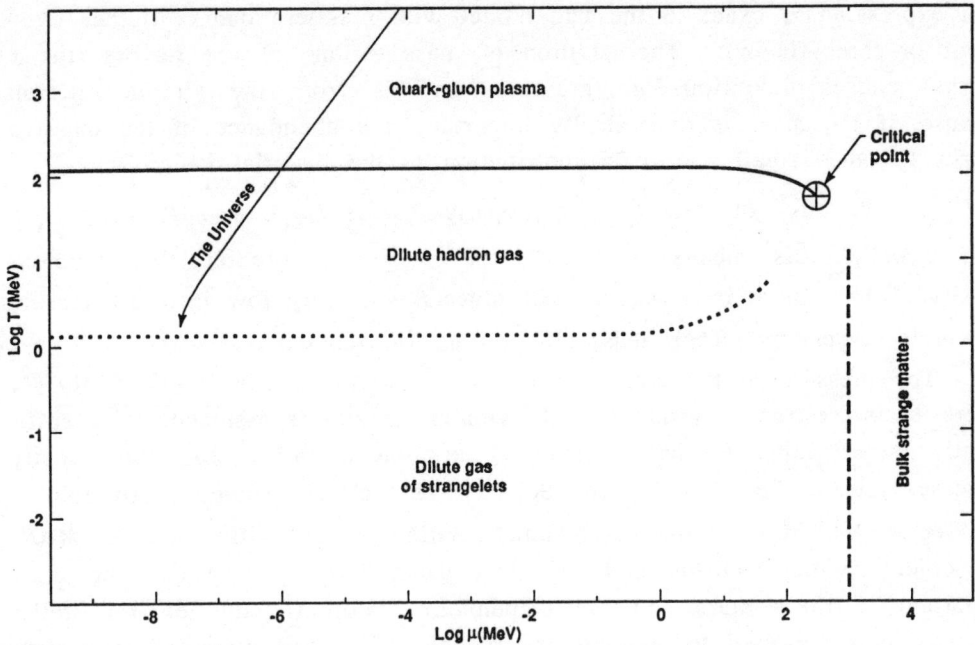

Fig. 3 As for figure 1, but for the strange matter picture.

u and d quarks to s quarks. Very small lumps of strange matter ($A \leq 10$) are not expected to be stable because of shell effects [12]; in particular, we know that the Λ hyperon, the $A=1$ strangelet, is not stable.

The phase diagram in the strange matter picture is quite different [1], as shown in Figure 3. The presumed absolute stability of strange matter ensures that most of the region of the plane that would otherwise be hadronic either comprises bulk strange matter or a gas of strangelets. In the gas of strangelets, most of the volume of the system carries no baryon number, which is concentrated into isolated lumps of strange matter.

IV. Strange Matter and Strange Stars

If the strange matter hypothesis is correct, neutron stars are metastable with respect to stars made of strange matter. This in turn means that the objects known to astronomers as neutron stars are probably made of strange matter, not of neutron matter, and should be called "strange stars"[14,15,16]. The properties of strange stars are discussed here.

These objects have extremely simple structures, because the zero-temperature equation of state is, to high accuracy, $P = \frac{1}{3}(\rho - 4B)$.

This expression is exact in the bag model with massless quarks (either two-flavor or three-flavor). The addition of mass to one of the flavors (the s quarks) causes deviations no greater than 4% from the simple relation because, if the mass is dynamically important, the abundance of the massive quarks becomes small and their contribution to the material insignificant.

This equation of state has the property that as $P \to 0$, $\rho \to 4B$. For $B = (145\ MeV)^4$ this means $\rho = 4 \times 10^{14} g\ cm^{-3}$, slightly greater than nuclear density. Thus, there is a sequence of objects with very low internal pressure and uniform density. Their mass (M)–radius (R) relation is $M \propto R^3$.

The pressure at the center of one of these objects is $P_c = (2\pi G/3)\rho^2 R^2$, where G is Newton's constant and Newtonian gravity is assumed. For sufficiently large radius R the pressure P_c approaches $4B/3$ and the density increases toward the center of the object. This effect becomes noticeable at $R \approx 5 km$, $M \approx 0.1\ M_\odot$ and the mass–radius relation is very different from $M \alpha R^3$ for objects with $R \approx 10 km$, and $M \approx 1 M_\odot$ Relativistic gravity also becomes important in these stars and the Oppenheimer–Volkoff equation for stellar structure must be used to compute the models. The full mass–radius relation is shown in Figure 4. The sequence terminates at the limit of dynamical stability, known as the Chandrasekhar limit. The dynamical stability of relativistic stars is discussed fully in [17].

Figure 4 also shows some well-known mass–radius relations for neutron stars, which are computed for a variety of different nuclear matter equations of state: MF is a mean field theory calculation; TI is a tensor–interaction model; BJ is a Bethe–Johnson model, which includes hyperons; R is a pure neutron model with a soft–core interaction; π is the R model with pion condensate. These models were reviewed by Baym and Pethick [3,4]. These mass–radius relations are very different from that for strange stars, and the difference arises entirely because, for nuclear matter, $\rho \to 0$ as $P \to 0$. One would hope that such a large, qualitative difference could be exploited to discover the truth regarding the strange matter hypothesis.

Nature has not been kind here. All neutron/strange stars for which masses have been determined have masses near 1.4 $M_\odot$ where the two models of compact stars have very similar radii. Should a very low–mass compact star be discovered, the two pictures would be distinguishable.

The fact that strange matter is absolutely stable raises the possibility that strange stars are made exclusively of strange matter, and that the surface of the star is exposed quark matter. Early discussions of strange stars presumed that this would be the case, and some interesting consequences for the appearance of these objects were found. However, as

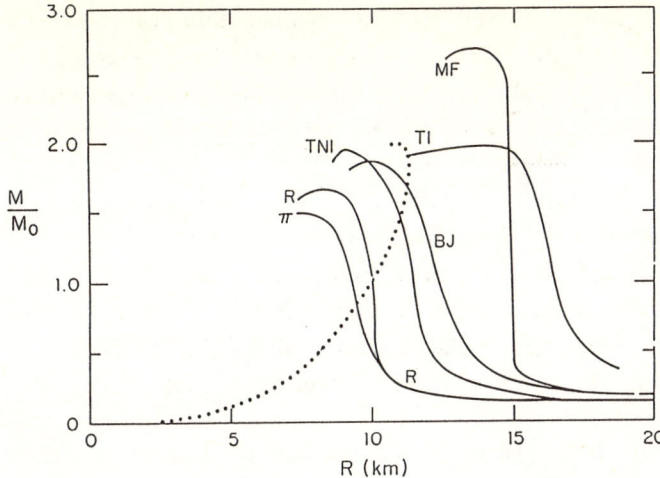

Fig. 3   As for figure 1, but for the strange matter picture.

we show below, there is also the strong possibility that the surface of a strange star is made of the same material as the surface of a neutron star.

A bare strange surface has very unusual properties [16]. The thickness of the "quark surface" is $\sim 1\,fm$; the integrity of this surface is ensured by the strong force. The electrons are held to the quark matter electrostatically, and the thickness of the "electron surface" is several hundred fermis; the electric field in this region is $\sim 5 \times 10^{17}\,V\,cm^{-1}$. Since neither component is held in place gravitationally, the traditional "Eddington Limit" to the luminosity that a static surface may emit does not apply, and these objects may (in principle) have photon luminosities much greater than $10^{38}\,erg\,s^{-1}$.

A strange matter surface will have a low emissivity for x-ray photons[16]. This conclusion is reached by calculating the dispersion relation for photons in strange matter. The result is much like the dispersion relation for photons in an electron plasma, but with characteristic "plasma frequency" $\omega_p = (8\pi\alpha/3)\,N_u^2/\rho_u$ (where α is the fine structure constant, N_u the number density of up quarks, ρ_u the energy density of up quarks). For typical parameters, $\omega_p \approx 19\,MeV$. This means that the surface of a bare strange star is highly reflective in the x-ray region and has a low emissivity. The emissivity has not yet been calculated.

There is a further consequence of the electrical properties of this surface. The very high electric field in the electron surface will exert a strong outward force on an ion. Clearly, a certain amount of normal ionic material can be supported by this electric field. It turns out that a crust of mass up to $\sim 5 \times 10^{28}\,g$ may be supported, with density at the inner edge up to

$4 \times 10^{11} g\ cm^{-3}$. This upper limit is set by the requirement that nuclear reactions between the crust and the strange matter must be prevented, or else the ions at the base of the crust would be converted to strange matter. This requirement is satisfied if: (a) there are no free neutrons in the crust [i.e. there is no "neutron drip" (13)]; (b) there is a "gap" between the ions at the base of the crust and the quark surface in which a Coulomb barrier prevents direct reactions between the ions and the strange matter.

This thin layer is identical to the "outer crust" of a neutron star[16]. For this reason, a strange star with a crust is not different from a neutron star in regard to its photon emissivity. Furthermore, since the crust is held onto the star by gravitation, this new surface is subject to the Eddington limit.

It seems likely that this latter view of the surface of a strange star is more realistic. The universe is a "dirty" environment, and certainly supernova remnants contain a lot of material that may accrete onto the surface of a newly formed strange star and make a crust. Hence, we are once again driven to conclude that a strange star is very similar to a neutron star in its observable properties.

Radio pulsars are observed to have periods that steadily increase. This is attributed to the loss of angular momentum by magnetic dipole radiation. In some pulsars small "glitches" in this smooth spin-down are occasionally observed. In a glitch, the period abruptly (in less than a day) decreases; over the next 40–80 days most of this decrease is lost as the pulsar appears to "heal" back toward its original spin-down curve.

A model has been developed for this phenomenon involving the behavior of superfluid neutrons in the inner crust of a neutron star [18]. There is no equivalent for this model involving strange stars. It is not clear how seriously the lack of a model for glitches should be taken; this may reflect only lack of imagination on our part.

A variety of "routes" from neutron matter to strange matter have been suggested [16]. These include conversion via two-flavor quark matter, clustering of lambdas, kaon condensates, direct "burning", and seeding from the outside. The uncertainties in each of these are so large that estimates of conversion rates cannot be made with confidence. It is possible, if unlikely, that neutron stars will not convert to strange stars, even if the strange matter hypothesis is correct.

However, once there is a seed of strange matter inside a neutron star it is possible to calculate the rate of growth [19]. The strange matter front absorbs neutrons, liberating u and d quarks into the strange matter. Weak equilibrium is then reestablished by the diffusion of strange quarks and by the weak interactions. The rate of progress of this front has a strong inverse temperature dependence. If this conversion happens just after the supernova explosion one expects a neutrino signature of 10^{52} ergs over a period between minutes and hours. This conversion can happen in later stages of neutron star evolution. If it happens in an active pulsar, a macroglitch will be observed because of the change in moment of inertia. An old defunct pulsar will convert even faster, and a gamma–ray burst will be its signature.

V. A Brief Review of Elementary Cosmology

For the purposes of discussing the cosmic quark–hadron phase transition it will be necessary to review some elements of physical cosmology. (There are several excellent texts on physical cosmology, e.g., [20].) This is most easily accomplished by combining some general considerations with a Newtonian analysis. The Newtonian analysis yields correct, exact expressions which describe the rate of expansion of the universe. The full analysis based on the General Theory of Relativity, which is not discussed here, is essential in order to devise meaningful global expressions which describe the universe.

The universe today is observed to be expanding. The the extent that observers can determine, the universe satisfies Hubble's Law of uniform expansion:

$$v = \dot{r} = H_o r \; , \tag{44}$$

where r is the proper vector distance between the observer and a distant galaxy, and v is the measured recession velocity of the galaxy. Note that if one shifts one's "center" from $r = O$ to $r = r_o$, the law is preserved; this is the property of *homogeneity*. Furthermore, the expansion is *isotropic*.

In addition to the expansion, the universe is filled with thermal radiation at $T \approx 2 \cdot 735K$. This radiation is isotropic to a high degree.

Modern physical cosmology is largely based on the assumption that the universe is homogeneous and isotropic, and that Einstein's General Theory of Relativity applies. We shall now see what consequences that has for a small, spherical "patch" of the universe.

This "patch" has proper radius $r = r_p(t)$, where t is the time; particles within this patch have proper coordinates $r = r(t)$. It is convenient to label each particle within this patch by the coordinate position it has at some time $t = t_o$, i.e. $x \equiv r(t_o)$. It is conventional to choose t_o to refer to the present epoch, so that t_o is the age of the universe. The coordinate x is time–independent, and is known as the "comoving coordinate". Uniform expansion can be expressed using an expansion factor $a(t)$,

$$r(t) = a(t)x, \qquad (45)$$

where $a(t=t_o)=1$.

The recession velocity is

$$v = \dot{a}(t)x = \left(\frac{\dot{a}}{a}\right)r = Hr . \qquad (46)$$

The expansion coefficient, when evaluated at $t=t_o$, is referred to as the "Hubble Constant" H_o.

Observational cosmologists measure the *redshift* z of photons which are emitted by atomic transitions at some time $t<t_o$ (thus fixing the emission wavelength λ_ε) and received at the telescope at t_o with wavelength λ, with definition $1+z \equiv \lambda/\lambda_\varepsilon$. It is easy to show in this model universe that:

$$1+z = a(t_o)/a(t) = a(t)^{-1} . \qquad (47)$$

The rate of change of the density of some conserved object can readily be expressed using H. Examples of conserved quantities include baryon number and entropy; for reasons that will become clear in section VII we look at the entropy density, s, which satisfies the equation of continuity

$$\frac{\partial s}{\partial t} + \nabla \cdot (sv) = 0 . \qquad (48)$$

Since the universe is homogeneous, $(v \cdot \nabla) = 0$. Equation (46) shows that $\nabla \cdot v = 3H$, and hence we have

$$\frac{\partial s}{\partial t} = -3Hs . \qquad (49)$$

We can now describe the evolution of the universe by obtaining a dynamical equation for $a(t)$. Consider the motion of a test particle at the edge of the local patch. It feels a gravitational acceleration due to the material inside the patch, but no acceleration due to the remainder of the matter in the universe (this is a consequence of spherical symmetry, and is rigorously true in General Relativity). Thus, we have:

$$\ddot{r}_p = -\frac{4\pi}{3}G\rho r_p , \qquad (50)$$

where ρ is the density of mass and energy inside the patch, and G is Newton's constant. Using equation (46) this becomes:

$$\ddot{a} = -\frac{4\pi}{3}G\rho a . \qquad (51)$$

In order to make further progress we must have some knowledge of ρ, and how it behaves with expansion. At the present epoch the matter density greatly exceeds the energy density, so we know that $\rho = \rho_o a^{-3}$, where ρ_o is the mean density of the universe at $t = t_o$. The universe is said to be matter dominated at the present epoch. Inserting the expression of ρ into equation (51) and integrating once yields:

$$\left(\frac{\dot{a}}{a}\right)^2 = \frac{8\pi G \rho}{3a^3} - \left(\frac{1}{aR}\right)^2 , \qquad (52)$$

where R^{-2} is an integration constant. The General Relativistic analysis reveals that $(aR)^{-1}$ is the curvature of the space–like hypersurface at fixed cosmic time [20].

Precise measurement of ρ_o and of H_o can, in principle, determine R^{-2}, which in turn fixes the geometry of the universe. It is conventional to use the ratio:

$$\Omega_o \equiv \frac{8\pi G \rho_o}{3 H_o^2} . \qquad (53)$$

If $\Omega_o > 1$, then $R^{-2} > 0$ 0 i.e. the spatial hypersurfaces are closed. Further integration of equation (52) reveals that the universe will reach a maximum expansion, after which it will collapse back onto itself. If $\Omega_o < 1$, then $R^{-2} < 0$ i.e. the spatial hypersurfaces are open, and the universe will expand forever, with $\dot{a} \to$ constant as $t \to \infty$. The special case $\Omega_o = 1$ has $R^{-2} = 0$, i.e. the spatial hypersurfaces are flat. The universe expands forever, but with $a \propto t^{2/3}$.

The observational situation with regard to Ω_o is unclear, but analyses of data lead astronomers to conclude that $0.3 \leq \Omega_o \leq 2$. For a review of this complicated subject see [20].

Theories based on the inflationary paradigm [21] predict that $R^{-2} = 0$, and hence that $\Omega_o = 1$. For this reason the geometrically flat models are taken very seriously. The composition of the matter that makes up the density ρ_o is a subject of considerable controversy, as we shall see below.

Note that the geometric term R^{-2} plays a significant role in equation (52) only at large a, which means at late times. At early times in the universe this term may be neglected i.e. all models resemble the flat model at sufficiently early times. Furthermore, the energy density of the photons (2.735K background) and neutrinos scale as $\rho \propto a^{-4}$; this means that at sufficiently early times (when $a \leq 10^{-4}$) the contribution of the matter to ρ may be neglected. Equation (52) becomes

$$\left(\frac{\dot{a}}{a}\right)^2 = \frac{8\pi G\rho}{3a^3} \;, \tag{54}$$

where ρ is the energy density. When most of the energy density is in the form of massless particles such as photons and neutrinos, $\rho \propto T^4$, where the constant of proportionality may readily be calculated, and involves the numbers of massless fermion and boson species. What is important about this result is that we can obtain a unique relationship between temperature T and time t. In addition, very high temperatures were found in the early universe, which establish good thermodynamic equilibria.

VI. The Cosmic Quark Hadron Phase Transition

The description of the cosmic quark–hadron phase transition is complicated by a variety of non–equilibrium effects. These include initial supercooling of the quark–gluon plasma below T_c followed by reheating to T_c, and the development of eddy currents in the quark phase. These departures from equilibrium are not expected to be large because the expansion of the universe is so very slow; recall from the introduction that the duration of this period is about 30μ sec.

The initial supercooling arose because, as the universe cooled to temperatures just below T_c, bubbles of the new phase did not appear immediately. This is because there is a positive free energy σ associated with the surface of the bubbles; this free energy is called the "surface tension". At $T = T_c$ the bulk thermodynamic potentials are equal $\left(P_H = P_Q\right)$, so there is no thermodynamic gain to creating a small bubble of hadron gas since the bubble has a positive surface contribution to the free energy. Only after some supercooling of the system to $T \leq T_c$ does the bulk thermodynamic advantage of the hadron phase compensate for the positive free energy of the bubble surface.

The supercooling has been analyzed using classical nucleation theory [6,7,22]. The degree of supercooling is found to be small, typically $\Delta T \sim 0.01\, T_c$, because the very slow rate of expansion (compared to the QCD timescale) permits the random formation of rare, large bubbles for which the surface term plays a smaller role. This same analysis yields an estimate of another physical quantity which is of great importance later, during nucleosynthesis. This quantity is the mean separation between nucleating bubbles at the end of the super–cooling phase,

$$\ell \approx 0.3 \frac{\sigma^{3/2} t}{T_c^{1/2} L} \;. \tag{55}$$

The dependence of ℓ on uncertain quantities, in particular on the surface tension σ, means that a good quantitative estimates of ℓ impossible to obtain. It is possible that good estimates of ℓ may be obtained using lattice calculations, but at this stage we can compute ℓ quantitatively. In the discussion below of nucleosynthesis, ℓ will be treated as an unknown parameter.

Once the small bubbles of hadron gas have nucleated, they grow rapidly, releasing latent heat, and reheating the remainder of the universe. This continues until the universe has heated back up to T_c, at which point the two phases are in equilibrium. Subsequent passage through the phase transition occurs through a sequence of equilibrium states—the process is (thermodynamically) reversible. The reheating process is complex and not fully understood; the first steps toward a good theory of this process are described in [23].

The period described above—supercooling followed by reheating—is brief, with duration $\sim 0.5\mu$ second. The ensuing period of phase equilibrium has some very interesting properties. The phase separation induces very interesting *peculiar* motions in the system, for two very different reasons that we now discuss. *Peculiar velocities* are velocities superimposed on top of the Hubble expansion described by equation (46).

During the ensuing period of near phase equilibrium, the overall expansion of the universe was manifested in a peculiar velocity field during the phase transition. Examining conditions in the "small patch" described above, there must be a velocity field $v(x)$ which, on a large scale, exhibits Hubble expansion. However, within each phase the entropy density is constant, since it is fixed by the thermodynamics. The velocity field must satisfy the equation of continuity, equation (48), whence we find $\nabla \cdot v = o$. This appears to contradict the assertion that there is large scale expansion. The apparent paradox is reconciled by noting that $\nabla \cdot v$ is singular at the phase boundary. All of the expansion of the universe occurs at the phase boundary. This is clearly very different from the normal Hubble Law described above.

The system is further complicated by gravitational instability. There is a substantial difference in energy density between the two phases, $\Delta \rho = \rho_Q - \rho_H \sim L$. Gravitational forces between different regions of the different phases are not in equilibrium. These forces give rise to accelerations and additional peculiar velocities. A crude estimate yields $\Delta v \sim H\ell$, which is of order the mean recession velocity between neighboring distinct regions of like phase.

This response of the system to the gravitational instability creates a very complicated velocity field, as large bubbles of one phase "plow through" the other phase. The new peculiar velocity has non-zero vorticity, and perhaps is best described as a system of "eddy currents", or even slow convection. The Reynolds number for the flow is ~100, too small for fully developed turbulence to arise. The significance of these eddy currents will be made clear in the next section.

VII. Baryon Transport During the Phase Transition

The eddy currents described above lead to the mixing of baryon number within each phase. We describe this mixing in terms of an effective diffusion coefficient $D \sim \Delta v \ell \sim H \ell^2$. Working again in the "small patch" we can construct a diffusion equation for the baryon number density n_B. In this patch the mean velocity of the baryons at any point x is:

$$v_B = H r - n_B^{-1} D \nabla n_B , \qquad (56)$$

where the first term is the Hubble expansion and the second is drift due to diffusion. Combining with the equation of continuity yields:

$$\frac{\partial n_B}{\partial t} = -3 H n_B - H (r \cdot \nabla) n_B + \nabla D \nabla n_B , \qquad (57)$$

where the terms on the right hand side represent expansion, Hubble drift and diffusion. This equation describes the evolution of n_B within a particular phase. It must be solved in each phase separately, with the two solutions satisfying the ratio R at the phase boundary. Note that this is a boundary condition on a moving surface.

Equation (57) assumes a much simpler form if rewritten in comoving coordinates, since the expansion is removed. The equation becomes:

$$\frac{\partial n_B}{\partial t} = \nabla^c \cdot D^c \nabla^c n_B^c , \qquad (58)$$

where superscript c denotes the use of comoving coordinates. This, of course, is the standard form of the diffusion equation. The "comoving diffusion coefficient" is related to the physical, or proper diffusion coefficient, by $D^c = a^{-2} D$.

A complete three dimensional treatment of this mixing is impracticable at present, and unwarranted until clearer understanding of the eddy currents emerge. An "equivalent sphere" model is shown in Figure 5. The sphere is centered at a point where a bubble of quark-gluon plasma will disappear. The full volume of the sphere represents a patch of volume $\sim \ell^3$. The phase boundary is always a spherical shell, which starts at the edge of

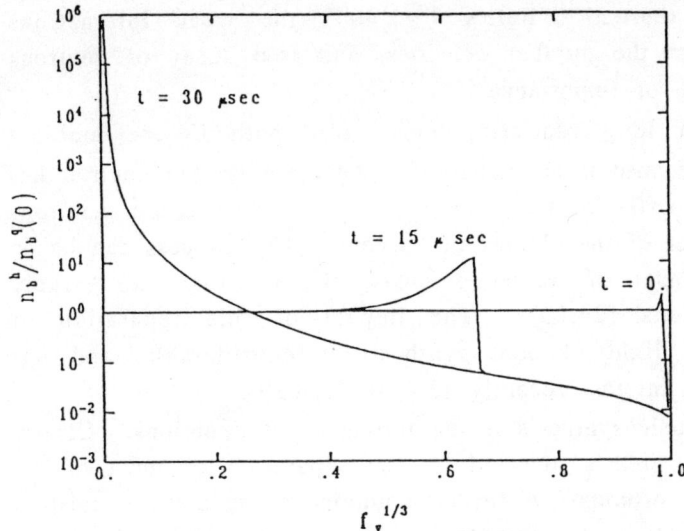

Fig. 5 The development of a baryon number inhomogeneity. Profiles of baryon number versus radius at t = 0 μs (after Tc is reached), t = 15 μs and t = 30 μs are shown.

the sphere and moves toward the center. The proper diffusion coefficient in the quark phase is, in this example, $D = 4 \times 10^9 \, cm^2 s^{-1}$.

The inhomogeneity develops as the phase boundary moves inward. Chemical equilibrium sweeps baryon number ahead of the boundary. Diffusion sweeps the excess baryon number deep into the quark–gluon plasma. As the boundary moves inwards, the amplitude of the inhomogeneity grows rapidly. The high central peak in this figure probably would not occur in a more realistic calculation, because in the inner core the phase boundary moves faster than the speed of sound, which means that quasi-equilibrium cannot be maintained.

This model shows that the solubility ratio at the phase boundary, together with mixing, leads to a very substantial inhomogeneity in the distribution of the baryon number.

VIII. Light Element Synthesis

The inhomogeneities described above remain frozen in comoving coordinates until $t \sim 1s$, $T \sim 1 \, MeV$. At this point there is a very interesting development. Neutrons and protons are now the only baryons in the system, and for $T > 1 \, MeV$ the neutrons and protons are strongly coupled by weak interactions with the thermal neutrinos. The ratio of number densities of neutrons and protons is thermal. Below $T \sim 1\ MeV$, the weak interactions become too slow to affect the number densities, and free decay of neutrons becomes the only reaction of importance.

The neutron has a long scattering mean free path in the ambient plasma of e^+, e^- and γ, because it is uncharged. Accordingly, the neutron has a much larger diffusion coefficient than the proton. For this reason, neutrons begin to diffuse rapidly out of the clumps of baryons. This process can be so efficient that, by the onset of nucleosynthesis, the neutrons are largely separated from the protons [24,25]. The impact of this separation of neutrons and protons on light element synthesis is considerable, and has been the subject of great interest recently [6,7,26,27,28,29].

The first stage of nucleosynthesis is the formation of deuterons. Clearly this important reaction is greatly affected by the separation of most of the neutrons from most of the protons. A further complication is that, as most of the neutrons are consumed in the proton rich regions, neutrons begin to diffuse back into the these cores and sustain further nucleosynthesis.

A full description of these processes involves solving the coupled, non-linear differential equations that describe the nuclear reaction rates of importance and include spatial diffusion of neutrons. These processes have been simulated by my colleagues and I by differencing the equations on a sixteen-zone representation of the initial baryon number distribution [30]. These models consist of a dense core of a high baryon number surrounded by a low density spherical mantle. The baryon number density contrast is R, and the volume fraction of the core is f_V.

Figure 6 illustrates the time history of the comoving neutron density for an optimum spherical fluctuation configuration with $\Omega_b = 1.0, R = 10^6, f_V^{1/3} = 0.25$ and r=50m. Up to $t \sim 10$ sec there is significant diffusion of neutrons into the low-density zones. This is evidenced by the increased neutron density in the low-density zones by as much as four orders of magnitude and the decreased neutron density for the high-density zones by as much as an order of magnitude. This diffusion eventually leads to decreased ^{4}He production in the high-density regions and enhanced ^{2}H production in the low-density regions. The low-density zones come into neutron-diffusion equilibrium by $t \sim 50$ sec, which is about the time that nucleosynthesis begins in the inner most high-density zone. By t=100 sec, the loss of neutrons by nuclear reactions in the high-density zone regions causes the neutron density in all of the zones to be nearly equal. However the further nuclear reactions then will continue to deplete neutrons from the high-density zones by as much as eighteen orders of magnitude. At this point neutrons begin to diffuse back into the high-density regions [28].

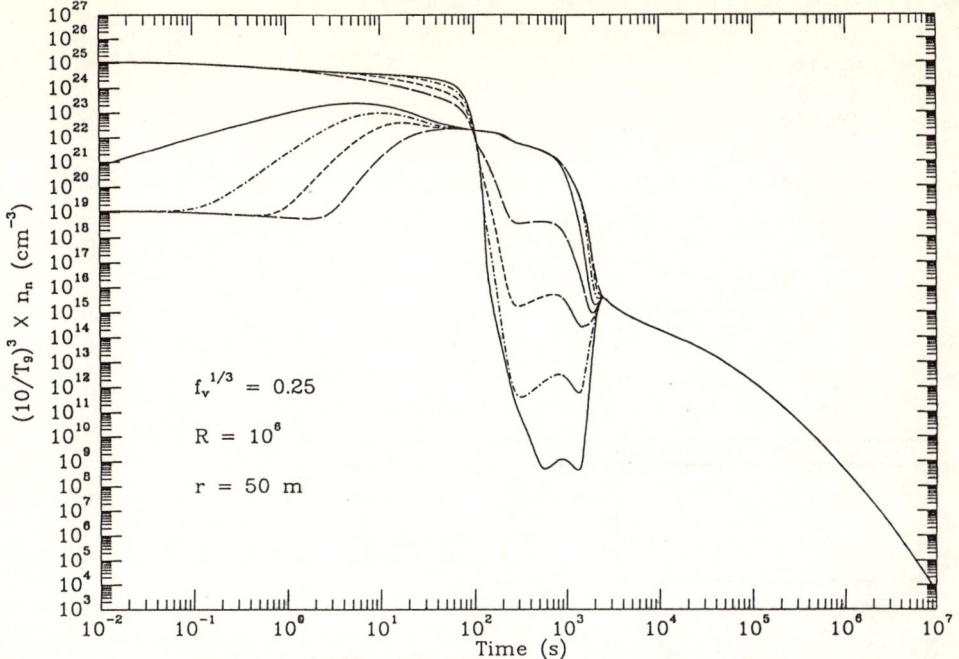

Fig. 6 Comoving neutron density as a function of time for eight different zones in a calculation with $R=10^6, f_v^{1/3}=0.25, r=50m, and \Omega_b=1.0$. The solid line corresponds to the inner most zones in high- and low-density regions. The long dashed lines are for the outer-most zones in each region.

Between t~200 to 1000 sec the back diffusion of neutrons is roughly cancelled by the neutron capture reactions and the neutron density in the different zones changes. Between 1000 and 200 sec a neutron diffusion equilibrium is obtained in all zones. Beyond 2000 sec most of the original big-bang neutrons would have decayed. However, some neutrons remain due to production of neutrons, primarily by the $D(D,n)^3He$ reaction.

The results from the optimum condensed sphere calculation for $\Omega_b=1.0, f_v^{1/3}=0.25, R=10^6$ are compared with the observed light element abundance constraints on Fig. 7. For the present purposes we take the following conservative limits on the observed primordial abundances from Boesgaard and Steigman [31]:

$$Y_p = 0.239 \pm .015; 1 \times 10^{-5} \leq D/H \leq 2 \times 10^{-4}; 1 \times 10^{-5} \leq {}^3He/H \leq 2\ 10^{-5}.$$

For ^7Li we take the results of the recent analysis of Mathews et al. [32] which showed that due to the uncertainties of galactic chemical evolution and

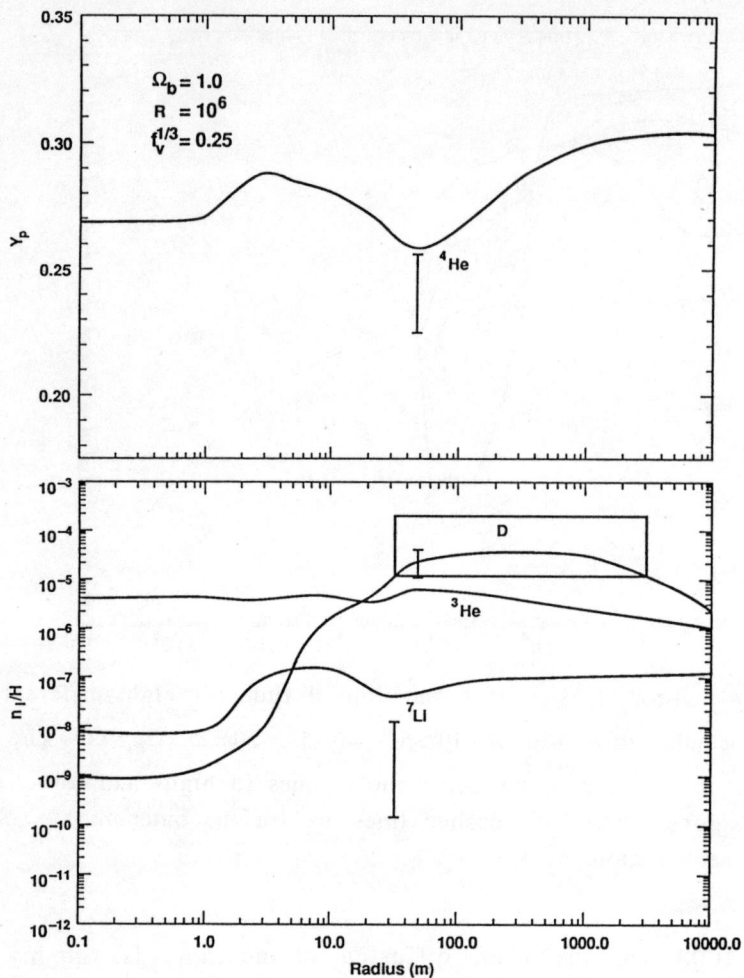

Fig. 7 A best fit $\Omega_b = 1.0$ inhomogeneous nucleosynthesis model with a spherical high-density fluctuation. Results are plotted as a function of separation radius, r.

main sequence destruction, together with a scarcity of data, the range of possible primordial ^{7}Li abundances which are consistent with observation

Contrary to our previous results, the only light element constraint which can be satisfied with $\Omega_b = 1.0$ in these models is deuterium. As indicated by the box on Fig. 7, this constraint is satisfied for a broad range of fluctuation separation radii, r~30 to 3000 m. The other constraints are not satisfied even with these conservative limits and optimum conditions. The calculation does come tantalizingly close to the limits, however, for separation radii ~30-50m.

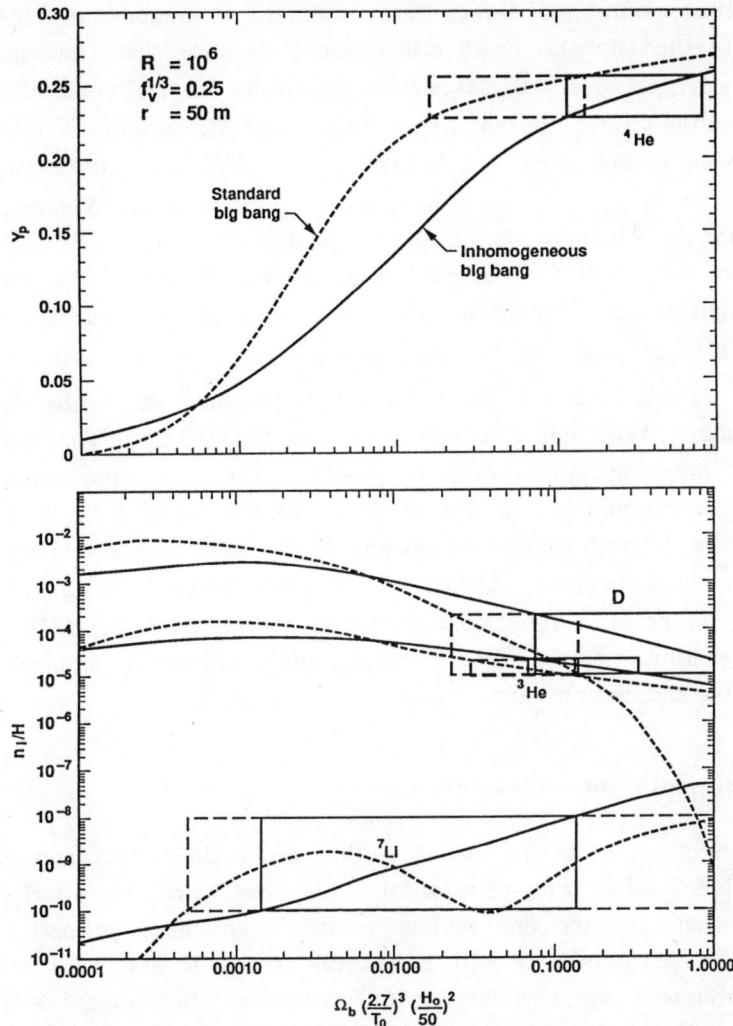

Fig. 8 Calculated light-element abundance yields as a function of Ω_b. The solid lines are for an inhomogeneous big bang with $R=10^6, f_v^{1/3}=0.25,$ and $r=50m$ and a spherical fluctuation shape. The dashed lines are for the standard homogeneous big bang. The boxes denote values consistent with the observed abundances of light elements.

To obtain a clearer picture of the possible limits on Ω_b for optimum inhomogeneous nucleosynthesis conditions we show on Fig. 8 the calculated primordial abundances as a function of Ω_b for an inhomogeneous big-bang with $R=10^{-6}, f_v^{1/3}=0.25$ and $r-50m$. The values of Ω_b which are consistent with the adopted constraints are shown as boxes on this figure. Also shown for comparison are the results and limits from standard homogeneous big-bang nucleosynthesis (dashed lines). From this figure it is clear that although the standard big-bang limits are $0.02 \leq \Omega_b \leq 0.15$ (primarily based upon the deuterium constraint), the limits based upon ^{4}He and deuterium in this particular inhomogeneous model allow Ω_b as high as 0.8. The ^{3}He constraint is somewhat more uncertain due to the production of ^{3}He in stars [31] and also due to the fact that the ^{3}He production does not depend very sensitively on Ω_b. However, from the adopted observed limits, we would place an upper limit of $\Omega_b \leq 0.4$ which is also consistent with the constraint derived if an upper limit of $Y_p \leq 0.24$ is adopted. It has been pointed out [33] that a better constraint than using ^{3}He is that of $(^3\text{He} + \text{D})/\text{H} \leq 10^{-4}$, since deuterium is converted to ^{3}He in stars. With this constraint values of $\Omega_b > 0.2$ up to $\Omega_b > 1.0$ are consistent in this inhomogeneous big-bang model. The ^{7}Li abundance constraint would be more restrictive. In this work we would obtain a limit of $\Omega_b \leq 0.15$. However, the hydrodynamic expansion of the high-baryon-density regions during nucleosynthesis [34] and/or an enhanced back-diffusion of neutrons [28] is likely to cause a significant reduction of the synthesized ^{7}Li by the $^7\text{Li}(p,\alpha)^4\text{He}$ reaction. Hence, we prefer the more conservative upper limits to Ω_b based upon the helium and deuterium abundances.

IX. The Primordial Lithium Abundance

In the previous section, I briefly remarked that my collaborators and I have generally concluded that the primordial ^{7}Li abundance is poorly determined by observations. Since this review volume is intended primarily for nuclear and particle physicists, it will be useful if the reader were to know the chain of inferences upon which these "primordial abundances" are based. I have singled out lithium (over the helium isotopes, and the deuterium) because it has been the most controversial in the past three years.

The "conventional wisdom "[31,33] regarding the priomordial lithium abundance is based upon work such as that of Spite and Spite [35]. These astronomers took the reasonable approach that if they looked at very old stars, the mix of elements they found would likely be representative of the primordial material left over from the big-bang. They selected their stars using two criteria. First, they chose only stars belonging to the halo of the Milky Way (the halo is described in Section X). Since the halo is believed to contain only objects which are at least as old as the galaxy, this is a logical way to select old stars. Second, they estimated the iron abundances of the stars, and chose only those stars with low iron abundances. In these stars they searched spectroscopically for the absorption line of neutral lithium, and determined $^7\text{Li}/^1\text{H}$ ratios using theoretical models of the atmospheres.

When their results for the number ratios of lithium to hydrogen are plotted versus surface temperature of the star, an interesting curve results. For stars of surface temperature between 6000K and ~5600K, the number ratio is ~10^{-10}, while for stars between ~5500K and ~4800K the abundances fall off; several stars showed no detectable lithium, meaning the number ratio is less than ~10^{-11}. There were no stars in their sample hotter than 6000K or cooler than 4800K.

The conventional interpretation of these data, first put forward by the original authors, is that the stars between 6000K and 5600K have their primordial lithium abundance, and the lower abundances seen in the cooler stars reflect slow depletion of lithium due to nuclear buring just below the atmosphere. To support this interpretation, one must believe that there was no depletion at all in the hotter stars.

This low estimate of the primordial lithium abundance would, if adopted, eliminate from further consideration the models discussed in VIII above. For this reason, it is important to look into the assumptions upon which the conclusion is based.

The "conventional wisdom "[31,33] regarding the priomordial lithium abundance is based upon work such as that of Spite and Spite [35]. These astronomers took the reasonable approach that if they looked at very old stars, the mix of elements they found would likely be representative of the primordial material left over from the big-bang. They selected their stars using two criteria. First, they chose only stars belonging to the halo of the Milky Way (the halo is described in Section X). Since the halo is believed to contain only objects which are at least as old as the galaxy, this is a logical way to select old stars. Second, they estimated the iron abundances of the stars, and chose only those stars with low iron abundances. In these stars they searched spectroscopically for the absorption line of neutral lithium, and determined $^7\text{Li}/^1\text{H}$ ratios using theoretical models of the atmospheres.

When their results for the number ratios of lithium to hydrogen are plotted versus surface temperature of the star, an interesting curve results. For stars of surface temperature between 6000K and ~5600K, the number ratio is ~10^{-10}, while for stars between ~5500K and ~4800K the abundances fall off; several stars showed no detectable lithium, meaning the number ratio is less than ~10^{-11}. There were no stars in their sample hotter than 6000K or cooler than 4800K.

The conventional interpretation of these data, first put forward by the original authors, is that the stars between 6000K and 5600K have their primordial lithium abundance, and the lower abundances seen in the cooler stars reflect slow depletion of lithium due to nuclear buring just below the atmosphere. To support this interpretation, one must believe that there was no depletion at all in the hotter stars.

This low estimate of the primordial lithium abundance would, if adopted, eliminate from further consideration the models discussed in VIII above. For this reason, it is important to look into the assumptions upon which the conclusion is based.

The reason for all of the uncertainty in this discussion is the possibility that the lithium abundance in the atmosphere of an old star does not reflect the abundance in the gas out of which the star formed. This is because of nuclear reactions. One can avoid the possibility of nuclear reactions by looking for lithium in the gas between the stars. If one can find some gas with relatively low iron content, and measure the lithium abundance in that gas, one can perhaps infer the primordial lithium abundance.

This seemed to be possible with the advent of the recent supernova in the Large Magellanic Cloud (SN1987A). The importance of SN1987A in this regard was that it was very bright; the low abundance of lithium means that only very weak absorption lines are produced, which can only be detected against bright sources of light.

Three different attempts were made to detect lithium absorption in the interstellar gas of the Large Magellanic Cloud, using SN1987A as a background source [37,38,39]. The gas in the Magellanic Clouds has lower iron abundance then the gas in our galaxy, so it is a reasonable place to study primordial lithium. Only upper limits were obtained. Sahu et al [38] inferred that the primordial lithium abundance was $\leq 1.6 \times 10^{-10}$. Malaney and I [39], using similar data, concluded that the upper limit was more like 4.4×10^{-9}!

The difference between these two upper limits arises because there are many correction factors which must be applied to the data before an abundance is obtained. To summarize, there are:

(1) The absorption line measured is in neutral lithium, whereas most of the lithium is singly ionized. The ratio of the two is very uncertain, since the ionization is produced by starlight in a region of the ultraviolet for which there is little data.

(2) Very likely most of the lithium in the interstellar medium is in solid grains (dust), and not in the gas phase at all. The correction for this is only weakly constrained by the data.

(3) Some lithium may be in LiH molecules.

(4) The hydrogen column density (which goes into the lithium to hydrogen ratio) is not measured for the same gas as the lithium column density! The hydrogen column density is measured by using the 21cm emission line,

which is measured with a radio telescope. The wide beam of the radio telescope averages over much more material than the line of sight to SN1987A.

It is unlikely that the many difficulties involved in the interpretation of the lithium abundances will be resolved soon. It would seem wise to avoid making decisions in cosmology based upon these measurements.

X. The Search for Baryonic Dark Matter

There is considerable interest in astronomy and cosmology in a variety of dark matter problems. There are many dark matter problems around, where we have strong evidence for the presence of unseen matter, and are presently unable to determine the quantity of unseen matter, or the constituents of the unseen matter. Among the more interesting of these problems are:

(1) What is the total mass of comets in the solar system? Estimates of this mass range from 7 earth masses to 10^3 earth masses [40].

(2) What objects comprise the unseen matter in the disk of the Milky Way, our galaxy? There is evidence that the surface mass density in the disk exceeds the mass of all the gas and the stars that we see directly [41].

(3) What objects comprise the mass in the halo of the Milky Way? This question will be discussed at length in this section.

(4) What is the mean density of the universe? Is $\Omega_o = 1$? If $\Omega_o = 1$, what makes up most of the matter? This question was discussed already in Section VIII.

While the questions addressed in (4) above are the most profound, the question (3) is well defined and addressable by experiment. All dark matter experiments which currently are proposed or underway are experimental searches for the dark matter in the halo of our galaxy. In the remainder of this section I will describe the evidence that this matter exists, and then how to search for one particular form of this matter. Specifically, I will describe a proposed experimental search for baryonic dark matter.

(a) The Dark Halo Around the Milky Way

The stars and the interstellar gas in our galaxy are confined to a thin disk. The gas disk is about 100 parsecs thick, the stellar disk is about 500 parsecs thick in the vicinity of the sun. The sun is about 8kpc (8000 parsecs) from the center of the galaxy, where the disk is fatter and merges into a sphere of radius ~1 kpc called the spheroid or bulge. The outer edge of the disk is not well defined by observation, but certainly extends out to about 20 kpc.

The disk is rotating about its axis. The rotation velocity is readily measured, using one of a variety of spectral lines (the hyperfine line at 21 cm of the atomic hydrogen in the interstellar medium is the most useful; also used are the rotational line of the CO molecule at 3 mm, and a variety of visible band atomic transitions seen in the atmospheres of stars). A remarkable fact emerges from these investigations (reviewed in [42]): the rotation velocity is about 220 km s^{-1}, and does not vary with radius, for radii greater than about 2 kpc! Studies of other disk galaxies have indicated that the phenomenon of "flat rotation curves" is the norm, not the exception.

The flat rotation curve implies that, out to 20 kpc (the limit to which measurements have been made), the total mass interior to that radius is increasing. (If most of the mass in the galaxy were near the center, we would expect to see $v \alpha R^{-1/2}$, as in Kepler's Law) Clearly, for our galaxy and other disk galaxies the mean density at radius R scales as $\rho \alpha R^{-2}$, and the mass interior to R scales at M α R! Most of the mass is in the outer regions of the galaxy!

Putting in numbers for our galaxy, one obtains

$$M_{HALO} \approx 3 \times 10^{11} \, M_\odot \left(\frac{R}{20 kpc}\right). \tag{59}$$

This total is remarkable in that M_{HALO} exceeds the total mass of the disk by at least a factor of ten! Very little else is known about the halo except that it is probably spherical (or at least nearly spherical) and that it probably does not extend beyond R~50kpc!

Clearly the halo is not made of normal stars - it would be seen by astronomers! (There are some stars seen in the halo, such as those discussed in Section IV, but they contribute very little to the total mass.) It cannot be made up of ordinary hydrogen and helium gas, either; if the gas were cool astronomers would detect the characteristic emission lines of atomic hydrogen, and if the gas were hot, astronomers would detect x-ray emission from the hot gas. Among the various candidate objects that are under investigation are:

(1) Massive neutrinos - if the μ or τ neutrino had a mass approximately 50 eV the halo could be made up of these objects.

(2) Axions - low mass (~10^{-4} eV) fundamental particles which are formed in the early universe.

(3) Wimps/Cosmions. Wimps (Weekly interacting massive particles) are fundamental particles of mass $\geq 5 GeV$ which are formed in the early universe.

All of these candidate objects are speculative, involving some extrapolation beyond known physics. There is no experimental evidence for any of (1), (2) or (3). Nevertheless, finding one of these candidates would have enormous consequences for cosmology and particle physics, and several experiments are underway; for a review see [43].

Another candidate for the dark matter is

(4) Brown dwarfs ($0.08\ M_\odot > M > 10^{-3}\ M_\odot$) and planets ($M < 10^{-3}\ M_\odot \approx M_{Jupiter}$)

Brown dwarfs and planets are made up of hydrogen, helium and the other elements in the standard cosmic ratios, and are not in the least speculative. Furthermore, since they contain no internal heat sources, they become quite invisible after 10^{10} years. They are viable candidates for the dark matter in the halo of our galaxy.

Brown dwarfs and planets have not been widely discussed as candidates for the dark matter because of the arguments, based on primordial nucleosynthesis, discussed in Section VIII. It was believed until recently that the upper limit to the amount of baryonic matter in the universe was lower than the total mass contained in galactic halos (and in clusters of galaxies). The discussion in Section VIII clearly shows that the quantity of baryonic matter in the universe could be much larger than previously supposed, and there is no longer any argument against brown dwarfs and planets as candidates for the dark matter in the halo.

Before going further, it would be useful to describe what brown dwarfs and planets are, and what the distinction is between them. Brown dwarfs are often thought of as "failed stars", for the following reason. When a cloud of gas condenses and collapses to form a star, it heats up as the gas is compressed. Heat is radiated from the surface, allowing further collapse. This process continues until the central temperature is high enough to ignite thermonuclear burning of hydrogen to helium. A balance between the heat generated by burning and the heat lost from the surface is established, and the collapse stops.

Condensations of mass $<0.08 M_\odot$ never get hot enough to ignite thermonuclear burning. They continue collapsing and loosing heat until their electrons form a degenerate Fermi sea, and they are supported by electron degeneracy pressure. At this point they start cooling slowly, and already after 10^7 years it would be difficult to detect a nearby brown dwarf by its radiation. After 10^{10} years these objects are effectively invisible.

Objects which are supported by electron degeneracy pressure have a simple relationship between their mass and their radius, $R \propto M^{-1/3}$, which extends in this case down to $M \approx 10^{-3}\ M_\odot$. In objects of even lower mass the electrons are mostly bound into atoms and molecules, and the equation of state is quite different from the degenerate gas equation of state. In this, the planetary range, the radius is an increasing function of mass. Over the whole mass range, there is a maximum of $R = 7 \times 10^9$ cm at about the mass of Jupiter.

(b) Search for Brown Dwarfs and Planets

Our group at Lawrence Livermore National Laboratory is developing an experiment to test the hypothesis that the mass in the galactic halo is made up of brown dwarfs or planets. Presently this group comprises T. Axelrod, D. Bennett, K. Cook, H.-S. Park, and myself. We may shortly be joined by C. Stubbs of UC Berkeley's Center for Particle Astrophysics. Our strategy is to look for "gravitational microlensing" of extragalactic stars by brown dwarfs and planets in the galactic halo. Another group led by M. Spiro at Saclay, France, is also pursuing this idea. The idea was originally proposed by Paczynski [43]. We expect to be sensitive to objects in the mass range $10^{-8} M_\odot \sim 0.1 M_\odot$.

The basis of this technique is the point mass gravitational lens. In Figure 9 we have the arrangement of observer, deflector of mass M, and star. Light rays are bent by the gravitational field of the deflector, according to Einstein's celebrated formula,

$$\zeta = \frac{2GM}{c^2 r} . \tag{60}$$

There are two solutions for light paths, resulting in two images of the background star. In the image plane these are located at r_1, and r_2, solutions of the equation.

$$r^2 - r_o r - R_o^2 = o , \tag{61}$$

where r_o is the location of the undeflected straight line from star to observer, and R_o is known as the "radius of the Einstein ring":

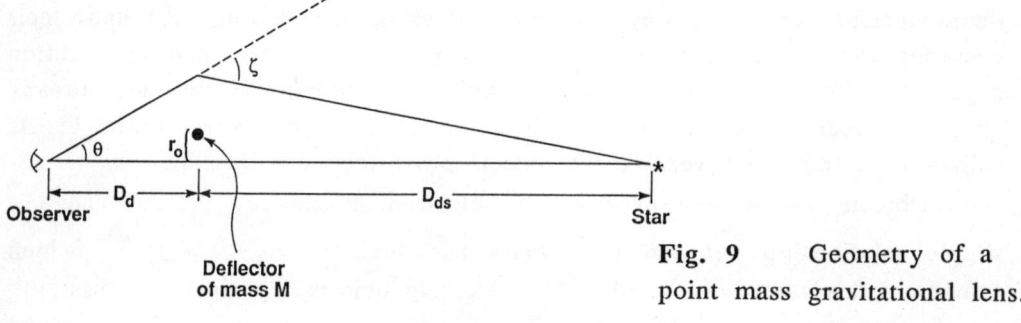

Fig. 9 Geometry of a point mass gravitational lens.

$$R_o{}^2 = \frac{2GM}{c^2} \frac{D_d \, D_{ds}}{D_d + D_{ds}} . \tag{62}$$

The two roots of equation (61) are of opposite sign, meaning that the images are located on opposite sides of the deflector. Note that when $r_0=0$ (i.e., the observer, deflector and star are aligned), the two images merge into a ring of radius R_0. Note further that if $D_{ds} >> D_d$, which will be the case for our experiment, $R_o{}^2$ is twice the product of the Schwarzschild radius of the deflector and the distance to the deflector.

It is not possible to detect this splitting into two images when the deflector is a planetary mass body in the halo. A solar mass deflector at 10 kpc has $R_o = 1.3 \times 10^{14}$ cm, and the angular separation of the images seen by the observer is only $\sim 4 \times 10^{-9}$ radians. This is much too small to be detected. However, the combined light of the two images is brighter than the undeflected star image, with amplification given by:

$$A = \frac{\left[r_o^2 + 2R_o^2\right]}{r_o \left[r_o^2 + 4R_o^2\right]^{1/2}} . \tag{63}$$

This equation has two important properties. Significant amplification occurs only when $r_0 \leq R_0$, and as $r_0 \to 0$, $A \to \infty$. Very significant amplification is possible.

It is not possible to observe a sample of extragalactic stars and recognize which are amplified and which are not. One must take advantage of the fact that all of the elements in this system (observer, deflector, star) are in relative motion. As they move, r_0 changes. A given system comes into near alignment and then moves out of alignment; as this happens the background stars appear to get brighter, and then get dimmer again. It is this transient amplification that is recognizable.

Since significant amplification occurs when $r_0 \leq R_0$, the duration of the event is $\Delta t \sim R_o / V$, where V is a typical relative velocity. We expect $V \sim 200$ km s^{-1}. This tells us the typical duration of the event $\Delta t \propto M^{1/2}$. It turns out that a 10^{-8} $M_\odot$ deflector has ~ 10 minute duration events, and a 0.08 $M_\odot$ deflector has a 21 day duration event. A sample of predicted amplification versus time curves are shown in Figure 10; these curves are parameterized by the minimum value of r_0/R_0 for the events.

There are many transient phenomena in astronomy, and it is important that we be able to distinguish a gravitational microlens event from some event that is intrinsic to the background star. In this regard, we note that there is a simple theory for the light curve of a microlens event, and the data must fit the theory to be accepted. Additionally, the gravitational lens is

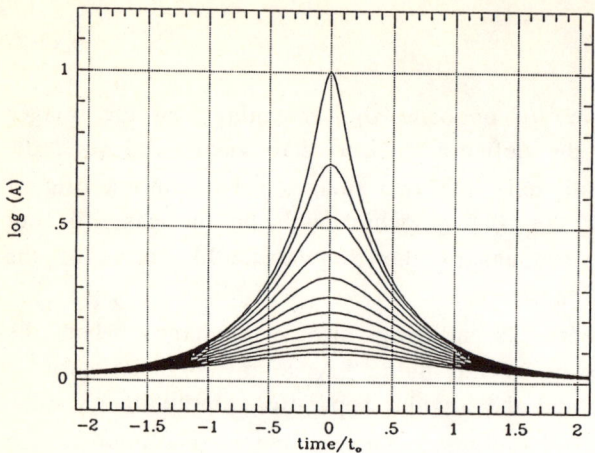

Fig. 10 Sample amplification versus time curves, for which, at closest approach, $r_0/R_0 = 0.1, 0.2,..., 1.2$.

achromatic, so the light curve should be identical in any wavelength band that is selected; it is important to collect data in at least two wavelength bands. Finally, all known stars which have occasional transient increases in luminosity can be recognized as "peculiar" by taking high dispersion spectra between outbursts. Hence, any star which undergoes a possible microlens event will be examined spectroscopically at a large telescope.

There is a good reason why this technique has not been exploited before: microlens events are rare. One can ask the question, what is the probability that a given extragalactic star is amplified at a given moment? This will be, approximately, the near number density of deflectors, times the effective cross-section, times the path length through the halo. The mean number density of deflectors is $3M_{HALO}/4\pi M R^3_{HALO}$ where M_{HALO} and R_{HALO} are the mass and radius of the halo, respectively. The cross-section is approximately πR_0^2, and the path length approximately R_{HALO}. The result is:

$$\text{Pr} \approx \frac{3GM_{HALO}}{c^2 R_{HALO}} \approx \left(\frac{V_{ROT}}{c}\right)^2, \qquad (64)$$

where V_{ROT} is the rotation velocity in the disk! $\text{Pr} \approx 10^{-6}$, which means that only one star in a million is amplified at any given time!

Of course, after an interval Δt the "one-in-a-million" star discussed in the previous paragraph is no longer amplified, but another star is. This enables us to define the necessary characteristics of a viable experiment.

(1) The duration of the experiment is greater than Δt.

(2)   (Number of stars) x (Duration of Experiment) > 10^6 Δt.

We do not know Δt ahead of time. Hence, we plan to make frequent measurements (every two minutes) on a small sample of stars, to look for short duration events, and less and less frequent measurements (up to every few days) on larger and larger samples of stars.

Up until now I have been talking about "extragalactic stars" as though there were no problem finding a suitable collection of these objects. To be useful for this experiment, the stars must be far enough away to be well outside our halo, but close enough that individual stars can readily be resolved at the telescope. The optimum sources of these stars are the two Magellanic Clouds, which are nearby, small irregular galaxies. These galaxies are very far south, and must be observed from the southern hemisphere. A sketch indicating the position of the Magellanic Clouds relative to the galaxy is shown in Figure 11.

In the vicinity of the central bar of the Large Magellanic Cloud there are about 10^5 stars per square degree brighter than magnitude 19.3. With about twenty square degrees of useful fields, there are about 2×10^6 stars available for the experiment.

Our plan to perform regular photometric observations on millions of stars is unprecedented. Previous photometric surveys have generally covered only hundreds of stars at a time. This experiment is made possible by several recent advances in semiconductor detector technology. Charge coupled devices (CCD's), which are becoming the detectors of choice in all areas of optical astronomy, are now available in very large formats, up to 1.6×10^7 pixels per device. This means that very large numbers of stars may be imaged simultaneously. Additionally, low cost computers of considerable

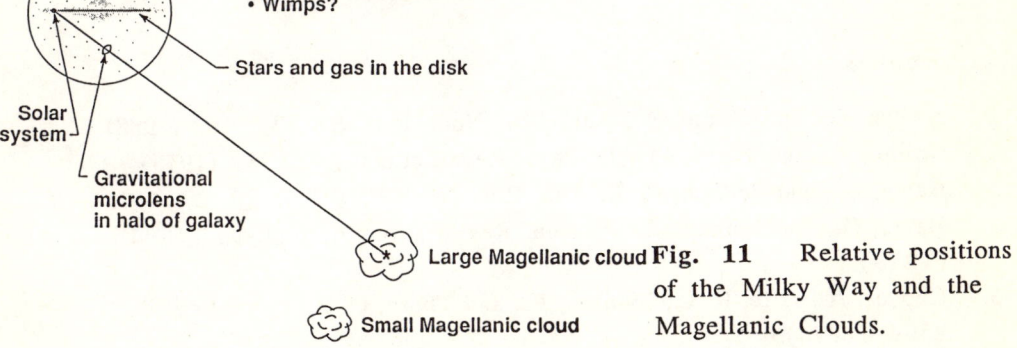

Fig. 11   Relative positions of the Milky Way and the Magellanic Clouds.

power are available, which can process the large volumes of data this experiment will produce at the rate the data are gathered.

The experiment must be performed on a dedicated telescope. This telescope need not be large, given the great quantrum efficiency of the CCD detectors. Our plan is to establish a telescope of aperture ~ 0.8 meter, with large format CCD detectros in the focal plane, and with a modest degree of automation. We believe that the duration of the experiment, once in operation, will be three to five years.

The Saclay group is taking a different approach. They plan to use the Schmidt telescope of the European Southern Observatory to take photographic plates of the Large Magellanic Cloud. The plates will be scanned and digitized by an automated plate measuring machine, and star brightness obtained from these data. The advantage of the plates is that one exposure covers the entire Magellanic Cloud, whereas our device covers at most 0.5 square degrees in one exposure. The disadvantages of plates are that they are non-linear, not reproducible, have low quantum efficiency, and have unpredictable defects. We have greater confidence in the linear CCD detectors than in the photographic plates. Time will tell which approach is superior.

In summary, we can use the gravitational microlens effect to perform a definitive search for evidence of bodies with masses in the brown dwarf/planet range.

XI. Conclusion

The quark-gluon plasma state has striking consequences for cosmology and astrophysics. In astrophysics, it appears that the most striking phenomenological conscequences arise when the strange matter hypothesisis is adopted, in which case neutron stars (most probably) all become strange stars. The conclusions in cosmology all depend upon the assumption that the quark-hadron phase transition is first order, in which event there may be significant consequences for light element nucleosynthesis.

References

1. Alcock, C., and Olinto, A., Ann. Rev. Nucl. Part. Sci., 38, 161 (1988)
2. Collins, J., and Perry, M. H., Phys. Rev. Letters, 34, 1353 (1975)
3. Baym, G., and Pethick, C. J., Ann. Rev. Nucl. Part. Sci., 25, 27 (1975)
4. Baym, G., and Pethick, C. J., Ann. Rev. Astron. Astrophys,. 17, 415 (1979)
5. Chodos, A., Jaffe, R. L., Johnson, K., and Thorn, C. B., Phys, Rev. D, 10, 2599 (1974)

6. Alcock, C., Fuller, G. M., and Mathews, G. J., Astrophys. J., 320, 439 (1987)
7. Fuller, G. M., Mathews, G. J., and Alcock, C., Phys. Rev. D, 37, 1380 (1988)
8. Brown, F. R., et al., Phys. Rev. Letters, 61, 2058 (1988)
9. Bacilieri, J. P., et al., Phys. Rev. Letters, 61, 1545 (1988)
10. Fukugita, M., preprint RIFP-781 (1989)
11. Freedman, B., and McLerran, L., Phys. Rev. D, 17, 1109 (1978)
12. Farhi, E., and Jaffe, R. L., Phys. Rev. D, 30, 2379 (1984)
13. Alcock, C., and Farhi, E., Phys. Rev. D, 32, 1273 (1985)
14. Witten, E., Phys. Rev. D., 30, 272 (1984)
15.. Haensel, P., Zdunik, J. L., and Schaeffer, R., Astron. Astrophys., 160, 121 (1986)
16. Alcock, C., Farhi, E., and Olinto, A., Astrophys. J., 310, 261 (1986)
17. Shapiro, S. L., and Teukolsky, S. A., "Black Holes, Neutron Stars, and White Dwarfs", (New York: Wiley; 1983)
18. Pines, D., and Alpar, M. A., Nature, 316, 27 (1985)
19. Olinto, A. V., Phys. Letters, 192, 71 (1987)
20. Kolb, E. W., and Turner, M. S., "The Early Universe" (Addison Wesley: New York; 1989)
21. Guth, A. H., Phys. Rev. D, 23, 347 (1981)
22. Kajantie, K., and Kurki-Suonio, H., Phys. Rev. D., 34, 1719 (1986)
23. Miller, J., and Pantano, O., Phys. Rev. D., 40, 1789 (1989)
24. Applegate, J. H., and Hogan, C. J., Phys. Rev. D., 30, 3037 (1985)
25. Applegate, J. H., Hogan, C. J., and Scherrer, R. J., Phys Rev. D, 35, 1151 (1987)
26. Kurki-Suonio, H., Phys. Rev. D, 37, 2104 (1988)
27. Kurki-Suonio, H., et al, Phys. Rev. D, 39, 1046 (1989)
28. Malaney, R. A., and Fowler, W. A., Astrophys. J., 333, 14 (1988)
29. Terasawa, N., and Sato, K., Phys. Rev. D, 39, 2893 (1989)
30. Mathews, G. J., Meyer, B., Alcock, C., and Fuller, G. M., to appear in Astrophys, J., (1990)
31. Boesgaard, A. M., and Steigman, G., Ann. Rev. Astron. Astrophys., 23, 319 (1985)
32. Mathews, G. J., Alcock, C., and Fuller, G. M., Astrophys. J., 349, 449 (1990)
33. Yang, J., et al., Astrophys. J., 281, 493 (1984)
34. Alcock, C., Dearborn, D., Fuller, G. M., Mathews, G. J., and Meyer, B., Phys. Rev. Letters, in press
35. Spite, M., and Spite, F., Astron. Astrophys., 115, 357 (1982)
36. Hobbs, L. M., and Pilachowski, C., Astrophys. J., 334, 734 (1988)

37. Vidal-Madjar, A., et al., Astron. Astrophys., 177, L17 (1987)
38. Sahu, K. C., Sahu, M., and Pottasch, S. R., Astron. Astrophys, 207, L1 (1988)
39. Malaney, R. A., and Alcock, C., Astrophys, J., 351, 31 (1990)
40. Hills, J. G., Astron. J., 86, 1730 (1981)
41. Bahcall, J. N., Astrophys. J., 287, 926 (1984)
42. Bahcall, J. N., Ann. Rev. Astron. Astrophys., 24, 577 (1986)
43. Sadoulet, B., in "The Quest for the Fundamental Constants in Cosmology" ed. J. Audouze and J. Tran Thanh Van (Editions Frontieres: Paris; 1990)

Subject Index

Additive quark model 143
Adiabatic fluctuation 272
Anticorrelation 224
Astrophysics 300
Asymptotic freedom 184

Bag model 237, 308
Bag pressure 189
Baryon–photon ratio 270
Big Bang, hot 265, 294
Big Bang 260, 281
Binning 125
Bjorken model 19
Boson interferometry 65
Brown dwarfs 336

Calorimeter, zero–degree 158
Cascade shower 121
Central region 16
Chaoticity parameter 70
Chemical potential 305
Chromodynamics 182, 233
Classical non-perturbation theory 203
Closed spaces 262
Coherent interactions 18
Cold quark matter 313
Collective behaviour 198
Color 238
Colorimeter 114
—, electromagnetic 121
—, hadron 122
Colour hydrodynamics 181
Confinement 234
Convolution 147
Correlation coefficients 69
Correlation, spin–flavor 238
Cosmological redshift 271
Cosmology 300
Coupling constant, "running" 184
Critical density 261
Critical energy 120
Critical temperature 189, 311
Cronin effect 143
Cross section, differential 6

Dark matter 264, 333
Debye screening 113, 199
Decay, two–body 129
Deceleration parameter 261

Deconfined quarks 108
Deconfinement 14
Decoupling
—, neutrino 277
—, radiation–matter 271
Density parameter 261
—, critical 261
—, energy 13
—, matter 321
—, nucleon 13
Differential cross section 6
Dileptons 223
Direct photons 37
Drell–Yan 143, 252
Dual parton model 20

E802 Experiment 24
Effect
—, Cronin 143
—, EMC 234
—, plasmon 194
—, seagull 135
Electromagnetic calorimeter 122
EMC effect 234
Energy density 13, 321
Energy stopping 155
Energy
—, critical 120
—, transverse 23
Entropy 112, 302
Equilibrium 301
—, thermal 112
Experiments
—, E802 24
—, NA34 27, 34
—, NA35 28
—, NA38 28, 36
—, WA80 25, 30
—, WA85 26

Finite temperature 190
Finite unbounded universe 263
Fireball model 161
Fluctuation 76
—, adiabatic 272
—, isocurvature 272
—, isothermal 272
Fragmentation region 16, 111
Freezing ratio 281

Geodesics 262
Geometric model 152
Glueball 237

Hadron–nucleon interactions 2
Hadron–nucleus interactions 2
Hadron calorimeter 122
Hadron gas 311
Hadron physics 233
Hadronic matter 13
Hadronization 220
Homogeneity 259, 319
Hubble constant 260, 320
Hydrodynamics, colour 181

Inclusive photons 37, 40
Incoherent interactions 18, 20
Infinite unbounded universe 263
Infrared divergence 195
Inhomogeneity 325
Interaction length 121
Interactions
—, coherent 18
—, hadron–nucleon 2
—, hadron–nucleus 2
—, heavy ion 1, 210
—, incoherent 18
—, proton–nucleus 140
—, strong 306
Interferometry 113
—, boson 65
Intermittency 76
Isocurvature fluctuation 272
Isothermal fluctuation 272
Isotropy 259, 319

J/Ψ production 113
J/Ψ suppresssion 42, 226

Landau model 20
Lattice calculations 196
Lepton pair 112, 143
Linear response theory 198
Lithium abundance 330
Lorentz transformation 126
Lund model 20

Magnetic moment 241
Magnetic spectrometer 124
Matter–dominated universe 265
Matter density 321
Matter, hadronic 13
Microwave radiation 267
Models
— additive quark 143
— bag 237
— Bjorken 19
— dual parton 20
— fireball 161
— geometric 152
— Landau 20
— Lund 20
— wounded nucleon 143
Moments, scaled factorial 76
Momentum resolution 124
Multiple scattering 120
Multiplicity 23, 114

NA34 Experiment 27, 34
NA35 Experiment 28
NA38 Experiment 28, 36
Neutrino 276
Neutrino decoupling 277
Neutrino degeneracy 295
Neutron stars 313
Non-abelian system 209
Nonequilibrium 322
Nuclear interaction length 121
Nucleon density 13
Nucleosynthesis 273, 281, 326

Olber's paradox 260
Open spaces 262
Oscillations 201

p–p collisions 133
Particle production 5
Particles, strange 87

Perturbation theory 190
Phase diagram 108, 306
Phase transition 187, 306
Photons 223
—, direct 37
—, inclusive 37, 40
Plasma oscillations 201
Plasmon effect 194
Polarized gluons 250
Pressure, bag 189
Primordial nucleosynthesis 273
Proton–nucleus interactions 140
Proton spin 244
Pseudorapidity 10, 110, 160

Quagma 252
Quantum chromodynamics 182, 233
Quarks, cold 313
Quarks, deconfined 108

Radiation–dominated universe 265
Radiation–matter decoupling 271
Radiation length 119
Radiation, microwave 267
Rapidity 8, 110
Redshift, cosmological 271
Region
—, central 16
—, fragmentation 16
Resolution, momentum 124

Scaled factorial moments 76
Screening, Debye 113, 199
Seagull effect 135
Signals 218
Space–time evolution 210
Spin–flavor correlation 238
Spin, proton 244
Strange matter 313, 315
Strange particles 87
Strange quarks 248

Strange stars 315
Strangelet 314
Strangeness enhancement 220
Strong interaction 306
Structures 272
Supercooling 322

Temperature 112
—, critical 189
—, finite 190
Thermal equilibrium 112
Thermodynamic properties 196
Thermodynamics 301
Transformation, Lorentz 126
Transparency 212
Transverse energy 23, 43, 139, 159
Transverse mass 3, 127
Transverse momentum 29, 48
Two–body decay 129

Ultrarelativistic heavy ion interactions 1, 210
Universe
—, finite unbounded 263
—, infinite unbounded 263
—, matter–dominated 265
—, radiation–dominated 265

Valence quarks 246

WA80 Experiment 25, 30
WA85 Experiment 26
World horizon 260
Wounded nucleon model 143

Yang–Mills plasma 203

Zero–degree calorimeter 158

B.N. Zakhariev, A.A. Suzko

Direct and Inverse Problems

Potentials in Quantum Scattering

1990. XIV, 223 pp. 42 figs. Softcover DM 48,– ISBN 3-540-52484-3

This textbook can almost be viewed as a „how-to" manual for solving quantum inverse problems, that is, for deriving the potential from spectra and/or scattering data. The formal exposition of inverse methods is paralleled by a discussion of the direct problem.
In part differential and finite-difference equations are presented side by side. A variety of solution methods is presented. Their common features and (dis)advantages are analyzed.
To foster a better understanding, the physical meaning of the mathematical quantities are discussed in detail. Wave confinement in continuum bound states, resonance and collective tunneling, and the spectral and phase equivalence of various interactions are some of the physical problems covered.

A. G. Sitenko

Scattering Theory

1990. Approx. 300 pp. 32 figs. (Springer Series in Nuclear and Particle Physics) Hardcover DM 88,– ISBN 3-540-51953-X

This mathematically rigorous introduction to nonrelativistic scattering theory addresses upper level undergraduates in physics. The relationship between the scattering matrix and physical observables is discussed in detail. Among the emphasized topics are the stationary formulation of the scattering problem, the inverse scattering problem, dispersion relations, three-particle bound states and their scattering, collisions of particles with spin and polarization phenomena. The analytical properties of the scattering matrix are discussed. Problems are included to help the reader to gain some experience and more expertise in scattering theory.

Springer-Verlag
Berlin
Heidelberg
New York
London
Paris
Tokyo
Hong Kong
Barcelona

J-M. Combes, A. Grossmann, P. Tchamitchian, Marseille (Eds.)

Wavelets

Time-Frequency Methods and Phase Space

Proceedings of the International Conference, Marseille, France, December 14-18, 1987

Inverse Problems and Theoretical Imaging

2nd rev. and enlarged ed. 1990. Approx. 350 pp. 98 figs. Softcover DM 98,– ISBN 3-540-53014-2

Time-frequency methods and phase space are well known to most physicists, engineers and mathematicians as is the traditional Fourier analysis. Recently the latter found for quite a few applications a competitor in the concept of wavelets. Crudely speaking a wavelet decomposition is an expansion of an arbitrary function into smooth localized contributions labeled by a scale and a position parameter.
This meeting brought together people exploring and applying these concepts in an interdisciplinary framework. The topics discussed range from purely mathematical aspects over signal analysis, seismic and acoustic applications via animal sonar systems to wavelets in computer vision.

P.C. Sabatier, Montpellier (Ed.)

Inverse Methods in Action

Proceedings of the Multicentennials Meeting on Inverse Problems, Montpellier, November 27th – December 1st, 1989

1990. XIV, 636 pp. 125 figs. Hardcover DM 138,– ISBN 3-540-51994-7

The basic idea of inverse methods is to extract from the evaluation of measured signals the details of the object emitting them. The applications range from physics and engineering to geology and medicine (tomography).
Although most contributions are rather theoretical in nature, this volume is of practical value to experimentalists and engineers and as well of interest to mathematicians. The review lectures and contributed papers are grouped into eight chapters dedicated to tomography, distributed parameter inverse problems, spectral and scattering inverse problems (exact theory), wave propagation and scattering (approximations); miscellaneous inverse problems and applications and inverse methods in nonlinear mathematics.

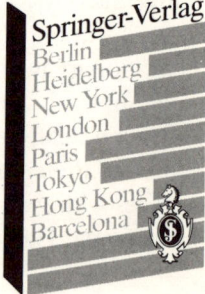

Springer-Verlag
Berlin
Heidelberg
New York
London
Paris
Tokyo
Hong Kong
Barcelona

Research Reports in Physics

The categories of camera-ready manuscripts (e.g., written in TeX; preferably hard plus soft copy) considered for publication in the **Research Reports** include:

1. Reports of meetings of particular interest that are devoted to a single topic (provided that the camera-ready manuscript is received within four weeks of the meeting's close!).
2. Preliminary drafts of original papers and monographs.
3. Seminar notes on topics of current interest.
4. Reviews of new fields.

Should a manuscript appear better suited to another series, consent will be sought from the author for its transfer to the other series.

Research Reports in Physics are divided into numerous subseries, e.g., nonlinear dynamics or nuclear and particle physics. Besides covering material of general interest, the series provides an opening for topics that are too specialized or controversial to be published within the traditional context. The implied small print runs make a consistent price structure impossible and will sometimes have to presuppose a financial contribution from the author (or a sponsor). In particular, in the case of proceedings the organizers are expected to place a bulk order and/or provide some funding.

Within **Research Reports** the timeliness of a manuscript is more important than its form, which may be unfinished or tentative. Thus in some instances, proofs may be merely outlined and results presented that will be published in full elsewhere later. Since the manuscripts are directly reproduced, the responsibility for form and content is mainly the author's, implying that special care has to be taken in the preparation of the manuscripts.

Springer-Verlag
Berlin Heidelberg New York
London Paris Tokyo Hong Kong

Research Reports in Physics

Manuscripts should be no less than 100 and no more than 400 pages in length. They are reproduced by a photographic process and must therefore be typed with extreme care. Corrections to the typescript should be made by pasting in the new text or painting out errors with white correction fluid. The typescript is reduced slightly in size during reproduction; the text on every page has to be kept within a frame of 16 × 25.4 cm ($6\frac{5}{16}$ × 10 inches). On request, the publisher will supply special stationary with the typing area outlined.

Editors or authors (of complete volumes) receive 5 complimentary copies and are free to use individual parts of the material in other publications later on.

All manuscripts, including proceedings, must contain a subject index. In the case of many-author books and proceedings an index of contributors is also required. Proceedings should also contain a list of participants, with complete addresses.

Our Instructions for the Preparation of Camera-Ready Manuscripts and further details are available on request.

Manuscripts (in English) or inquiries should be directed to

Dr. Ernst F. Hefter,
Physics Editorial 4,
Springer-Verlag, Tiergartenstrasse 17,
D-6900 Heidelberg, FRG,
(Tel. [0]6221-487495;
Telex 461723; Telefax 06221-43982).

Springer-Verlag
Berlin Heidelberg New York
London Paris Tokyo Hong Kong